recent advances in phytochemistry

volume 36

Phytochemistry in the Genomics and Post-Genomics Eras

RECENT ADVANCES IN PHYTOCHEMISTRY

Proceedings of the Phytochemical Society of North America
General Editor: John T. Romeo, *University of South Florida, Tampa, Florida*

Recent Volumes in the Series:

Volume 28 Genetic Engineering of Plant Secondary Metabolism
Proceedings of the Thirty-third Annual Meeting of the Phytochemical Society of North America, Pacific Grove, California, June-July, 1993

Volume 29 Phytochemistry of Medicinal Plants
Proceedings of the Thirty-fourth Annual Meeting of the Phytochemical Society of North America, Mexico City, Mexico, August, 1994

Volume 30 Phytochemical Diversity and Redundancy in Ecological Interactions
Proceedings of the Thirty-fifth Annual Meeting of the Phytochemical Society of North America, Sault Ste. Marie, Ontario, Canada, August, 1995

Volume 31 Functionality of Food Phytochemicals
Proceedings of the Thirty-sixth Annual Meeting of the Phytochemical Society of North America, New Orleans, Louisiana, August, 1996

Volume 32 Phytochemical Signals and Plant-Microbe Interactions
Proceedings of the Thirty-seventh Annual Meeting of the Phytochemical Society of North America, Noordwijkerhout, The Netherlands, April, 1997

Volume 33 Phytochemicals in Human Health Protection, Nutrition, and Plant Defense
Proceedings of the Thirty-eighth Annual Meeting of the Phytochemical Society of North America, Pullman, Washington, July, 1998

Volume 34 Evolution of Metabolic Pathways
Proceedings of the Thirty-ninth Annual Meeting of the Phytochemical Society of North America, Montreal, Quebec, Canada, July, 1999

Volume 35 Regulation of Phytochemicals by Molecular Techniques
Proceedings of the Fortieth Annual Meeting of the Phytochemical Society of North America, Beltsville, Maryland, June, 2000

Volume 36 Phytochemistry in the Genomics and Post-Genomics Eras
Proceedings of the Forty-first Annual Meeting of the Phytochemical Society of North America, Oklahoma City, Oklahoma, August, 2001

Cover design: Three-dimensional display of the photodiode array absorbance data obtained by HPLC/PDA/MS for a *M. truncatula* extract. The first dimension is HPLC retention time, second is wavelength, and third is absorbance (see Fig 3.5).

recent advances in phytochemistry

volume 36

Phytochemistry in the Genomics and Post-Genomics Eras

Edited by

John T. Romeo
University of South Florida
Tampa, Florida, USA

and

Richard A. Dixon
Samuel Roberts Noble Foundation
Ardmore, Oklahoma, USA

2002

PERGAMON
An Imprint of Elsevier Science

Amsterdam - Boston - London - New York - Oxford - Paris
San Diego - San Francisco - Singapore - Sydney - Tokyo

ELSEVIER SCIENCE Ltd
The Boulevard, Langford Lane
Kidlington, Oxford OX5 1GB, UK

© 2002 Elsevier Science Ltd. All rights reserved.

This work is protected under copyright by Elsevier Science, and the following terms and conditions apply to its use:

Photocopying
Single photocopies of single chapters may be made for personal use as allowed by national copyright laws. Permission of the Publisher and payment of a fee is required for all other photocopying, including multiple or systematic copying, copying for advertising or promotional purposes, resale, and all forms of document delivery. Special rates are available for educational institutions that wish to make photocopies for non-profit educational classroom use.

Permissions may be sought directly from Elsevier Science via their homepage (http://www.elsevier.com) by selecting 'Customer support' and then 'Permissions'. Alternatively you can send an e-mail to: permissions@elsevier.co.uk, or fax to: (+44) 1865 853333.

In the USA, users may clear permissions and make payments through the Copyright Clearance Center, Inc., 222 Rosewood Drive, Danvers, MA 01923, USA; phone: (+1) (978) 7508400, fax: (+1) (978) 7504744, and in the UK through the Copyright Licensing Agency Rapid Clearance Service (CLARCS), 90 Tottenham Court Road, London W1P 0LP, UK; phone: (+44) 207 631 5555; fax: (+44) 207 631 5500. Other countries may have a local reprographic rights agency for payments.

Derivative Works
Tables of contents may be reproduced for internal circulation, but permission of Elsevier Science is required for external resale or distribution of such material.
Permission of the Publisher is required for all other derivative works, including compilations and translations.

Electronic Storage or Usage
Permission of the Publisher is required to store or use electronically any material contained in this work, including any chapter or part of a chapter.

Except as outlined above, no part of this work may be reproduced, stored in a retrieval system or transmitted in any form or by any means, electronic, mechanical, photocopying, recording or otherwise, without prior written permission of the Publisher.
Address permissions requests to: Elsevier Science Global Rights Department, at the mail, fax and e-mail addresses noted above.

Notice
No responsibility is assumed by the Publisher for any injury and/or damage to persons or property as a matter of products liability, negligence or otherwise, or from any use or operation of any methods, products, instructions or ideas contained in the material herein. Because of rapid advances in the medical sciences, in particular, independent verification of diagnoses and drug dosages should be made.

Although all advertising material is expected to conform to ethical (medical) standards, inclusion in this publication does not constitute a guarantee or endorsement of the quality or value of such product or of the claims made of it by its manufacturer.

First edition 2002

Library of Congress Cataloging in Publication Data
A catalog record from the Library of Congress has been applied for.

British Library Cataloguing in Publication Data
A catalogue record from the British Library has been applied for.

ISBN: 0-08-044116-5

∞ The paper used in this publication meets the requirements of ANSI/NISO Z39.48-1992 (Permanence of Paper).
Printed in The Netherlands.

PREFACE

This volume presents a selection of papers originally contributed at the annual meeting of the Phytochemical Society of North America held at the Westin Hotel, Oklahoma City, Oklahoma, August 4-8, 2001. The title of the symposium, organized by Richard Dixon, was "Phytochemistry in the Genomics and Post-Genomics Eras". Approximately 150 scientists from North America and 12 other countries attended four days of presentations that centered on the role of phytochemistry in the context of the rapid developments in biology brought about by the application of large-scale genomics approaches. The papers in this volume represent a selection of the main symposium papers.

During the past quarter century, phytochemistry has taken something of a back seat to other branches of plant science, such as developmental biology and physiology. During the past ten years, the predominance of interest (and training of young scientists) in molecular biology, with its applications in the field of agricultural biotechnology, has tended to further obscure phytochemistry as an important discipline. It is, therefore, ironic that the new developments in molecular biology are now driving renewed interest in phytochemistry.

Genomics has fundamentally altered the way in which we view plant biology, by providing the possibility of taking a global view of cellular processes. The pre-genomics reductionist view ("I work on pathway X") that would once have been applauded as showing "focus" now seems somewhat outdated. The complete genome sequence of *Arabidopsis thaliana* is now available, soon to be followed by that of rice, with others such as *Medicago truncatula* in the pipeline. Large-scale sequencing programs are rapidly documenting the expressed genes in many other species, including wheat, soybean, corn, sugarcane, tomato, tobacco, and banana. The big question is what all these genes do. The science that addresses this question is termed functional genomics. Many companies and institutes engaged in large scale functional genomics programs now see the relative paucity of graduates trained in chemistry and biochemistry as a major limitation to future progress.

There are several approaches to addressing plant gene function on a large scale. In one, an attempt is made to sequence all expressed genes, from a range of developmentally or environmentally selected cDNA libraries (see the Chapters by Roje and Hanson and Lange et al.). Expressed sequence tags (ESTs) are compared to sequences in existing gene databases, from which function can sometimes be inferred (but always with caution, as the computer annotations are not always correct). In the case of *Arabidopsis*, availability of the whole genome sequence provides a complete listing of potential genes of interest (see Chapter by Feldmann et al.). If the function is not apparent, analysis of gene expression profiles using microarray technology ("DNA chips") might give clues as to function based on the developmental or environmental expression pattern of the candidate genes. Alternatively, various gene knock-out and mutational approaches can be utilized (see Chapters by Feldmann et al., Osbourn and Haralampidis, and Winkel-Shirley). However a gene becomes a candidate, it is generally necessary to confirm its function by expressing it in some heterologous expression system, such as *E. coli*, yeast, or insect cells (see the Chapter by Kutchan). Availability of substrates (possibly labeled, and with the correct stereochemistry, of course!) and a specific assay method are then essential. They always were, but the speed with which EST programs may lead to candidate genes for plant natural product biosynthesis now makes the biochemical assay of enzyme activities the potential rate limiting step.

Alternative approaches to functional genomics further expand the scale of experiments. Thus, several strategies now exist for creating DNA-tagged mutants in which genes may be randomly up-regulated or down-regulated, and the independent transgenic lines may then be directly screened for biochemical phenotype by phytochemical analysis (see Chapter by Xia et al.). Such an approach requires the establishment of high throughput "metabolic profiling" for analysis of gene function, since many thousands of lines may have to be screened before establishing a "hit" on a particular gene or pathway of interest. Phytochemical analysis, therefore, becomes the essential analytical tool for gene discovery. The currently applied techniques for metabolic profiling are well familiar to phytochemists- liquid or gas chromatography coupled to light absorption, fluorescence or mass detection (see Chapters by Sumner et al. and Trethewey). This new field of "metabolomics" or "metanomics" will facilitate a far greater understanding of how gene expression regulates the metabolic phenotype of the cell than has been hitherto possible. For phytochemists with an interest in computer data base technology, a major challenge will be to assemble databases of metabolic profiles that will be able to communicate with gene sequence, protein profile and gene expression databases (see Chapters by Mendes et al. and May).

The term "structural genomics" is used to describe how the primary sequence of amino acids in a protein relates to the function of that protein. Currently, the core of structural genomics is protein structure determination, primarily by X-ray crystallography, and the design of computer programs to predict protein fold structures for new proteins based on their amino acid sequences and structural principles derived from those proteins whose 3-dimensional structures have been determined. Plant natural product pathways are a unique source of information for the structural biologist in view of the almost endless catalytic diversity encountered in the various pathway enzymes, but based on a finite number of reaction types. Plants are combinatorial chemists par excellence, and understanding the principles that relate enzyme structure to function will open up unlimited possibilities for the

PREFACE

Plants are combinatorial chemists par excellence, and understanding the principles that relate enzyme structure to function will open up unlimited possibilities for the rational design of new enzymes to generate novel biologically active natural products (see Chapter by Noel et al.).

Understanding the molecular genetics of plant natural product pathways facilitates the engineering of these pathways for plant improvement and human benefit (see Chapter by Halkier et al). This chapter focuses attention on the need to start applying genomics technology to "exotic" species for development of "biotech crops" with novel natural products and with improved pest resistance and increased nutritional value.

Looking back to ten years ago, it is impossible to imagine the progress that has been made in understanding the blueprints of life embodied in genome sequences. In Vol 35 of this series, Virginia Walbot looked ahead to the day when it would be possible to monitor changes of metabolites in living organisms using biosensors. The increased attention to the "metabolome" as the chemical phenotype of the cell, particularly relevant to plants with their complex chemistries, will hopefully drive the development of new technologies that will make *in situ* analysis of multiple metabolites a reality. Phytochemistry truly has a great future in the genomics and post-genomics eras.

We greatly enjoyed working with the authors of these chapters. JTR expresses special gratitude to Darrin King for his expertise and patience in the final preparation of this volume.

Richard A Dixon
Samuel Roberts Noble Foundation

John T. Romeo
University of South Florida

CONTENTS

1. Bioinformatics and Computational Biology for Plant Functional Genomics.. 1
   Pedro Mendes, Alberto de la Fuente, and Stefan Hoops

2. A Genomics Approach to Plant One-Carbon Metabolism......................... 15
   Sanja Roje and Andrew D. Hanson

3. Metabolomics: A Developing and Integral Component in Functional Genomic Studies of *Medicago Truncatula* 31
   Lloyd W. Sumner, Anthony L. Duran, David V. Huhman, and Joel T. Smith

4. Metabolite Profiling: From Metabolic Engineering to Functional Genomics. 63
   Richard N. Trethewey

5. Triterpenoid Saponin Biosynthesis In Plants... 81
   Anne E. Osbourn and Kosmas Haralampidis

6. A Mutational Approach to Dissection of Flavonoid Biosynthesis in *Arabidopsis*.. 95
   Brenda Winkel-Shirley

7. Biopanning by Activation Tagging.. 111
   Yiji Xia, Justin Borevitz, Jack W. Blount, Richard A. Dixon, and Chris Lamb

8. Functional Genomics of Cytochromes P450 in Plants............................ 125
   Kenneth A. Feldmann, Sunghwa Choe, Hobang Kim, and Joon-Hyun Park

9. Functional Genomics Approaches to Unravel Essential Oil Biosynthesis...... 145
   Bernd Markus Lange and Raymond E.B. Ketchum

10. Sequence-Based Approaches to Alkaloid Biosynthesis Gene Identification... 163
    Toni M. Kutchan

CONTENTS

11. An Integrated Approach To *Medicago* Functional Genomics 179
    Gregory D. May

12. Structurally Guided Alteration of Biosynthesis in Plant Type III
    Polyketide Synthases 197
    Joseph P. Noel, Joseph M. Jez, Michael B. Austin, Marianne E. Bowman,
    and Jean-Luc Ferrer

13. The Role Of Cytochromes P450 in Biosynthesis and Evolution of
    Glucosinolates 223
    Barbara Ann Halkier, Carsten Hørslev Hansen, Michael Dalgaard Mikkelsen,
    Peter Naur, and Ute Wittstock

Index... 249

Chapter One

BIOINFORMATICS AND COMPUTATIONAL BIOLOGY FOR PLANT FUNCTIONAL GENOMICS

Pedro Mendes,[*] Alberto de la Fuente, and Stefan Hoops

*Virginia Bioinformatics Institute,
Virginia Polytechnic and State University,
1880 Pratt Drive,
Blacksburg, 24061 Virginia, U.S.A.*

[*] *Author for correspondence, e-mail: mendes@vt.edu*

Introduction...	2
Functional Genomics and Systems Biology..	2
Biochemistry = Functional Genomics?..	4
Computational Functional Genomics...	7
Summary...	10

INTRODUCTION

Genomics is rapidly changing the way in which biological sciences proceed. We have moved away from a reductionist analysis of single molecular components (one enzyme, one gene) to global views of the entire cellular machinery. At first, genomics was an area of interest mostly to geneticists, as it catalogued complete DNA sequences of organisms. As the first complete genome sequences were determined, however, it became obvious that this inventory could only be completely deciphered with the help of all biological sciences. The problem is that a large number of genes in any genome cannot be identified by their sequence alone because they do not have sequence similarity with any other known gene, or if they do, the other gene is also of unknown function. The publication of the first eukaryotic genome, that of *Saccharomyces cerevisiae*,[1] perhaps the organism with best known biochemistry, revealed a staggering 40% of genes to which no function could be assigned by sequence similarity. Based on this, Oliver called for a systematic approach to the discovery of gene function,[2] which since became known as *functional genomics*. We believe that functional genomics will be a unifying force pulling all biological sciences together. Here, we outline reasons why we think that functional genomics is equivalent to high-throughput biochemistry. The barriers between genetics and biochemistry have already fallen, and it is fair to say that genetics cannot ignore the underlying biochemistry and vice-versa.

A major characteristic of functional genomics is that it produces data in massive amounts, more than can be dealt with manually by a single human. Computers are absolutely needed to collect, organize, and interpret the results of experiments. Bioinformatics is the activity that organizes the data in electronic format and facilitates finding patterns that imply links among cellular components. These data mining processes are means of generating hypotheses and formulating models that explain the observed phenomena. Computational biology is the activity of formalizing these models in a computable framework to use them for prediction, and to test the consistency of the hypotheses with prior observations. Phytochemistry has much to gain from these computational approaches, and it is becoming dependent on them. Our laboratory is actively researching bioinformatics and computational biology, specifically in relation to phytochemistry, and we briefly describe these activities here.

FUNCTIONAL GENOMICS AND SYSTEMS BIOLOGY

A well accepted meaning for the term *functional genomics* has been the research activities leading to the identification of the function of open reading frames (ORF) in complete genome sequences. The focus is on the particular ORFs that have no similarity to other sequences of known function and that, if removed from the genome, result in mutants without a phenotype. These ORFs are sometimes referred to as "orphans".[3] Just as it was recognized that bioinformatics (or more precisely sequence similarity) alone would not reveal the function of all genes,[2] it was also understood that the word "function" was poorly defined for these purposes. In terms of genome annotation, function is usually taken to mean the direct molecular activity of the gene product. Thus, the function of genes coding for enzymes would be the identity of their substrates and the nature of the reactions they catalyze.

However, knowing the enzymatic reaction does not necessarily reveal what cellular role the enzyme has. Even the classification of genes in terms of their cellular roles, pioneered by Riley,[4] and frequently used to summarize genome annotations, is not very useful,[5] and can even be misleading since most enzymes participate in more than one cellular process. For example, the protein kinases and phosphatases classified as part of signal transduction are obviously also part of energy metabolism. Others have pointed out the necessity of a holistic description of cellular interactions with the cell's environment in order to understand fully the function of gene products.[3,6] We strongly support this view, especially since, whether explicitly or not, one always uses a model for any interpretation. It is prudent to identify the model explicitly upfront, as this allows one to identify assumptions. Ultimately, understanding gene function means being able to locate the gene and its products in a biochemical network and to identify all their interactions with other genes, gene products, and the environment. We predict that model-based descriptions will become the only meaningful way to describe gene function objectively, but for now even identifying function in terms of molecular activity of enzymes is clearly a problem.

A topic relevant to this discussion of functional genomics is *systems biology*. Systems biology is rather an old field that was much in vogue in the sixties and seventies[7,8] and that is now becoming "hot".[9] Systems biology's basic proposition is that biological systems should be analyzed as a whole, rather than the parts in isolation. The motivation originally came from electrical and electronic engineering, which describes the global properties of circuits based on the properties and state of its elements (resistors, diodes, etc.). By analogy, in cellular biology, one would need to know the kinetic properties of enzymes and measure their levels as well as those of other molecules (small and large) to infer how the biochemical "circuits are wired". Unfortunately, in the days of the first appearance of systems biology, it was practical only to measure a small number of metabolites and enzymes. Combined with the extreme popularity of the more reductionist molecular biology and DNA technology, this reduced systems biology to a small set of researchers. Nevertheless, the theoretical elements that had germinated in those days were able to develop,[10-15] and today we have a nearly complete body of theory for biochemical systems. Now that functional genomics is here, with its high-throughput and massively parallel technologies able to measure thousands of molecular species, it is possible to characterize cells as systems in great detail. Systems biology has, thus, resurfaced,[9] and is now in place to make a greater impact as it progresses hand in hand with functional genomics, and expands its theoretical basis.

Another reason that systems biology did not succeed in the earlier days and is flourishing now is that the integration of large functional genomic data sets in a theoretical framework depends on the use of powerful computers, which were unavailable in the sixties. These are now ubiquitous, and much progress has been made in software designed for modeling biochemical systems. However, much is still to be achieved in computational biology in order to realize the full potential of functional genomics and systems biology.

BIOCHEMISTRY = FUNCTIONAL GENOMICS?

For many, the term functional genomics may still mean some form of large-scale molecular genetics. We disagree with that view and think it is much more: ironically, the word "functional" refers to biochemistry, as the function of genes is indeed biochemical. So far the most used technologies in functional genomics are those that focus on measurements of RNA. Two technologies are now common practice in plant biology: sequencing small pieces of cDNA (expressed sequence tags or ESTs) and gene expression profiling through the use of cDNA microarray [16] or DNA chips.[17] The former is a means of creating inventories of expressed genes associated with specific environmental conditions and tissues, while the latter is a way of measuring how much these genes are expressed in specific conditions. Microarrays and DNA chips are, thus, an excellent means to quantify the level of (nearly) all genes of a genome and can be a way of probing their function. However, since mRNA does little more than pass sequence information from chromosomes to ribosomes, it could be argued that it would be preferable to monitor proteins, the elements that actually carry out the functions.

Proteomics[18] fulfills this need with a combination of 2-dimensional polyacrylamide gel electrophoresis[19] and mass spectrometry,[20] and more recently with capillary electrophoresis and mass spectrometry.[21] Although *a priori* it could be conceived that monitoring mRNA or proteins would be two alternative ways of obtaining similar results, both theory and experiment contradict this. From a theoretical point of view, for the mRNA and protein concentrations to be strongly correlated, these molecules need to have equal (or proportional) rates of synthesis and degradation. Even though their rates of synthesis may be proportional (if the number of ribosomes is large), their rates of degradation are unlikely to be – thus, it should not be expected that changes in mRNA and protein concentration will be correlated.[22] Indeed, lack of correlation has been observed in several experiments,[23-25] although in one other, some correlation seems to be present.[26] It is safe to say that in general there should not be much correlation between protein and mRNA changes. Measurements of proteins and mRNAs should, thus, be seen as complementary and together form a broader picture of cellular function.

Following the above rationale, it seems logical to extend the argument to the level of metabolites. An even stronger case could be made on the basis that metabolites are closer to phenotype than proteins or mRNA (and in some cases the concentration of a single metabolite is the phenotype). Here, we show, using simple models, that correlation between protein and metabolites cannot also be expected, at least in general. The model depicted in Fig. 1.1 consists of the expression of one gene and the reaction that its enzyme catalyzes, and we examine how the system components respond to external perturbations. The framework of metabolic control analysis[10, 27, 28] and its extension to multi-level reaction networks[29] is a convenient way to study such perturbation experiments (whether *in silico* or *in vivo*) and we have used it here. We present two extreme behaviors of this system, each corresponding to a specific set of parameter values (these sets correspond to two different genes). In both cases, we apply a perturbation to the rate of gene transcription, mimicking a wild type and a mutant. If there were correlation among the three biochemical levels, we should be able to see this in the relative changes in the mRNA, enzyme, and metabolite concentrations.

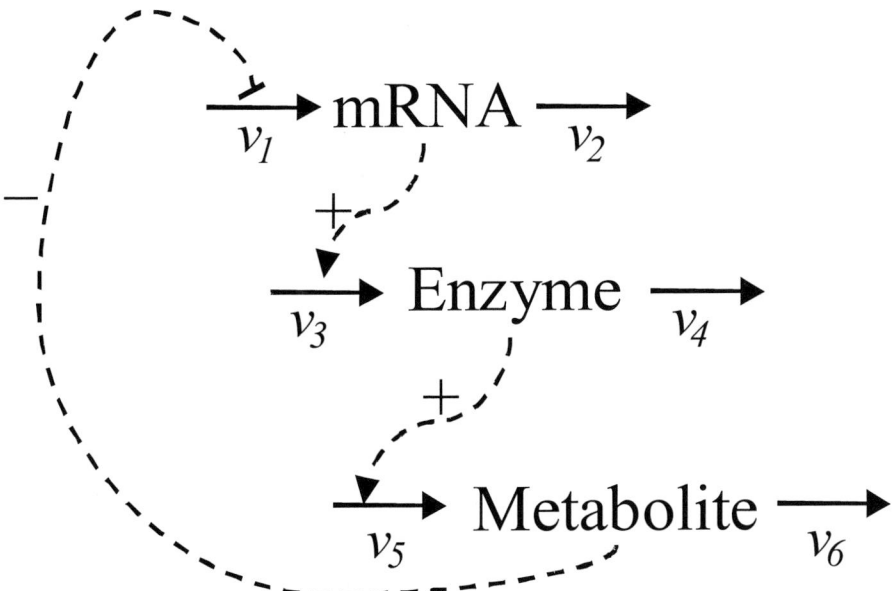

Figure 1.1: A simple model of gene expression. The model has been implemented and simulated in the program Gepasi.[41, 42] Nonlinear saturable functions chosen for the kinetics of all 6 reactions and parameter values chosen to produce extreme behaviors (details available from authors).

Table 1.1 presents values for the two extreme cases considered. In the first case (gene A), we observe that the mRNA concentration in the mutant increased by nearly 10-fold, while the metabolite concentration hardly changed. In the second, the mRNA level underwent only a 10% change, while the metabolite concentration increased 10-fold. These two examples demonstrate two limits of behavior, with most real cases lying in between. One cannot expect to always be able to characterize phenotype simply by measuring mRNA or protein levels. For gene A, we would assume a strong phenotype in the mutant, when in reality it was almost imperceptible; for gene B, we would assume there is no phenotype, while in reality there was a strong one. Obviously, for many genes, the problem may not be so serious, but not knowing when this is the case, we are suspicious of any study that measures mRNA or protein levels alone.

Table 1.1: Simulation of expression of two genes in wild type and mutant forms following the model of Fig. 1.1. The mutants were generated by changing values of their rate of transcription, which could correspond to point mutations in their upstream regulatory regions. Numeric values presented are steady-state concentrations relative to the wild type.

	Gene A		Gene B	
	wild type	mutant	wild type	Mutant
MRNA	1.000	9.911	1.000	1.090
Protein	1.000	9.911	1.000	1.090
Metabolite	1.000	1.008	1.000	10.09

Finally, we want to cast some doubt on a common assumption in metabolic studies, that the expression of genes is slower than metabolic reactions and, thus, that one can assume constant enzyme concentrations throughout the time course of an experiment. This also puts in question the complementary assumption, sometimes taken by geneticists, that metabolism is fast and, thus, gene expression experiments need not consider the metabolic level since it would be in fast equilibrium. In either case, it is only a matter of how slow expression needs to be for the assumptions to hold true. Again, by using a simple model, we observed (data not shown) that even if the gene expression time scale is some 3 orders of magnitude slower, both metabolism and gene expression change enough that both have to be taken into account (i.e., there is a component in the metabolite change that comes from the change in enzyme concentration, and there is a component of mRNA change that comes from the change in metabolite concentration, assuming that the transcription is sensitive to the metabolite). If the time scales were to be separated by 6 orders of magnitude, then one could indeed separate in time the metabolic and genetic responses. However, we note that this is likely to be a rather uncommon situation, as it would correspond to a metabolite with a half-life of 1 second and a protein with a half-life of 11 days, while the previous case corresponds to 1 second and 16 minutes, respectively. Without knowledge of the time scales of all the cellular components, one would be foolish to assume that all metabolites are very unstable and all proteins are very stable (we do know that mRNA is hardly ever stable!).

Given that functional genomics is about measuring functions of thousands of components, we propose that the temporal argument cannot be used to justify only studying mRNA, proteins, or metabolites alone. We feel strongly that functional genomics should measure all three classes of molecules, and we predict that metabolomics, the measurement of large numbers of small molecules, will become as important as transcriptomics and proteomics in the near future (the term has indeed already been used in a popular magazine[30]). Elsewhere in this volume, several mass spectrometry technologies are proposed for the use of metabolomics, while others propose the use of Fourier-transform infra-red spectrometry[31] and nuclear magnetic resonance.[32] We tend to favor the mass spectrometry approaches since they are able to resolve larger numbers of metabolites. In our minds, it is clear that functional genomics = biochemistry!

COMPUTATIONAL FUNCTIONAL GENOMICS

The above proposal to measure detailed molecular profiles implies high costs for data management. For the model plant *Arabidopsis thaliana*, we predict the existence of over 50,000 total molecular species. A time point in a detailed time course could, thus, occupy 400 kbytes, a time course with 20 time points, 8 Mbytes, and if these were determined in triplicate, 24 Mb. However large these numbers may seem, they are only the tip of the iceberg since they represent only one real number (concentration, concentration ratio, etc.) for each observation of a molecular species. In reality, for each of these numbers, there will be a much larger amount of raw data, this being defined as the immediate electronic result of an experimental measurement. For microarrays and 2D gels, raw data correspond to high-resolution images, for mass spectra, they correspond to pairs of vectors of reals (mass/charge ratios and ion intensities), and for chromatograms, also to pairs of reals (retention time and total ion intensities) or a matrix of reals (for each retention time, a vector of frequencies and intensities). All of these raw data types occupy multi-Mbyte of storage, and it is inconceivable that they could be managed in any *ad hoc* way.

Bioinformatics support is absolutely required in functional genomics, even for only a few experiments. Data management at this level is best done with industrial strength database management systems. Our preference is relational database management systems (RDBMS),[33] which store data in tables, allowing the tables to be connected (related) in explicit ways. RDBMS use the structured query language (SQL) for queries and data manipulation. While this is certainly more comprehensible than the machine code underlying all computer operations, one should not expect biologists to learn it in order to be able to manage their data. A more realistic solution is to construct graphical user interfaces that present the biologist with all the operations that can be done on the data. Such programs hide the SQL statements behind conveniently labeled buttons or menus, and still achieve the same objective. This is a rather standard mode of operation that has been extensively used in other bioinformatic applications, for example to manage sequencing experiments.

An issue that becomes very important for functional genomics is that there should be a flexible way for adding new methods of data analysis. This stems from the fact that there is still great activity in researching what algorithms will best reveal gene functions from the mass of data produced. In our collaboration with investigators at the Samuel Roberts Noble Foundation on functional genomics of the legume *Medicago truncatula*, we are designing an application programming interface (API) that insulates the details of any analysis method from the rest of the system. This allows us to develop the database management independently from the data analysis and, more importantly, to extend the data analysis subsequent to deploying the system. The benefits of this approach will be evident when a new algorithm is discovered that can be applied to the raw data. Then, we will be able to reprocess the data as soon as the new algorithm is implemented in our system and without any laborious data format conversions. We emphasize that this is possible only because we designed the system to be extendable from the start. Another design decision that we hope will pay off is that all data types (transcriptomics, proteomics, and metabolomics) will be included in the same database. At the small cost of a larger data model, we avoid the infamous problem of data integration,[34] since all our data will be in the same location, and the API will pass it directly to the analysis algorithms.

By the nature of its interdisciplinary nature, functional genomics is almost invariably a collaborative effort at distinct locations. A practical approach to the data management of such projects that we follow is to localize all data in one server and allow all collaborators to access it remotely, conveniently through a web browser without having to install specialized software. This also allows access by third parties, after public release of the data sets. In fact, these massive data sets cannot be published in the traditional way. One leaves the data out of journal articles and merely indicates there how to access the data directly from this system. Functional genomics is changing even the sociology of science!

Data management is only one aspect of computational functional genomics. Disregarding it invariably results in chaos, but by itself it contributes little to discovery. Data analysis, on the other hand, is where all the excitement will be, for this is how discoveries are revealed. Especially in the case of functional genomics, data analysis is a crucial step and one that is undergoing intense research. In early functional genomics experiments, analysis was limited to finding patterns in the data by application of hierarchical clustering algorithms,[35] and self-organizing maps[36] (a neural network-based method of grouping variables in a data set). More sophisticated means of relating genes have since appeared that seem to perform better.[37-39] All these methods belong to a class named unsupervised learning, where the main objective is to identify patterns in the data. A more traditional approach in the physical sciences has been to use supervised learning methods, where one calibrates the method with a known data set and uses that calibration to identify classes or quantify properties in the new data set. The latter class includes such diverse methods as measuring masses in scales to identifying molecules by their ultra-violet spectra. Arguments have been made that classes for functional genomics assignments should be derived from supervised learning methods rather than from *ad hoc* ways[40] (currently common practice).

Functional genomics, much like 19th century biology or 17th century astrophysics, is in a phase of data collection and pattern discovery. Unsupervised learning methods are merely a way of generating hypotheses that have to be confirmed with subsequent experiments. Supervised learning methods allow one to measure precisely the properties of interest, given that they have already been identified. Independently, if these methods come from statistics or machine learning and artificial intelligence, they are means to constructing models of the system with little assumptions. At a more advanced stage, it becomes important to move on to models that incorporate more assumptions, as it is these that will allow us to construct plausible models (mechanisms) for the processes. We believe that this also will be possible with functional genomics data, because the necessary body of theory for biochemical processes is in place already. We favor theoretical frameworks based on dynamics of chemical networks, although other complementary approaches are also available (for example, based on statistical mechanics or irreversible thermodynamics). Indeed, software support is already available, such as the metabolic simulator Gepasi,[41, 42] which includes sophisticated means for parameter estimation from experimental data.[43] However, we anticipate problems that originate from the scale of the systems being analyzed in functional genomics.[44] These are not only at the level of computation time and memory requirements, but also with the extent of data coverage in terms of exploring the possible behaviors of the system. The latter can be a severe problem that can be summarized as data-rich but information-poor, and can only be solved with more extensive experimentation. The former requires software to be written explicitly to deal with this large scale, something we are currently undertaking in our laboratory and in collaboration with colleagues at the European Media Lab in Heidelberg. We are rewriting our Gepasi simulator to take advantage of parallel processing and at the same time adding alternative methods of simulation and more analytical procedures; the new software to be named COPASI (for COmplex PAthway SImulator).

Finally, we draw attention to the problem of visualization of these very large data sets. This is a problem that lies more in the domains of psychology and computer science, but one that could result in immediate benefits to biochemical research if appropriately solved. This is an area of active research and, to some extent, generic to all sciences. Functional genomics has already introduced novel means of visualizing data,[45, 35] but more are needed. We propose that it is at the visualization level that data from transcriptomics, proteomics, and metabolomics should be fused. We are developing software, in collaboration with Phenomenome Discoveries, Inc., which will combine metabolomics and transcriptomics data sets through the use of metabolic maps and, thus, facilitate comparing these separate measurements in the context of cellular metabolism. This should be much more powerful than purely correlational methods, as it allows one to quickly relate metabolite changes to protein and mRNA changes and, thus, allows one to pinpoint the molecular function of an orphan gene in the metabolic map. For this purpose, we use the diagrams from the database system KEGG,[46] and color the circles (metabolites) and rectangles (enzymes) in these diagrams with colors that reflect the relative levels of the metabolites and genes. The same system can also be used to display relative levels of proteins. Other types of display are being considered to allow one to explore further sections of secondary metabolism for which there are no known pathways. We are also investigating methods to infer metabolic pathways from the metabolomics data. Although this is still in an early stage, it would be of great benefit to biochemistry in general and invaluable to phytochemistry.

SUMMARY

We propose that functional genomics is best carried out with a systems biology approach where detailed molecular measurements are made at the level of the transcriptome, proteome, and metabolome, and which are synthesized with integrative mathematical models. We discuss the computational challenges inherent in this proposal, outlining the infrastructure needed to support such research. Efforts currently ongoing in our laboratory for data management, visualization, and analysis are briefly described. Research to use functional genomics data to infer novel secondary metabolic networks is under way.

ACKNOWLEDGEMENTS

We thank Dr. Jacky Snoep for collaborating in the time scale analysis of gene expression and metabolism, Douglas Kell for helpful discussions, and the Commonwealth of Virginia for financial support.

REFERENCES

1. GOFFEAU, A., BARRELL, B.G., BUSSEY, H., DAVIS, R.W., DUJON, B., FELDMANN, H., GALIBERT, F., HOHEISEL, J.D., JACQ, C., JOHNSTON, M., LOUIS, E.J., MEWES, H.W., MURAKAMI, Y., PHILIPPSEN, P., TETTELIN, H., OLIVER, S.G., Life with 6000 genes, *Science*, 1996, **274**, 546-567.
2. OLIVER, S.G., From DNA sequence to biological function, *Nature*, 1996, **379**, 597-600.
3. CASARI, G., DE DARUVAR, A., SANDER, C., SCHNEIDER, R., Bioinformatics and the discovery of gene function, *Trends Genet.*, 1996, **12**, 244-245.
4. RILEY, M., Functions of the gene products of *Escherichia coli*, *Microbiol. Rev.*, 1993, **57**, 862-952.
5. SOMERVILLE, C., SOMERVILLE, S., Plant functional genomics, *Science*, 1999, **285**, 380-383.
6. SMITH, T.F., Functional genomics - bioinformatics is ready for the challenge, *Trends Genet.*, 1998, **14**, 291-293.
7. VON BERTALANFFY, L., Basic concepts in quantitative biology of metabolism. In: Quantitative Biology of Metabolism - First International Symposium (O. Kinneand A. Locker, eds.), Helgoland. 1964, (vol 9), pp. 5-37.
8. ROSEN, R., Dynamical System Theory in Biology, Wiley-Interscience, New York. 1970.
9. KITANO, H., Perspectives on systems biology, *New Gener. Comput.*, 2000, **18**, 199-216.
10. KACSER, H., BURNS, J.A., The control of flux, *Symp. Soc. Exp. Biol.*, 1973, **27**, 65-104.
11. SAVAGEAU, M.A., Biochemical Systems Analysis, Addison-Wesley, Reading, MA. 1976, 379 p.
12. HEINRICH, R., RAPOPORT, S.M., RAPOPORT, T.A., Metabolic regulation and mathematical models, *Progr. Biophys. Mol. Biol.*, 1977, **32**, 1-82.
13. REICH, J.G., SEL'KOV, E.E., Energy Metabolism of the Cell. A Theoretical Treatise, Academic Press, London. 1981, 345 p.
14. HAYASHI, K., SAKAMOTO, N., Dynamic Analysis of Enzyme Systems. An Introduction, Springer-Verlag, Berlin. 1986.
15. KOSHLAND, D.E., Switches, thresholds and ultrasensitivity, *Trends Biochem. Sci.*, 1987, **12**, 225-229.
16. SCHENA, M., SHALON, D., DAVIS, R.W., BROWN, P.O., Quantitative monitoring of gene expression patterns with a complementary DNA microarray, *Science*, 1995, **270**, 467-470.
17. LOCKHART, D.J., DONG, H.L., BYRNE, M.C., FOLLIETTE, M.T., GALLO, M.V., CHEE, M.S., MITTMANN, M., WANG, C.W., KOBAYASHI, M., HORTON, H., BROWN, E.L., Expression monitoring by hybridization to high-density oligonucleotide arrays, *Nature Biotechnol.*, 1996, **14**, 1675-1680.

18. WILKINS, M.R., PASQUALI, C., APPEL, R.D., OU, K., GOLAZ, O., SANCHEZ, J.C., YAN, J.X., GOOLEY, A.A., HUGHES, G., HUMPHERY-SMITH, I., WILLIAMS, K.L., HOCHSTRASSER, D.F., From proteins to proteomes: large scale protein identification by two-dimensional electrophoresis and amino acid analysis, *Bio/technology*, 1996, **14**, 61-65.
19. O'FARREL, P.H., High resolution two-dimensional electrophoresis of proteins, *J. Biol. Chem.*, 1975, **250**, 4007-4021.
20. HUMPHERY-SMITH, I., CORDWELL, S.J., BLACKSTOCK, W.P., Proteome research: complementarity and limitations with respect to the RNA and DNA worlds, *Electrophoresis*, 1997, **18**, 1217-1242.
21. AEBERSOLD, R., DUCRET, A., FIGEYS, D., GU, M., ZHANG, Y.N., WATTS, J., CE-ESI-MS/MS: A microanalytical technique for probing physiological function, *Abstracts of Papers of the American Chemical Society*, 1997, **213**, 201.
22. HATZIMANIKATIS, V., CHOE, L.H., LEE, K.H., Proteomics: theoretical and experimental considerations, *Biotechnol. Prog.*, 1999, **15**, 312-318.
23. HAYNES, P.A., GYGI, S.P., FIGEYS, D., AEBERSOLD, R., Proteome analysis: biological assay or data archive?, *Electrophoresis*, 1998, **19**, 1862-1871.
24. GYGI, S.P., ROCHON, Y., FRANZA, B.R., AEBERSOLD, R., Correlation between protein and mRNA abundance in yeast, *Mol. Cell. Biol.*, 1999, **19**, 1720-1730.
25. IDEKER, T., THORSSON, V., RANISH, J.A., CHRISTMAS, R., BUHLER, J., ENG, J.K., BUMGARNER, R., GOODLETT, D.R., AEBERSOLD, R., HOOD, L., Integrated genomic and proteomic analyses of a systematically perturbed metabolic network, *Science*, 2001, **292**, 929-934.
26. FUTCHER, B., LATTER, G.I., MONARDO, P., MCLAUGHLIN, C.S., GARRELS, J.I., A sampling of the yeast proteome, *Mol. Cell. Biol.*, 1999, **19**, 7357-7368.
27. HEINRICH, R., RAPOPORT, T.A., A linear steady-state treatment of enzymatic chains. General properties, control and effector strength, *Eur. J. Biochem.*, 1974, **42**, 89-95.
28. FELL, D.A., Understanding the Control of Metabolism, Portland Press, London. 1996.
29. HOFMEYR, J.H., WESTERHOFF, H.V., Building the cellular puzzle: control in multi-level reaction networks, *J. Theoret. Biol.*, 2001, **208**, 261-285.
30. ANONYMOUS, Metabolomics - Array of hope. In: The Economist, 2001.
31. OLIVER, S.G., WINSON, M.K., KELL, D.B., BAGANZ, F., Systematic functional analysis of the yeast genome, *Trends Biotechnol.*, 1998, **16**, 373-378.
32. RAAMSDONK, L.M., TEUSINK, B., BROADHURST, D., ZHANG, N., HAYES, A., WALSH, M.C., BERDEN, J.A., BRINDLE, K.M., KELL, D.B., ROWLAND, J.J., WESTERHOFF, H.V., VAN DAM, K., OLIVER, S.G., A functional genomics strategy that uses metabolome data to reveal the phenotype of silent mutations, *Nature Biotechnol.*, 2001, **19**, 45-50.
33. CODD, E.F., A relational model of data for large shared data banks, *Commun. ACM*, 1970, **13**, 377-387.
34. DAVIDSON, S.B., OVERTON, C., BUNEMAN, P., Challenges in integrating biological data sources, *J. Comp. Biol.*, 1995, **2**, 557-572.
35. EISEN, M.B., SPELLMAN, P.T., BROWN, P.O., BOTSTEIN, D., Cluster analysis and display of genome-wide expression patterns, *Proc. Natl. Acad. Sci. USA*, 1998, **95**, 14863-14868.

36. TAMAYO, P., SLONIM, D., MESIROV, J., ZHU, Q., KITAREEWAN, S., DMITROVSKY, E., LANDER, E.S., GOLUB, T.R., Interpreting patterns of gene expression with self-organizing maps: methods and application to hematopoietic differentiation, *Proc. Natl. Acad. Sci. USA*, 1999, **96**, 2907-2912.
37. HILSENBECK, S.G., FRIEDRICHS, W.E., SCHIFF, R., O'CONNELL, P., HANSEN, R.K., OSBORNE, C.K., FUQUA, S.A., Statistical analysis of array expression data as applied to the problem of tamoxifen resistance, *J. Natl. Cancer Inst.*, 1999, **91**, 453-459.
38. ALTER, O., BROWN, P.O., BOTSTEIN, D., Singular value decomposition for genome-wide expression data processing and modeling, *Proc. Natl. Acad. Sci. USA*, 2000, **97**, 10101-10106.
39. HOLTER, N.S., MITRA, M., MARITAN, A., CIEPLAK, M., BANAVAR, J.R., FEDOROFF, N.V., Fundamental patterns underlying gene expression profiles: simplicity from complexity, *Proc. Natl. Acad. Sci. USA*, 2000, **97**, 8409-8414.
40. KELL, D.B., KING, R.D., On the optimization of classes for the assignment of unidentified reading frames in functional genomics programmes: the need for machine learning, *Trends Biotechnol.*, 2000, **18**, 93-98.
41. MENDES, P., GEPASI: a software package for modelling the dynamics, steady states and control of biochemical and other systems, *Comput. Appl. Biosci.*, 1993, **9**, 563-571.
42. MENDES, P., Biochemistry by numbers: simulation of biochemical pathways with Gepasi 3, *Trends Biochem. Sci.*, 1997, **22**, 361-363.
43. MENDES, P., KELL, D.B., Non-linear optimization of biochemical pathways: applications to metabolic engineering and parameter estimation, *Bioinformatics*, 1998, **14**, 869-883.
44. MENDES, P., Modeling large scale biological systems from functional genomic data: parameter estimation. In: Foundations of Systems Biology (H. Kitano, ed.) MIT Press, Cambridge, MA. 2001, pp. 165-186.
45. WEINSTEIN, J.N., MYERS, T.G., O'CONNOR, P.M., FRIEND, S.H., FORNACE, A.J., JR., KOHN, K.W., FOJO, T., BATES, S.E., RUBINSTEIN, L.V., ANDERSON, N.L., BUOLAMWINI, J.K., VAN OSDOL, W.W., MONKS, A.P., SCUDIERO, D.A., SAUSVILLE, E.A., ZAHAREVITZ, D.W., BUNOW, B., VISWANADHAN, V.N., JOHNSON, G.S., WITTES, R.E., PAULL, K.D., An information-intensive approach to the molecular pharmacology of cancer, *Science*, 1997, **275**, 343-349.
46. OGATA, H., GOTO, S., FUJIBUCHI, W., KANEHISA, M., Computation with the KEGG pathway database, *BioSystems*, 1998, **47**, 119-128.

Chapter Two

A GENOMICS APPROACH TO PLANT ONE-CARBON METABOLISM

Sanja Roje and Andrew D. Hanson*

*Horticultural Sciences Department,
University of Florida,
Gainesville, Florida 32611-0690*

Author for correspondence, e-mail: adha@mail.ifas.ufl.edu

Introduction	16
Overview of C_1 Metabolism	16
Pre- and Post-Genomics Maps of C_1 Metabolism	18
Characterization of MTHFR, a Known C_1 Enzyme	19
Genomics-Driven Discovery of New C_1 Enzymes	22
Metabolism of 5-Formyl-THF	22
Enzymes of Formate Metabolism	22
10-Formyl-THF Deformylase	23
S-Formylglutathione Hydrolase	23
Formamidase	23
Sarcosine Oxidase	24
Towards Understanding of the SMM Cycle	24
Summary	26

INTRODUCTION

The network of one-carbon (C_1) reactions is crucial to primary and secondary metabolism in plants because it supplies the C_1 units needed to synthesize proteins, nucleic acids, pantothenate, and a great variety of methylated molecules.[1,2] However, this network is hard to study with the tools of classical biochemistry and genetics; its enzymes may be of low abundance and/or exist as several isoforms, mutants are on the whole lacking, and its key intermediates - C_1 substituted folates, S-adenosylmethionine (AdoMet) and S-adenosylhomocysteine (AdoHcy) - are labile and hard to quantify.[2,3] Fortunately, the availability of genome data, EST collections, and reverse genetic technologies now make it easier to characterize known C_1 enzymes, to discover new ones, and to develop and test hypotheses about compartmentation and fluxes. Concatenating these new resources has started to provide a clearer, and excitingly unexpected, picture of plant C_1 metabolism. This chapter briefly summarizes our current understanding of plant C_1 metabolism, emphasizing recent advances driven by new technologies of the genomics era.

OVERVIEW OF C_1 METABOLISM

C_1 units required for various reactions of primary and secondary metabolism are carried by the coenzyme tetrahydrofolate (THF), enzymatically attached to its N_5 or N_{10} positions, or bridged between the two (Fig. 2.1A). These C_1 units range in oxidation level from formyl (most oxidized) through methenyl and methylene to methyl (most reduced). Specific C_1 derivatives of THF are generated by the metabolism of formate, glycine, serine, and other molecules, and can be enzymatically interconverted (Fig. 2.1B). These C_1 derivatives are used in anabolic reactions generating purines, formylmethionyl-tRNA in organelles, thymidylate, pantothenate, and methionine (Fig. 2.1B). The largest anabolic flux is the use of 5-methyl-THF to convert homocysteine to methionine. Methionine can then be incorporated into proteins or converted to AdoMet, the donor for numerous methylation reactions. A unique auxiliary feature of plant C_1 metabolism is the apparently futile S-methylmethionine (SMM) cycle, in which AdoMet is used to methylate methionine to SMM, which serves to methylate homocysteine, thereby regenerating methionine.[4]

C_1 pathways are particularly active in tissues that produce methylated compounds such as lignin, alkaloids, and betaines because the C_1 demands for these physiologically and economically important secondary metabolites can dwarf those of primary metabolism. As a clue to the levels at which C_1 pathways might be operating in tissues with active secondary metabolism, consider lignin synthesis, which has a huge demand for C_1 units to make methoxy groups.[5-7] Analysis of EST abundance data[6] shows that in differentiating pine xylem, almost 8% of the mRNAs

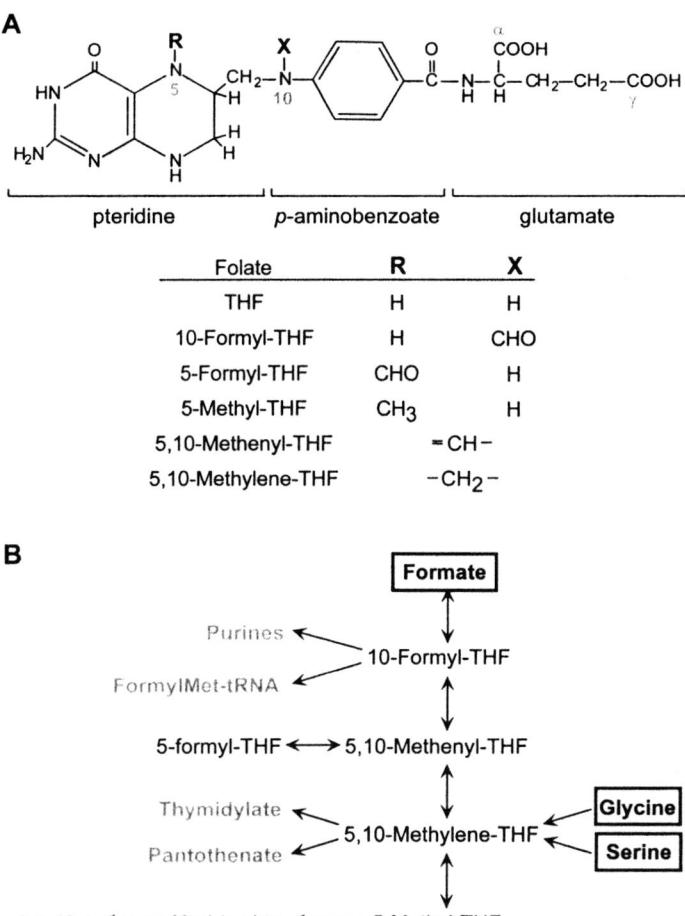

Fig. 2.1: Tetrahydrofolate and its role in C_1 interconversions and transfers. **A.** Chemical structures of THF and its C_1-substituted derivatives. **B.** Major C_1 unit interconversions and transfers involving THF derivatives. Sources of C_1 units are shown in boxes and their metabolic fates in gray.

putatively encode enzymes of C_1 metabolism, while just 3.5% code for the well-known enzymes of glycolysis plus the TCA cycle (Fig. 2.2). C_1 transfers are also absolutely central to the massive photorespiratory fluxes that occur in C_3 plants.[8,9]

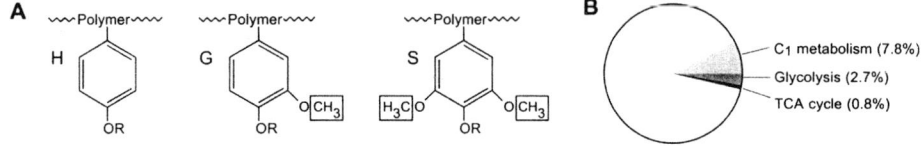

Fig. 2.2: **A.** Chemical structure of the three types of lignin monomer units: H, *p*-hydroxyphenyl-; G, guaiacyl; and S, syringyl. Note the methyl groups (boxed) in the methoxy moieties of the G and S monomers. **B.** Pie chart showing the proportions of pine xylem ESTs that putatively encode enzymes of C_1 metabolism, glycolysis, and the TCA cycle.

PRE- AND POST-GENOMICS MAPS OF C_1 METABOLISM

Before the advent of public EST and whole-genome sequencing projects, the metabolic map of plant C_1 metabolism had many gaps: various C_1 enzymes were known to exist but had not been cloned or characterized; certain C_1 enzymes known from other organisms had never been investigated; and the subcellular compartmentation of a number of enzymes had not been addressed. The existing gaps in the map were, by default, commonly filled in based on presumed parallels with the C_1 metabolism of mammals or fungi.[1,2] However, distinctive aspects of the plant C_1 map were to be expected, because plant C_1 metabolism clearly has unique features. First, plants have photorespiration. In essence, this is the recycling of two molecules of phosphoglycolate, produced by the oxygenase activity of Rubisco, to give one molecule of phosphoglycerate plus CO_2.[8,9] The photorespiratory pathway has two folate-dependent reactions, both mitochondrial, catalyzed by serine hydroxymethyltransferase (SHMT) and the glycine decarboxylase complex (GDC). The flux through the SHMT and GDC reactions is enormous – the CO_2 released is typically ≈25% of the amount fixed by Rubisco.[10] Other specific features of plant C_1 metabolism include its huge demand for C_1 units (as methyl groups) to support the synthesis of methyl-rich secondary products such as lignins, betaines, and alkaloids,[3,11] the SMM cycle,[4] and the ability of plants to generate formate from source(s) other than the folate C_1 pool.[11]

The availability of data from the public sequencing projects has enabled the use of nucleotide sequence information from genomes and ESTs to complement biochemical approaches, and to assemble a new, clearer working map of plant C_1

metabolism (Fig. 2.3). It is evident that before genomics only about half of the enzymes of the plant C_1 network were cloned; these correspond to the gray arrows in Fig. 2.3A. Because most enzymes of C_1 metabolism are highly conserved, homology with bacterial, yeast, or animal sequences helped identify DNA sequences for a number of biochemically known plant enzymes that had not been cloned (the black arrows in Fig. 2.3A). Homology searches also suggested the presence of several enzymes for which there was no biochemical evidence (the empty arrows in Fig. 2.3A); some of them were entirely unsuspected. The putative subcellular localization of the C_1 reactions inferred from a combination of biochemical and genomics evidence is laid out in Fig. 2.3B. Note that some C_1 reactions appear to occur in parallel in cytosol, mitochondria, and plastids, whereas others appear to be confined to a single compartment, which implies transport of metabolites between the cytosol and the organelles. Specific examples of recent advances in C_1 metabolism that were either facilitated or entirely made possible by the availability of genomics data are described in the sections that follow.

CHARACTERIZATION OF MTHFR, A KNOWN C_1 ENZYME

Methylenetetrahydrofolate reductase (MTHFR) catalyzes the NAD(P)H-dependent reduction of 5,10-methylenetetrahydrofolate (CH_2-THF) to 5-methyltetrahydrofolate (CH_3-THF). CH_3-THF then serves as a methyl donor for the synthesis of methionine. The MTHFR proteins and genes from mammalian liver and E. coli have been characterized,[12-15] and MTHFR genes have been identified in S. cerevisiae[16] and other organisms. The MTHFR of E. coli (MetF) is a homotetramer of 33-kDa subunits that prefers NADH as reductant,[12] whereas mammalian MTHFRs are homodimers of 77-kDa subunits that prefer NADPH and are allosterically inhibited by AdoMet.[13,14] Mammalian MTHFRs have a two-domain structure: the amino-terminal domain shows 30% sequence identity to E. coli MetF, and is catalytic; the carboxyterminal domain has been implicated in AdoMet-mediated inhibition of enzyme activity.[13,14]

The MTHFRs of Arabidopsis and maize have recently been cloned by genomics-based approaches, based on homology with the enzymes from other organisms.[17] Like mammalian MTHFRs, the plant enzymes were found to be homodimers of two-domain subunits that are homologous to the mammalian enzymes throughout both domains. However, when the recombinant plant proteins were expressed in yeast, they were found to differ radically from the mammalian MTHFRs in both their pyridine nucleotide preference and their regulatory properties: plant enzymes prefer NADH to NADPH, and they are insensitive to AdoMet.[17]

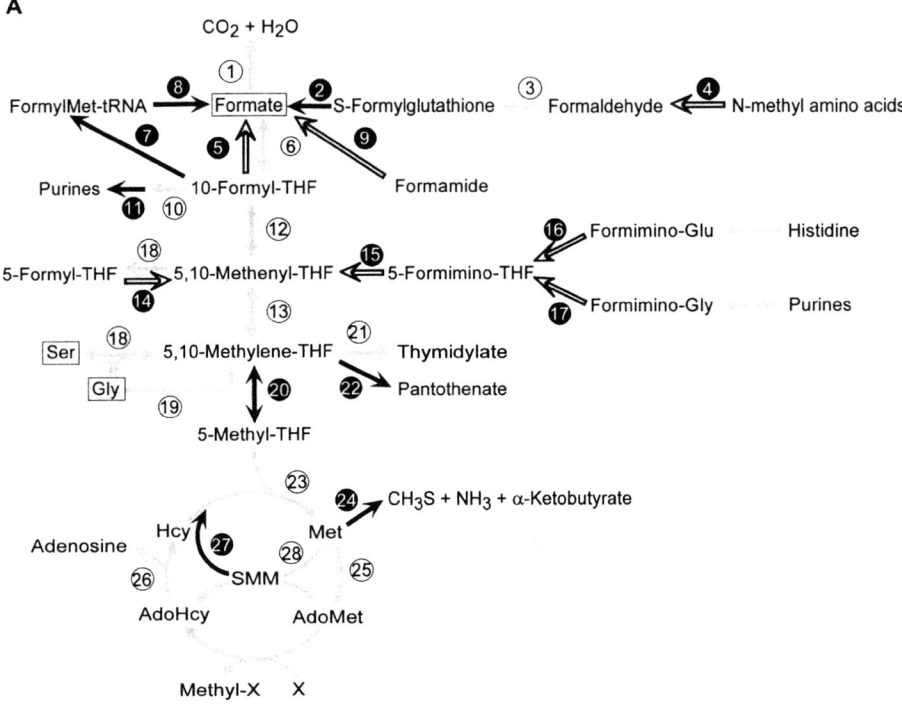
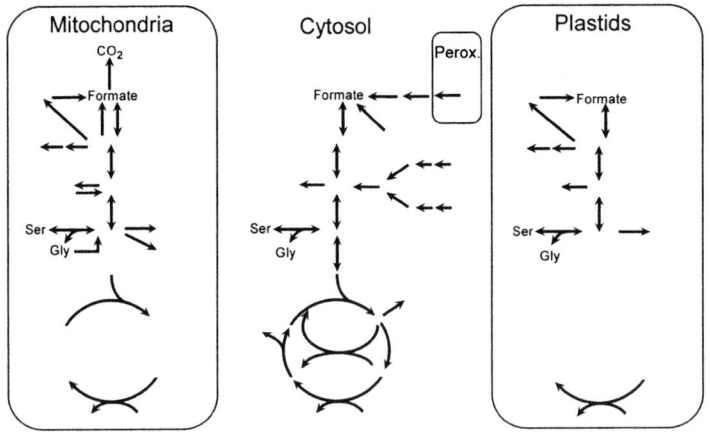

Fig. 2.3: The plant C_1 metabolism network. **A.** Current metabolic map of the C_1 metabolism, as deduced form biochemical and genomic data. Main sources of C_1 units are boxed. Grey arrows, enzymes cloned and characterized prior to the genomics era; black arrows, enzymes that were anticipated based on biochemical evidence, but for which genes had not previously been identified; empty arrows, genes that apparently encode enzymes – some of them unexpected – that have not been investigated in plants for which there is genomic evidence. Met, methionine; Hcy, homocysteine. Enzymes: 1, formate dehydrogenase; 2, S-formylglutathione hydrolase; 3, formaldehyde dehydrogenase; 4, sarcosine oxidase; 5, 10-formyl-THF deformylase; 6, 10-formyl-THF synthetase; 7, Met-tRNA transformylase; 8, polypeptide deformylase; 9, formamidase; 10, GAR transformylase; 11, AICAR transformylase; 12 & 13, bifunctional 5,10-methenyl-THF cyclohydrolase/5,10-methylene-THF dehydrogenase; 14, 5-formyl-THF cycloligase; 15, 5-formimino-THF cyclodeaminase; 16, glutamate formiminotransferase; 17, glycine formiminotransferase; 18, SHMT; 19, GDC; 20, 5,10-methylene-THF reductase; 21, thymidylate synthase/dihydrofolate reductase; 22, ketopantanoate hydroxymethyltransferase; 23, Met synthase; 24, Met γ-lyase; 25, AdoMet synthetase; 26, AdoHcy hydrolase; 27, Hcy methyltransferase; 28, Met methyltransferase. **B.** A simplified scheme of the inferred subcellular compartmentation of the C_1 reactions. The arrows indicate enzyme reactions as in Fig. 2.3A; the numbers are omitted for simplicity.

These differences in the properties of plant and mammalian MTHFRs have important regulatory implications. CH_2-THF, the substrate for the MTHFR reaction, is directly required for the synthesis of thymidylate, and can supply 10-formyltetrahydrofolate for the synthesis of purines.[1,2] It is, therefore, important that its partitioning among the different anabolic reactions be regulated. The NADPH-dependent MTHFR reaction in mammalian liver is physiologically irreversible due to the high cytosolic NADPH/NADP ratio and the large standard free energy change for the reduction of CH_2-THF.[18,19] This reaction, therefore, has the potential to deplete the cytosolic pool of CH_2-THF.[1,20] AdoMet sensitivity of the mammalian enzyme is considered to be a key regulatory feature that prevents such depletion.[1,20] The MTHFR reaction in plant cytosol is NADH-dependent, and is consequently likely to be reversible due to the high cytosolic NAD/NADH ratio, obviating the need for regulation by AdoMet.[17] MTHFRs provide a cautionary example of how enzymes sharing a high degree of homology can differ in critical regulatory properties that cannot be predicted from their primary structure.

GENOMICS-DRIVEN DISCOVERY OF NEW C_1 ENZYMES

In this section, we summarize information about five C_1 enzymes involved in metabolism of 5-formyl-THF (5-CHO-THF) and formate, whose existence was predicted from genomics data. 5-Formyltetrahydrofolate cycloligase (5-FCL) and S-formylglutathione hydrolase have since been shown to catalyze the anticipated reactions; the other three enzymes (10-formyl-THF deformylase, formamidase, and sarcosine oxidase) are still putative.

Metabolism of 5-Formyl-THF

5-CHO-THF is formed by a side-reaction of serine hydroxymethyltransferase (SHMT) in the presence of glycine.[21] This C_1 metabolite has been found in all organisms investigated so far, including plants.[22] 5-CHO-THF is not a C_1 donor, but it strongly inhibits SHMT and other C_1 enzymes, and may have a regulatory role in animals.[23] The only enzyme known to destroy 5-CHO-THF is 5-CHO-THF cycloligase (5-FCL), which catalyzes its irreversible conversion to 5,10-methenyl-THF, returning it to the metabolically active C_1 pool.[21,22] The presence of 5-FCL in plants was, therefore, to be anticipated, but biochemical evidence for it was lacking.

The *Arabidopsis* genome contains a single-copy gene homologous to 5-FCLs from other organisms. EST database searches indicated that this gene was expressed in plant tissues at a modest level. The presence of a putative signal peptide suggested mitochondrial localization. Cloning and overexpression of the putative *Arabidopsis* 5-FCL in *E. coli* has confirmed that it has a high level of 5-FCL activity, with kinetic properties intermediate between the bacterial and the mammalian enzymes (S. Roje, A.D. Hanson, and M.T. Janave, unpublished). It will be interesting to learn if this enzyme has acquired plant-specific roles. Does it, for example, allow steady operation of mitochondrial SHMT by removing 5-CHO-THF that is generated during the high photorespiratory flux of glycine?

Enzymes of Formate Metabolism

Formate metabolism is the most obscure sector of the plant C_1 network, and the one where the genome has yielded the most surprises.[3] Plants have formate pools that are important sources of C_1 units,[24-27] and plants also have formate dehydrogenase activity.[28-31] However, the origin of formate is not clear. It may come from non-enzymatic decarboxylation of glyoxylate in leaves in the light,[27,32] but this cannot be the source of formate in the dark or in non-green tissues. Data from the *Arabidopsis* genome sequencing project suggest four novel possibilities, discussed *seriatim* below.

10-Formyl-THF Deformylase

Two *Arabidopsis* genes appear to encode homologs of bacterial 10-formyl-THF deformylase (PurU). This enzyme irreversibly hydrolyzes 10-formyl-THF, releasing formate. In bacteria, PurU is a major source of formate and also has a regulatory function.[33,34] It is activated by methionine and inhibited by glycine, allowing cells to jettison THF-bound formate when methionine is abundant but glycine is not.[33,34] This activity has not been reported in plants or other eukaryotes. Since PurU provides formate in bacteria, and formate dehydrogenase is mitochondrial in plants,[28-30] it is notable that both PurU homologs from *Arabidopsis* appear to have mitochondrial transit peptides. If plants indeed have mitochondrial deformylases, do these enzymes drive a one-way flux out of C_1-folate pools into formate, which is oxidized to CO_2?

S-Formylglutathione Hydrolase

S-Formylglutathione hydrolase, which catalyzes the irreversible hydrolysis of *S*-formylglutathione into glutathione and formate, has recently been cloned from *Arabidopsis* by a genomics approach (S. Roje and A.D. Hanson, unpublished). Like other organisms, plants have NAD-linked formaldehyde dehydrogenase,[35] which acts on *S*-hydroxymethylglutathione and yields the thioester *S*-formylglutathione. The hydrolase and dehydrogenase are hypothesized to be essential to formaldehyde detoxification in all organisms.[36] Detoxifying formaldehyde may be especially significant in plants due to the high photorespiratory flux through 5,10-methylene-THF, because the latter is in chemical equilibrium with - and can release - formaldehyde.[37] A combined action of these two enzymes could, thus, be a significant source of formate.

Formamidase

Two *Arabidopsis* genes specify homologs of formamidase, which cleaves formamide to formate and NH_3. It is not clear how plants make formamide, but possibilities include cyanide hydratase activity, or a bacterial-type histidine degradation enzyme (HutG), which releases formamide.[38] These are the most enigmatic of the putative C_1 enzymes. Both proteins are uncannily good matches to formamidase from a methylotrophic bacterium and to related microbial enzymes but do not match other GenBank entries. They are, thus, excellent candidates for formamidases, perhaps with acetamidase activity.[39]

Sarcosine Oxidase

Formaldehyde might be derived from spontaneous dissociation of CH_2-THF[37] or from oxidation of methanol released by pectin demethylation.[40] However, the *Arabidopsis* genome contains a putative sarcosine oxidase gene, suggesting that formaldehyde could also come from oxidative demethylation of sarcosine (*N*-methylglycine). Sarcosine occurs in plants[41,42] and – intriguingly – feeding it to leaves induces formate dehydrogenase.[28] Radiotracer data show that some plants can oxidize methyl groups,[43] and in animals this occurs via a pathway in which sarcosine is an intermediate.[44] Moreover, sarcosine oxidases or sarcosine oxidase-like proteins from other organisms act on various *N*-methyl amino acids and imino acids such as proline and pipecolate,[45] all of which occur in plants.[46] The protein specified by the *Arabidopsis* sarcosine oxidase gene appears to be peroxisomal, as in mammals.[45,47]

TOWARDS UNDERSTANDING THE SMM CYCLE

SMM synthesis is mediated by the enzyme methionine *S*-methyltransferase (MMT) through the essentially irreversible, AdoMet-mediated methylation of methionine.[48-50] Both MMT and SMM are unique to plants.[48,50] The opposite reaction, in which SMM is used to methylate homocysteine to yield two molecules of methionine, is catalyzed by the enzyme homocysteine *S*-methyltransferase (HMT).[48] Unlike MMT, HMTs also occur in bacteria, yeast, and mammals, enabling them to catabolize SMM of plant origin, and providing an alternative to the methionine synthase reaction as a means to methylate homocysteine. Plant MMT and HMT reactions, together with those catalyzed by AdoMet synthetase and AdoHcy hydrolase, constitute the SMM cycle (Fig. 2.4).[4]

The function of the SMM cycle is not yet fully understood. Its established role, involving separation of the MMT and HMT reactions in time and space, is the long-distance transport of methyl groups and reduced sulfur. SMM is a major constituent of the phloem sap in wheat and most probably many other species; after synthesis in the leaves, it moves in the phloem to sink organs, where it is reconverted to methionine.[51] A hypothetical role that requires the complete SMM cycle to turn in individual plant organs is the maintenance of the pool of free methionine in the event of an overshoot in conversion of methionine to AdoMet.[4] Other hypothetical roles include avoidance of homocysteine toxicity, and ensuring that AdoMet, which is chirally unstable in physiological conditions,[52] does not accumulate to high levels and thereby increase the rate of racemization to the inactive *R,S* diastereomer.

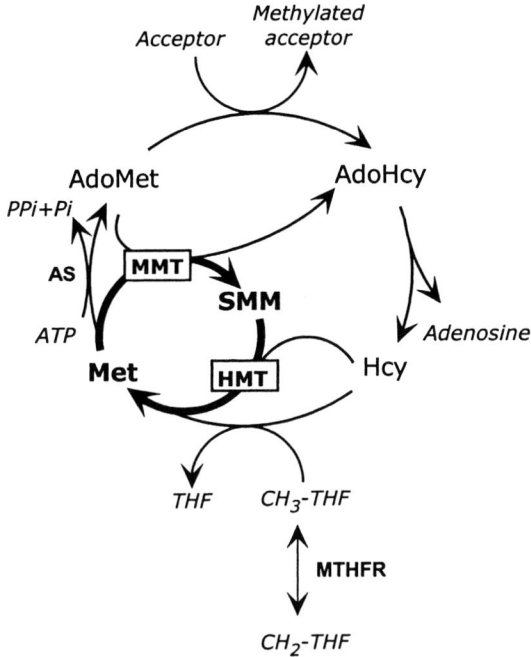

Fig. 2.4: The *S*-methylmethionine cycle and its interaction with the activated methyl cycle. The SMM cycle operates within the activated methyl cycle, and in effect short-circuits it. The reactions mediated by MMT and HMT are shown in bold. THF, tetrahydrofolate; CH_2-THF, 5,10-methylenetetrahydrofolate, CH_3-THF, 5-methyltetrahydrofolate, AS, AdoMet synthetase.

Both MMTs and HMTs have recently been cloned and characterized.[51,53,54] Homology searches of *Arabidopsis* and maize EST collections and the *Arabidopsis* genome identified three different HMT-coding sequences from *Arabidopsis* (AtHMT-1 to 3), and four from maize (ZmHMT-1 to 4).[53] Phylogenetic analysis of these sequences indicated that they comprise two subfamilies that pre-date the separation of monocots and dicots. One subfamily contains AtHMT-1 and ZmHMT-1; the other contains the rest.[53] The three HMTs from *Arabidopsis* were cloned using ESTs, and the recombinant enzymes were expressed in *E. coli*.[53,54] The analysis of their biochemical properties revealed that AtHMT-1 is strongly inhibited by methionine, whereas the other two HMTs (which are members of a phylogenetically separate subfamily) are methionine-insensitive. This finding raised the possibility that methionine sensitivity of HMTs is important for the control of flux through the HMT reaction and the SMM cycle, and that different HMTs have different functions. Analysis of the expression levels of HMTs in *Arabidopsis* organs revealed that they are expressed throughout the plant, with different members prevalent in different organs. Interestingly, seeds, which have the highest methionine content of all organs, predominantly express the methionine-insensitive AtHMT-2.[54]

MMT has also recently been cloned, and shown to be encoded by a single gene in *Arabidopsis* and maize.[51] As with HMTs, it was found to be expressed throughout the plant, supporting the hypothesis that a complete SMM cycle operates within all plant organs. Radiotracer evidence subsequently confirmed that the complete cycle indeed turns in all organs investigated.[54]

In order to estimate the flux through the SMM cycle and to explore its function, a computer model of methionine metabolism in mature *Arabidopsis* rosette leaves was developed based on data from radiotracer experiments and on metabolite contents. This model suggested that the cycle serves to stop accumulation of AdoMet, rather than to prevent depletion of free methionine, as proposed by Mudd and Datko.[54] Because plants lack the AdoMet feedbacks on MTHFR and AdoMet synthetase that regulate AdoMet pool size in other eucaryotes, the SMM cycle may be the main mechanism whereby plants achieve short-term control of AdoMet level. MMT knockouts of maize and *Arabidopsis* recently became available, and these can now be used to further investigate the role of the SMM cycle, and to test the predictions of the model.

SUMMARY

The network of C_1 reactions is essential to both primary and secondary metabolism in plants, but the area of C_1 metabolism in plants is underdeveloped because it is hard to study with the tools of classical biochemistry and genetics. The availability of data from public sequencing projects has opened up new perspectives for investigation of C_1 metabolism. The nucleotide sequence information from genomes and ESTs has made it possible to catalog the C_1 enzymes that appear to be present or absent in plants, to deduce their probable subcellular localization, and, thus, to assemble a new map of C_1 metabolism. Sequences of enzymes whose presence was known or anticipated from biochemical evidence became available (MTHFR, 5-FCL, HMTs); these enzymes were subsequently cloned by ordering cognate ESTs, or by PCR-based techniques. Good homologs of several C_1 enzymes that were not previously known to exist in plants were discovered during the cataloging (10-formyl-THF deformylase, formamidase, sarcosine oxidase). Should these prove to have the anticipated activity, their discovery could be the most significant contribution of genomics to C_1 metabolism, since the cloning and characterization of these enzymes may help solve the long-time mystery of the origin of formate in plants.

ACKNOWLEDGEMENTS

The authors' research summarized in this article was supported in part by the Florida Agricultural Experiment Station, by an endowment from the C.V. Griffin, Sr. Foundation, and by grants from NSF (to A.D.H.), and approved for publication as Journal Series no. N-02119.

REFERENCES

1. APPLING, D.R., Compartmentation of folate-mediated one-carbon metabolism in eukaryotes, *FASEB J.*, 1991, **5**, 2645-2651.
2. COSSINS, E.A., CHEN, L., Folates and one-carbon metabolism in plants and fungi, *Phytochemistry*, 1997, **45**, 437-452.
3. HANSON, A.D., GAGE D.A., SHACHAR-HILL, Y., Plant one-carbon metabolism and its engineering, *Trends Plant Sci.*, 2000, **5**, 206-213.
4. MUDD, S.H., DATKO, A.H., The S-methylmethionine cycle in *Lemna paucicostata*, *Plant Physiol.*, 1990, **93**, 623-630.
5. WHETTEN, R., SEDEROFF, R., Lignin biosynthesis, *Plant Cell*, 1995, **7**, 1001-1013.
6. ALLONA, I., QUINN, M., SHOOP, E., SWOPE, K., CYR, S.S., CARLIS, J., RIEDL, J., RETZEL, E., CAMPBELL, M.M., SEDEROFF, R., WHETTEN, R.W., Analysis of xylem formation in pine by cDNA sequencing, *Proc. Natl. Acad. Sci. USA.*, 1998, **95**, 9693-9698.
7. STERKY, F., REGAN, S., KARLSSON, J., HERTZBERG, M., ROHDE, A., HOLMBERG, A., AMINI, B., BHALERAO, R., LARSSON, M., VILLARROEL, R., VAN MONTAGU, M., SANDBERG, G., OLSSON, O., TEERI, T.T., BOERJAN, W., GUSTAFSSON, P., UHLEN, M., SUNDBERG, B., LUNDEBERG, J., Gene discovery in the wood-forming tissues of poplar: Analysis of 5,692 expressed sequence tags, *Proc. Natl. Acad. Sci. USA.*, 1998, **95**, 13330-13335.
8. DOUCE, R., NEUBURGER, M., Biochemical dissection of photorespiration, *Curr. Opin. Plant Biol.*, 1999, **2**, 214-222.
9. WINGLER, A., LEA, P.J., QUICK, W.P., LEEGOOD, R.C., Photorespiration - metabolic pathways and their role in stress protection, *Proc. Roy. Soc. Lond. Ser. B.*, 2000, **355**, 1517-1529.
10. OLIVER, D.J., The glycine decarboxylase complex from plant mitochondria, *Annu. Rev. Plant Physiol. Plant Mol. Biol.*, 1994, **45**, 323-337.
11. HANSON, A.D., ROJE, S., One-carbon metabolism in higher plants, *Annu. Rev. Plant Physiol. Plant Mol. Biol.*, 2001, **52**, 119-137.
12. SHEPPARD, C., TRIMMER, E., MATTHEWS, R.G., Purification and properties of NADH-dependent 5, 10-methylenetetrahydrofolate reductase (MetF) from *Escherichia coli*, *J. Bacteriol.*, 1999, **181**, 718-725.

13. MATTHEWS, R.G., SHEPPARD, C., GOULDING, C., Methylenetetrahydrofolate reductase and methionine synthase: biochemistry and molecular biology, *Eur. J. Pediatr.*, 1998, **157**, S54-S59.
14. GOYETTE, P., SUMNER, J. S., MILOS, R., DUNCAN, A.M.V., ROSENBLATT, D.S., MATTHEWS, R.G., ROZEN, R., Human methylenetetrahydrofolate reductase: isolation of cDNA, mapping and mutation identification, *Nat. Genet.*, 1994, **7**, 195-200.
15. GOYETTE, P., PAI, A., MILOS, R., FROSST, P., TRAN, P., CHEN, Z., CHAN, M.,.ROZEN, R., Gene structure of human and mouse methylenetetrahydrofolate reductase (MTHFR), *Mamm. Genome*, 1998, **9**, 652-656.
16. RAYMOND, R.K., KASTANOS, E.K., APPLING, D.R., *Saccharomyces cerevisiae* expresses two genes encoding isozymes of methylenetetrahydrofolate reductase, *Arch. Biochem. Biophys.*, 1999, **372**, 300-308.
17. ROJE, S., WANG, H., MCNEIL, S.D., RAYMOND, R.K., APPLING, D.R., SHACHAR-HILL, Y., BOHNERT, H.J., HANSON, A.D., Isolation, characterization, and functional expression of cDNAs encoding NADH-dependent methylenetetrahydrofolate reductase from higher plants, *J. Biol. Chem.*, 1999, **274**, 36089-36096.
18. VANONI, M.A., MATTHEWS, R.G., Kinetic isotope effects on the oxidation of reduced nicotinamide adenine dinucleotide phosphate by the flavoprotein methylenetetrahydrofolate reductase, *Biochemistry*, 1984, **23**, 5272-5279.
19. WOHLFARTH, G., DIEKERT, G., *Arch. Microbiol*, 1991, **155**, 378-381.
20. BAGLEY, P.J., SELHUB, J., A common mutation in the methylenetetrahydrofolate reductase gene is associated with an accumulation of formylated tetrahydrofolates in red blood cells, *Proc. Natl. Acad. Sci. USA.*, 1998, **95**, 13217-13220.
21. STOVER, P., SCHIRCH, V., The metabolic role of leucovorin, *Trends Biochem. Sci.*, 1993, **18**, 102-106.
22. COSSINS, E.A., The fascinating world of folate and one-carbon metabolism, *Can. J. Bot.*, 2000, **78**, 691-708.
23. GIRGIS, S., SUH, J.R., JOLIVET, J., STOVER, P.J., 5-Formyltetrahydrofolate regulates homocysteine remethylation in human neuroblastoma, *J. Biol.Chem.*, 1997, **272**, 4729-4734.
24. COSSINS, E.A., One-carbon metabolism. In: The Biochemistry of Plants, Vol. 2 (P. K. Stumpf and E. E. Conn, eds.), Academic Press, New York. 1980, pp. 365-418.
25. AMORY, A.M., CRESSWELL, C.F., Role of formate in the photorespiratory metabolism of *Themeda triandra* Forssk, *J. Plant Physiol.*, 1986, **124**, 247-255.
26. AMORY, A.M., FORD, L., PAMMENTER, N.W., CRESSWELL, C.F., The use of 3-amino-1,2,4-triazole to investigate the short-term effects of oxygen toxicity on carbon assimilation by *Pisum sativum* seedlings, *Plant Cell Env.*, 1992, **15**, 655-663.
27. WINGLER, A., LEA, P.J., LEEGOOD, R.C., Photorespiratory metabolism of glyoxylate and formate in glycine-accumulating mutants of barley and *Amaranthus edulis*, *Planta*, 1999, **207**, 518-526.
28. HOURTON-CABASSA, C., AMBARD-BRETTEVILLE, F., MOREAU, F., DAVY DE VIRVILLE, J., REMY, R., COLAS DES FRANCS-SMALL, C., Stress induction of mitochondrial formate dehydrogenase in potato leaves, *Plant Physiol.*, 1998, **116**, 627-635.

29. OLSON, B.J.S.C., SKAVDAHL, M., RAMBERG, H., OSTERMAN, J.C., MARKWELL, J., Formate dehydrogenase in *Arabidopsis thaliana*: Characterization and possible targeting to the chloroplast, *Plant Sci.*, 2000, **159**, 205-212.
30. LI, R., ZIOLA, B., KING, J., Purification and characterization of formate dehydrogenase from *Arabidopsis thaliana*, *J. Plant Physiol.*, 2000, **157**, 161-167.
31. SUZUKI, K., ITAI, R., SUZUKI, K., NAKANISHI, H., NISHIZAWA, N.K., YOSHIMURA, E., MORI, S., Formate dehydrogenase, an enzyme of anaerobic metabolism, is induced by iron deficiency in barley roots, *Plant Physiol.*, 1998, **116**, 725-732.
32. BRISSON, L.F., ZELITCH, I., HAVIR, E.A., Manipulation of catalase levels produces altered photosynthesis in transgenic tobacco plants, *Plant Physiol.*, 1998, **116**, 259-269.
33. NAGY, P.L., MAROLEWSKI, A., BENKOVIC, S.J., ZALKIN, H., Formyltetrahydrofolate hydrolase, a regulatory enzyme that functions to balance pools of tetrahydrofolate and one-carbon tetrahydrofolate adducts in *Escherichia coli*, *J. Bacteriol.*, 1995, **177**, 1292-1298.
34. NAGY, P.L., MCCORKLE, G.M., ZALKIN, H., purU, A source of formate for purT-dependent phosphoribosyl-*N*-formylglycinamide synthesis, *J. Bacteriol.*, 1993, **175**, 7066-7073.
35. MARTINEZ, M.C., ACHKOR, H., PERSSON, B., FERNANDEZ, M.R., SHAFQAT, J., FARRES, J., JORNVALL, H., PARES, X., *Arabidopsis* formaldehyde dehydrogenase. Molecular properties of plant class III alcohol dehydrogenase provide further insights into the origins, structure and function of plant class P and liver class I alcohol dehydrogenases, *Eur. J. Biochem.*, 1996, **241**, 849-857.
36. BARBER, R.D., DONOHUE, T.J., Function of a glutathione-dependent formaldehyde dehydrogenase in *Rhodobacter sphaeroides* formaldehyde oxidation and assimilation, *Biochemistry*, 1998, **37**, 530-537.
37. KALLEN, R.G., JENCKS, W.P., The mechanism of the condensation of formaldehyde with tetrahydrofolic acid, *J. Biol. Chem.*, 1966, **241**, 5851-5863.
38. MCFALL, E., NEWMAN, E.B., Amino acids as carbon sources. In: *Escherichia coli and Salmonella* - Cellular and Molecular Biology, Vol. 1. (F.C. Neidhardt et al., eds.), ASM Press, Washington DC. 1996, pp. 358-379.
39. WYBORN, N.R., SCHERR, D.J., JONES, C.W., Purification, properties and heterologous expression of formamidase from *Methylophilus methylotrophus*, *Microbiology*, 1994, **140**, 191-195.
40. FALL, R., BENSON, A.A., Leaf methanol - the simplest natural product from plants, *Trends Plant Sci.*, 1996, **9**, 296-301.
41. WAINWRIGHT, T., SLACK, P.T., LONG, D.E., *N*-Nitrosodimethylamine precursors in malt, *IARC Sci. Publ.*, 1982, **41**, 71-80.
42. HUNT, S. The non-protein amino acids. In: Chemistry and Biochemistry of the Amino Acids (G. C. Barrett, ed.), Chapman & Hall, London. 1985, pp. 55-138.
43. MAZELIS, M. Catabolism of sulfur-containing amino acids. In: Sulfur Nutrition and Assimilation in Higher Plants (L. De Kok, I. Stulen, H. Rennenberg, C. Brunold and W. Rauser, eds.), SPB Academic Publishing, The Hague. 1993, pp. 95-108.
44. BALAGHI, M., HORNE, D.W., WAGNER, C., Hepatic one-carbon metabolism in early folate-deficiency in rats, *Biochem. J.*, 1993, **291**, 145-149.

45. REUBER, B.E., KARL, C., REIMANN, S.A., MIHALIK, S.J., DODT, G. Cloning and functional expression of a mammalian gene for a peroxisomal sarcosine oxidase, *J. Biol. Chem.*, 1997, **272**, 6766-6776.
46. KOYAMA, Y., OHMORI, H. Nucleotide sequence of the *Escherichia coli solA* gene encoding a sarcosine oxidase-like protein and characterization of its product, *Gene*, 1996, **181**, 179-183.
47. CHIKAYAMA, M., OHSUMI, M., YOKOTA, S., Enzyme cytochemical localization of sarcosine oxidase activity in the liver and kidney of several mammals, *Histochem. Cell Biol.*, 2000, **113**, 489-495.
48. GIOVANELLI, J., MUDD, S.H., DATKO, A.H., Sulfur amino acids in plants. In: The Biochemistry of Plants, Vol 5 (Miflin, B.J., ed.), Academic Press, New York. 1980, pp. 453-505.
49. JAMES, F., NOLTE, K.D., HANSON A.D., Purification and properties of S-adenosyl-L-methionine:L-methionine S-methyltransferase from *Wollastonia biflora* leaves, *J. Biol. Chem.*, 1995, **270**, 22343-22350.
50. PAQUET, L., LAFONTAINE, P.J., SAINI, H.S., JAMES, F., HANSON, A.D. Evidence en faveur de la présence du 3-diméthylsulfoniopropionate chez une large gamme d'Angiospermes, *Can. J. Bot.*, 1995, **73**, 1889-1896.
51. BOURGIS, F., ROJE, S., NUCCIO, M.L., FISHER, D.B., TARCZYNSKI, M.C., LI, C., HERSCHBACH, C., RENNENBERG, H., PIMENTA, M.J., SHEN, T.L., GAGE, D.A., HANSON, A.D., S-Methylmethionine plays a major role in phloem sulfur transport and is synthesized by a novel type of methyltransferase, *Plant Cell*, 1999, **11**, 1465-1498.
52. HOFFMAN, J.L., 1986. Chromatographic analysis of the chiral and covalent instability of S-adenosyl-L-methionine, *Biochemistry*, 1986, **25**, 4444-4449.
53. RANOCHA, P., BOURGIS, F., ZIEMAK, M.J., RHODES, D., GAGE, D.A., HANSON, A.D., Characterization and functional expression of cDNAs encoding methionine-sensitive and -insensitive homocysteine S-methyltransferases from *Arabidopsis*, *J. Biol. Chem.*, 2000, **275**, 15962-15968.
54. RANOCHA, P., MCNEIL, S.D., ZIEMAK, M.J., LI, C., TARCZYNSKI, M.C., HANSON, A.D., The S-methylmethionine cycle in angiosperms: Ubiquity, antiquity and activity, *Plant J.*, 2001, **25**, 575-584.

Chapter Three

METABOLOMICS: A DEVELOPING AND INTEGRAL COMPONENT IN FUNCTIONAL GENOMIC STUDIES OF *MEDICAGO TRUNCATULA*

Lloyd W. Sumner,[1]* Anthony L. Duran,[1] David V. Huhman,[1] Joel T. Smith[2]

[1]*Plant Biology Division,*
The Samuel Roberts Noble Foundation,
Ardmore, OK 73401

[2]*Southeastern Oklahoma State University,*
Durant, OK 74701

*Author for correspondence, e-mail: lwsumner@noble.org

Introduction..32
    Medicago truncatula: A Model Legume.. 32
    Integrated Functional Genomics: Transcriptomics, Proteomics, and
        Metabolomics...32
Current Challenges of Metabolomics... 35
    Chemical Complexity... 35
    Analytical and Biological Variance... 35
    Dynamic Range... 36
Metabolic Profiling Approach: Sequential Extraction and Parallel Analyses 37
Analytical Tools for Metabolic Profiling.. 38
    Mass Spectrometry.. 38
    Gas Chromatography-Mass Spectrometry...39
    High Performance Liquid Chromatography-Mass Spectrometry..............41
    Tandem Mass Spectrometry (MS/MS).. 44
    High Resolution Accurate Mass Measurements................................. 48
    Capillary Electrophoresis... 50
Bioinformatics and Statistical Processing of Metabolome Data.......................52
    Principal Component Analysis (PCA)... 53
    Self Organizing Maps (SOMs)... 53
    Hierarchical Cluster Analysis (HCA)... 54
Summary... 55

INTRODUCTION

Medicago truncatula, a Model Legume

Legumes are important agricultural and commercial crops characterized by root nodules formed as a result of the symbiotic relationship with nitrogen-fixing *rhizobia*. *Medicago truncatula,* a close relative of alfalfa *(Medicago sativa)*, has been chosen as a model legume because of its prolific nature, small diploid genome ($\sim 5 \times 10^8$ bp), self-fertilization, ease of genetic transformation, and rapid generation time.[1-4] *Medicago truncatula* is the subject of a functional genomics project to understand better the biological processes associated with legumes and their interaction with the environment.[5] In this chapter, we provide an overview of functional genomics and focus on metabolomic approaches currently being pursued for *Medicago truncatula*.

Integrated Functional Genomics: Transcriptomics, Proteomics, and Metabolomics

The biological sciences have greatly advanced through the large scale physical mapping and sequencing of over twenty genomes including humans.[6] These efforts have yielded genomic "parts lists" or lists of genes that govern the biochemical processes of life. Although these "parts lists" are of great utility, they often do not yield information concerning gene relationships or function. A significant number of gene functions can be inferred through similarity matching to gene sequences of known function determined through traditional empirical methods. This approach has been utilized in most species that have known genomes; however, there still remain a large number of predicted open reading frames (ORFs) that have no assigned function.[7] As a result, many studies are now being directed toward the determination of the functional aspects of large numbers of genes or even total genomes. These studies are generally classified as functional genomics and serve as powerful discovery tools for understanding gene function.[8] The ability to monitor large numbers of genes and gene products simultaneously allows the study of how whole organisms or systems response to genetic or environmental stimuli. These global studies are increasingly being referred to as systems biology.[9]

Functional genomics seeks to determine gene function through the correlation of genes and gene products. (Fig.3.1) These reverse genetic approaches are pursued by altering gene expression through genetic perturbation, followed by monitoring of the expression products to infer function. Genetic perturbation can be achieved by random or directed gene activation tagging, gene transfer, transient gene silencing, or transient gene over expression. Once expression has been altered, mRNA, protein and/or small molecule metabolite levels are quantified through various profiling

METABOLOMICS

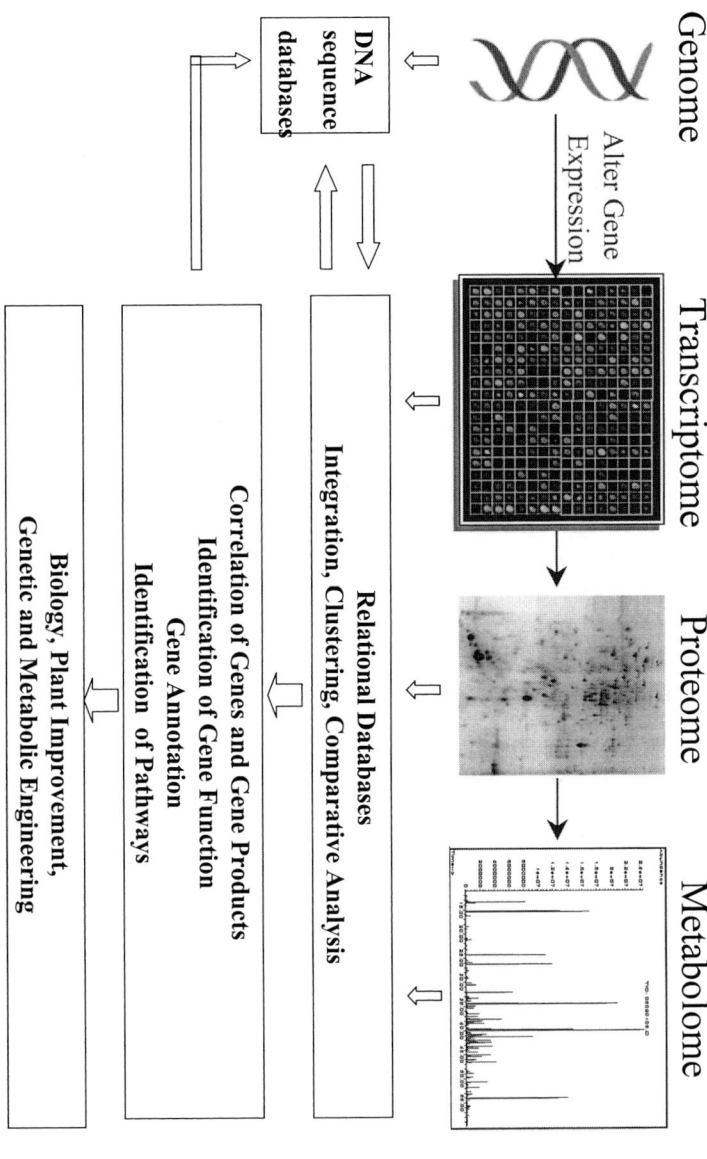

Figure 3.1: An integrated functional genomic approach monitors quantitative and qualitative differences in the transcriptome, proteome, and metabolome as a means to study gene function and cellular responses to external stimuli. The large amount of information contained within the profile data is deposited into relational databases where it can be correlated, compared, and interrogated by bioinformatic tools to yield a better understanding of biology. Profile data are also emerging as unique means of annotating genome data. Confirmation of putative proteins through proteome analysis is one example.

approaches. Simultaneous profiling of large numbers of gene expression products has been termed transcriptomics, proteomics,[10,11] and metabolomics.[12-14] The quantitative and qualitative gene expression profiling information can be stored in relational databases so that it can be further interrogated, correlated, clustered, and compared to understand gene function. This correlated information can be used to better understand biology through the assembly of pathways. Once gene function has been elucidated, this information can be used to improve plants through genetic and metabolic engineering. Global profiling is also useful for studying cellular and systems responses to environmental challenges. Environmental challenges might include light, temperature, radiation, pathogens, biotic or abiotic elicitors. We propose that maximum biological information is obtained when an integrated functional genomics approach is employed that collects data across multiple levels of gene expression; however, many projects monitor gene expression at a single level, *i.e.*, transcriptome, proteome, or metabolome, due to obvious limitations of time and/or resources.

The technological platforms being used to approach large scale profiling are rapidly evolving and maturing. Transcriptome profiling is being approached primarily through cDNA macro/micro arrays[15] and/or SAGE (serial analysis of gene expression).[16] Traditional protein profiling has been performed by using two-dimensional polyacrylamide gel electrophoresis (2-DE),[17-19] and this approach is greatly enhanced when coupled with modern mass spectrometry.[11,20-24] On the horizon, there are developing proteome technologies based on multidimensional chromatography coupled to mass spectrometry (LC/LC/MS/MS),[25,26] accurate mass tags for unique differentiation of large numbers of peptides,[27,28] and isotope-coded affinity tags.[29] The tools for profiling the metabolome are emerging and will be discussed in detail below.

There are advantages to profiling the metabolome as a means to understanding gene function. Foremost, metabolites represent the end products of gene expression and profiling them yields a more definitive view of function.[13,14,30] Quantitative and qualitative measurements of large numbers of cellular metabolites yield a broad view of the biochemical status or biochemical phenotype of an organism that can be directly linked to function.[12] For example, Roessner *et al.* have used metabolic profiling to correlate increased glucose and fructose phosphate levels to sucrose catabolism in transgenic potato lines.[31] Metabolic profiling also offers unique opportunities to study regulation and signaling under the control of small molecules (*i.e.*, metabolites). Quite often, signaling and regulation are transparent at the transcriptome and/or proteome level. Finally, metabolomics offers the unbiased ability to differentiate phenotypes and genotypes based on metabolite levels that may or may not produce visible phenotypes/genotypes.[31,32]

Profiling of the transcriptome and proteome has received some criticism in its ability to predict function. The transcriptome has a singular purpose, which is to transport a message for a protein that is to be translated; however, increases in the mRNA do not always correlate to increases in protein levels.[33] Once translated, a protein may or may not be in its enzymatic active form. Due to these factors, changes in transcriptome or proteome do not always correspond to direct alterations in activity or phenotype [see Mendes *et al*, this volume]. Another consideration when profiling the transcriptome and proteome is that most modern techniques identify the mRNA and protein through similarity or database matching, thus, identification is based primarily on the quality of the match and represents an indirect approach. If database information or function is not known, then transcript or protein profiling yields only limited information. Although the above arguments are true in many instances, there are countless instances in which transcriptome and proteome profiling have correctly identified function, and we, therefore, continue to support an integrated approach. An integrated approach provides multiple levels of confidence.

CURRENT CHALLENGES OF METABOLOMICS

Chemical Complexity

Profiling the metabolome is challenging. The genome and transcriptome basically consist of four nucleotides with similar chemical properties; therefore, a global profiling approach is reasonably achieved. The proteome is substantially more complex, but is still composed of a limited set of twenty-two primary amino acids. Although more complex, 2-DE can differentiate a large number of proteins in a single analysis, with several thousand being routine and 10,000 representing the upper boundary.[19] When one surveys the metabolome, the chemical complexity is significantly greater. The chemical properties of metabolites range from ionic inorganic species to hydrophilic carbohydrates and sophisticated secondary natural products to hydrophobic lipids. The chemical diversity and complexity of the metabolome make it extremely challenging to profile ALL of the metabolome simultaneously. Currently, we do not believe that a single analytical technique provides the ability to profile all of the metabolome. In this chapter, we outline many of the approaches to surmounting the obstacle of chemical complexity.

Analytical and Biological Variance

Analytical variance is defined as the coefficient of variance or relative standard deviation that is directly related to the experimental approach. This variance does differ in accordance with the technology platform being used and is indeterminate in origin. Biological variance is also indeterminate in origin and arises from quantitative variations in metabolite levels among plants of the same species grown under identical or as near as possible identical conditions. Biological variations typically exceed analytical variations. Recently, Roessner and coworkers reported that the biological variability exceeded the analytical variability of GC/MS by a factor of ten.[34] These large biological variations are major limitations on the "resolution" of the metabolomics approach. One way to reduce biological variance is to pool tissues. This tactic helps minimize random variations through statistical averaging; however, many variations in metabolite levels often have biological significance and result from functional differentiation of tissues. Pooling tissue can also result in undesirable dilution of site or tissue specific up/down-regulated metabolites. An alternative is to start with homogeneous tissue such as cell cultures, but this has obvious restrictions on the ability to study intact plants. The bottom line is that sampling is important and strategies need to be incorporated to minimize variations.

Dynamic Range

A technological challenge encountered in metabolomics is dynamic range. Dynamic range defines the concentration boundaries of an analytical determination over which the instrumental response is linear to the analyte concentration. The dynamic range of many techniques can be severely limited by the sample matrix or the presence of interfering and competing compounds. This is one of the most difficult issues to address in metabolomics. Most analytical mass spectrometric methods have dynamic ranges of 10^4 to 10^6 for individual components; however, this range is commonly and significantly reduced by the presence of other chemical components. In other words, the presence of some excessive metabolites can cause significant or severe chemical interferences that limit the range in which other metabolites may be successfully profiled. For example, large abundances of primary metabolites such as sugars often interfere with our ability to profile secondary metabolites such as flavonoids by GC/MS. The positive aspect of this dilemma is that many of the highly expressed metabolites are often unique in differing tissues or organisms, thus providing an exclusive basis for differentiation. These exclusive compounds are often referred to as biomarkers, and one could view metabolomics as an array of biomarkers.

Interfering or competing analytes can often lower performance and/or bias MS profiling techniques. For example, it is difficult to profile oligosaccharides in the presence of peptides or amino acids by LC/MS. The reason is that amino acids have higher proton affinities than oligosaccharides and, therefore, yield higher abundances of charged species necessary for mass measurement. Another problem in electrospray ionization mass spectrometry (ESI/MS) is salts. Levels of ionic

METABOLOMICS

species above >10^{-4} M are known to reduce the ionization efficiency in ESI/MS and significantly interfere with profiling all species.[35] Different analytical approaches have been developed to improve dynamic range and to minimize complications. These are discussed below.

METABOLIC PROFILING APPROACH: SEQUENTIAL EXTRACTION AND PARALLEL ANALYSES

Our approach to addressing the chemical complexity and dynamic range of the metabolome employs sequential extraction followed by parallel analyses. This approach is outlined (Fig.3.2), and is designed to segregate the metabolome into more manageable subclasses with similar chemical properties. The subclasses are subjected to parallel analytical profiling techniques to record metabolite profile information. Segregation of the subclasses helps minimize chemical interferences, while parallel analyses help visualize a greater portion of the metabolome. Our technological approaches to multidimensional parallel profiling center around mass spectrometry, due to its enhanced sensitivity and specificity. Specific methods such as GC/MS, LC/MS, and CE are matched with the target subclass to achieve the best performance.

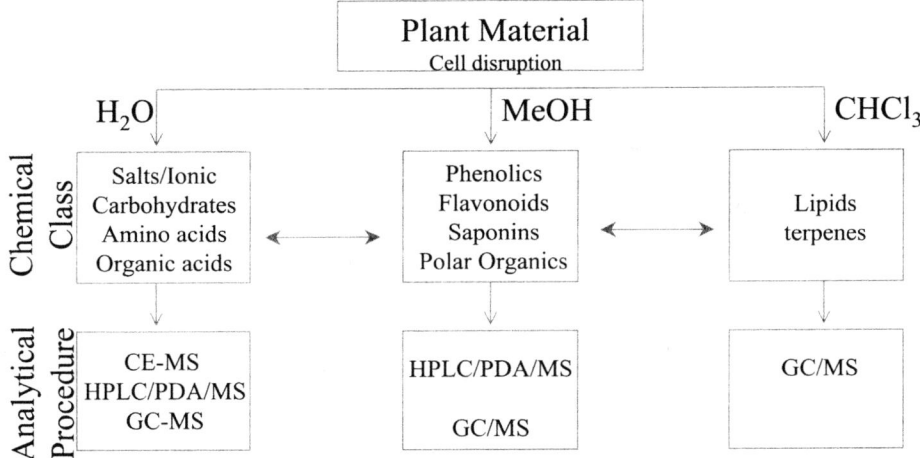

Figure 3.2: Our approach to surmounting the metabolome obstacles of chemical complexity and dynamic range employs sequential extraction followed by parallel analyses. Segregation of the metabolome into subclasses helps minimize chemical interferences, while parallel analyses help to visualize a greater portion of the metabolome.

Metabolic profiling of plant tissues is also complicated by changes in metabolite composition and concentration that can occur due to the presence of a large variety of enzymes. This necessitates rapid harvesting of plant materials to minimize degradation of materials. Tissues being profiled are harvested, immediately frozen in liquid nitrogen, lyophilized, and stored at $-80°C$ until extracted or processed. Dry tissue can then be sequentially extracted based on solvent properties such as dielectric constant. Combined solvents can also be used for extraction such that metabolites selectively partition between solvents. Extracts can be further fractionated by using solid phase extraction (SPE) if needed. Additional fractionation, however, moves away from the objective of global profiling, but may be necessary to get a more detailed view of the metabolome.

ANALYTICAL TOOLS FOR METABOLIC PROFILING

Mass Spectrometry (MS)

There are multiple tools and technologies being developed to assist in metabolic profiling. Many of the approaches contain mass selective detection that is incorporated due to its enhanced sensitivity and selectivity. Several good overviews covering the general principles of mass spectrometry have been presented;[36-38] however, a quick synopsis here may be beneficial.

Mass spectrometers discriminate or mass measure molecules based on a mass-to-charge ratio (m/z) of positive or negative gas phase ionic species. Ionization is an important process in MS because all compounds do not ionize equally. Ionization efficiencies vary with chemical structure and result in preferential or competitive ionization during the analyses of complex mixtures. Competitive ionization makes it difficult to detect one chemical compound in the presence of another having a greater ionization efficiency. Multidimensional analytical approaches that include separation prior to ionization reduce problems of competitive ionization. Examples will be provided below for both gas chromatography (GC) and high performance liquid chromatography (HPLC) coupled to MS. Even poor separations that yield coeluting components are beneficial as coeluting compounds are generally more similar in chemical nature and have less difference in ionization efficiencies. The ionization process also dictates the coupling capabilities of the mass spectrometer to separation platforms. For example, GC/MS is performed with electron or chemical ionization, whereas HPLC/MS is commonly performed with electrospray ionization.

Gas Chromatography – Mass Spectrometry

Gas chromatography coupled to mass spectrometry (GC/MS) is emerging as a powerful tool for profiling large numbers of primary metabolites,[12,34,39,40] and we are incorporating this approach into our program. The favorable attributes of GC/MS include high reproducibility (low analytical variance), standardized technique, and high separation efficiencies. High separation efficiencies allow for the separation of complex mixtures, and are achieved with long (30 to 60 m) capillary columns (internal diameters 75 to 320 m). GC/MS is commonly used in conjunction with electron ionization and requires the analyte to be volatile, thermally stable, and energetically stable. Many important biological analytes are polar and nonvolatile; therefore, they must be first chemically modified or derivatized prior to GC/MS analysis.

We are using GC/MS for profiling primary metabolites in *M. truncatula*. This approach allows for the simultaneous profiling of approximately 300 to 500 components, including amino acids, organic acids, monosaccharides, disaccharides, alcohols, and aromatic amines.[12,31] A typical GC/MS profile of a *M. truncatula* root extract is shown (Fig.3.3). The figure illustrates the naturally occurring relative abundances of metabolites visualized by GC/MS, while also providing an expanded view of the profiles revealing the large amount of information contained within the data.

A large number of primary metabolites can be readily identified because most of these compounds are commercially available. Standard compounds are derivatized, co-chromatographed, and the data are deposited into databases. Unknown metabolites are identified by matching chromatographic retention times and mass spectra to that of known compounds in the databases.[41,42] Mass spectral identification is performed by matching target spectra with commercial libraries such as The National Institute of Standards and Technology (NIST) library or custom libraries constructed in-house by using authentic standards. There are several computer algorithms that automate the process of database searching and identification.[43-45] An example includes the Automated Mass Spectral Deconvolution and Identification Software (AMDIS) provided with many Hewlett Packard GC/MS instruments.[46] We exploit both custom and commercial libraries for metabolite identifications. By using this approach we have identified a large number (~130 currently) of primary metabolites in *M. truncatula* (Fig.3.4). This method has also been used to compare the profiles of various *M. truncatula* tissues (data not shown).

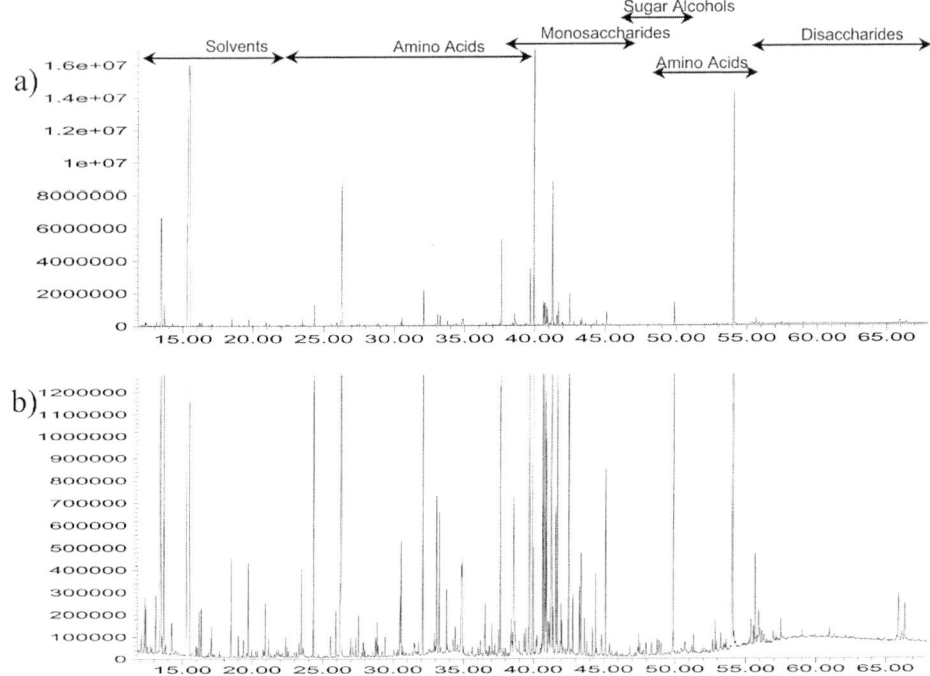

Figure 3.3: GC/MS metabolic profiles of a polar *M. truncatula* root extract that illustrates the elution regions of various metabolite classes. The (a) normalized chromatogram is dominated by several peaks, but the (b) expanded view of the same root profile reveals a substantially larger amount of information not apparent at first glance.

The primary limitation associated with GC/MS is the need for derivatization. Derivatization introduces additional complexity to the system and is not 100% efficient. Inefficient reactions result in the presence of multiple derivatized forms of the same compound. For example, we can detect three different derivatization products of the amino acid asparagine (mw = 132) in *M. truncatula* roots (Fig.3.4). These include asparagine, N,O-TMS (mw = 276), asparagine, N,N,O-TMS (mw = 348), and asparagine, N,N,N,O-TMS (mw = 420). Inefficiency of the derivation reactions also limits the lower concentration range of analytes that can be profiled. Finally, derivatization is not capable of achieving volatility for all compounds, such as many of the flavonoid glycosides. If derivatization is successful and the analyte is volatilized, it must still remain energetically stable enough to be detected. If the compound is not stable, it will fragment and molecular weight information may be lost, thereby complicating identification.

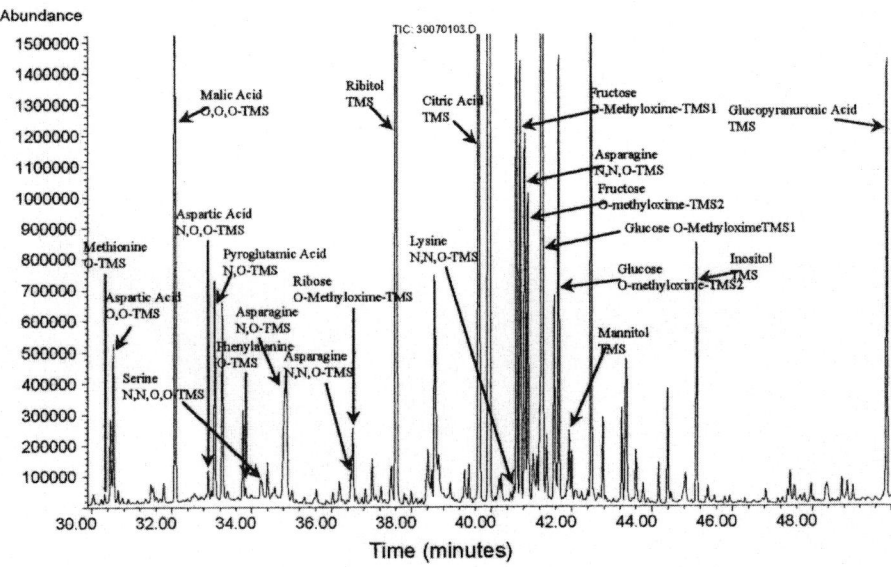

Figure 3.4: GC/MS metabolic profile of a polar *M. truncatula* root extract that provides the identification for many of the root components. Individual components are identified by matching their mass spectra to those in databases or by comparison with authentic samples. Using this approach we have identified a large number (>130 currently) of primary metabolites in *M. truncatula*.

High Performance Liquid Chromatography - Mass Spectrometry

HPLC is a universal separation technique that is capable of separating both volatiles and non-volatiles without the need for derivatization. We are developing methods that employ both on-line photodiode array (PDA) detection and mass selective detection, HPLC/PDA/MS. This approach also utilizes an ion-trap mass spectrometer that is capable of normal and tandem mass spectrometry.[47,48] Tandem mass spectrometry allows the isolation of compounds in the gas phase followed by controlled fragmentation to yield structural information.[49,50] The combination of these technologies, *i.e.*, HPLC/PDA/MS, yields a powerful tool for profiling *and* structural determinations.[51,52]

On-line photodioda array detection is most useful for the analysis of compounds containing chromophores, such as phenolic compounds including flavonoids, isoflavonoids, coumarins, and pterocarpans. An illustrative three-dimensional photodiode array display for a *Medicago truncatula* phenolic extract is provided (Fig.3.5). The three dimensional data consist of UV absorption spectra from 190 to 500 nm for each point along the chromatogram. The data can be rapidly previewed for unique absorption regions correlating to specific compounds

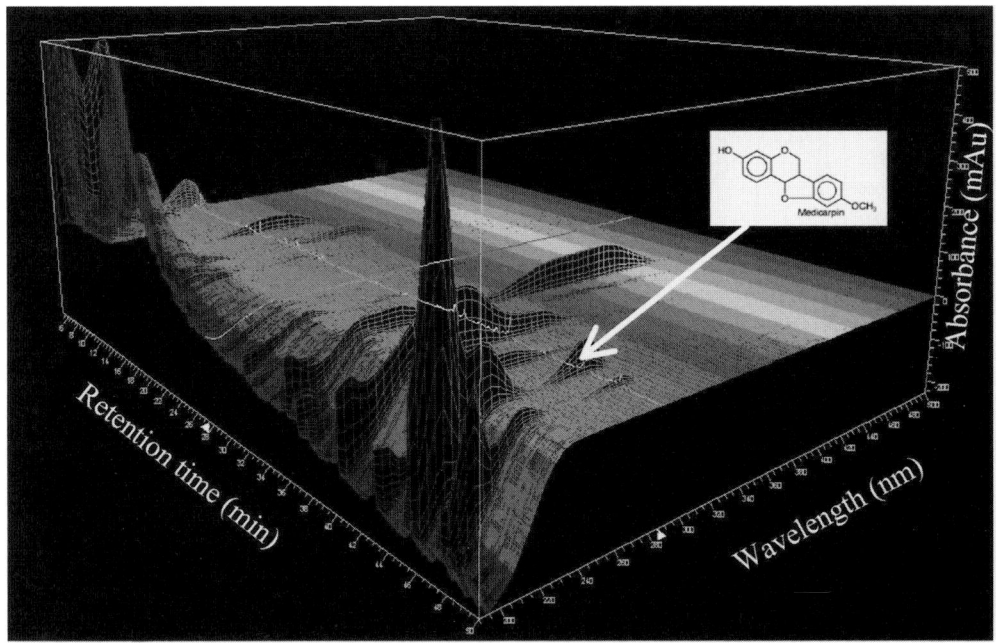

Figure 3.5: Three-dimensional display of the photodiode array absorbance data obtained by HPLC/PDA/MS for a *M. truncatula* extract. The first dimension is HPLC retention time, second is wavelength, and third is absorbance. The data can be rapidly previewed for specific absorbance regions characteristic of functional groups.

or functional groups. Independent chromatograms can also be constructed for each wavelength to increase the selectivity of the data. The UV data are complemented by the mass selective data. Illustrations of both types of data are provided (Fig.3.6). The chromatogram is generated from the ion abundances and mass spectra recorded in the negative-ion, electropsray ionization mode. The mass and UV spectra for the peak eluting at approximately 45 minutes are provided in the inserts and identify the eluting compound as medicarpin, known to have a λ_{max} at 287 nm and a molecular weight of 270. A negative-ion is observed for the deprotonated molecule at m/z 269 with a deprotonated dimer ion observed at m/z 539 confirming the molecular weight as 270.

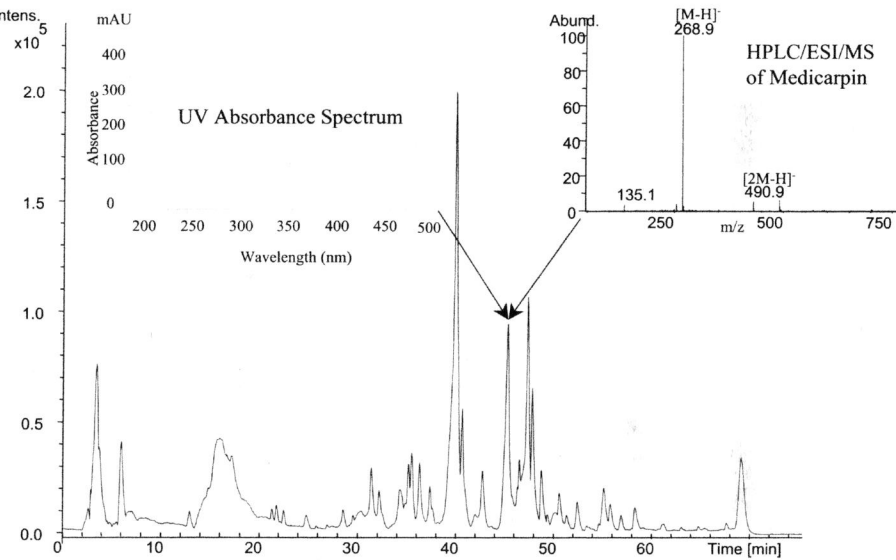

Figure 3.6: Multidimensional data obtained by HPLC/PDA/MS analysis of an alfalfa root extract. The HPLC retention time, the UV absorbance spectrum, and the mass spectrum readily identify the peak eluting at 45 minutes as medicarpin, a known phytoalexin in alfalfa.

The utility of mass selective detection is greatest when analyzing compounds that do not contain chromophores or when structural information is needed for chemical identification. Triterpene saponins contain very weak chromophores and have long been associated with a variety of biological activities including have long been associated with a variety of biological activities including allelopathy,[53] poor digestibility in ruminants,[54] deterrence to insect foraging,[55] and beneficial antifungal properties[56] (see Osbourn et al. this volume). Saponins also possess anti-inflammatory, cholesterol lowering, and anticancer properties.[57-59] Saponins isolated from the legume *Acacia victoriae* have been reported to trigger apoptosis in cancer cells.[60]

Saponins consist of triterpenoid or steroidal aglycones that are substituted with a varying number of sugar side chains. Unsubstituted, nonpolar aglycones are classified as sapogenins and two representative structures are included (Fig.3.7). Because glycoside conjugates are labile and nonvolatile, they must be ionized by using a lower energy technique such as electrospray ionization to retain their integrity during mass analysis. We have been profiling saponins in alfalfa (*Medicago sativa*) and *M. truncatula* by using HPLC coupled to an ESI ion-trap mass spectrometer to acquire normal and tandem mass spectra during profiling.[61] The mass spectra have also been used for structural characterization of *Medicago* saponins. A profile of an alfalfa extract is provided that illustrates both the enhanced sensitivity and selectivity of mass selective detection compared to UV detection at 206 nm (Fig.3.7).

HPLC/PDA/MS has also been used to compare the saponin profiles in multiple cultivars of alfalfa and *M. truncatula*. Comparative profiles are provided (Fig.3.8). It is interesting that these closely related legumes yielded different saponin profiles. The saponin profile of *M. truncatula* is more complex than alfalfa and may provide a richer source for mining putative pharmaceuticals.

Tandem Mass Spectrometry (MS/MS)

The analysis of many natural products including saponins is further hindered by the lack of commercial standards. This complicates the identification process and requires more powerful tools for structural elucidation. Tandem mass spectrometry (MS/MS) is useful in structural determinations and can be visualized as multiple mass spectrometers placed in tandem as illustrated (Fig.3.9). This technique performs gas-phase purification of a specified m/z value with the first mass spectrometer. This is achieved by allowing only the ion of interest to be transmitted, while simultaneously discriminating against (rejecting) all other ions. The transmitted ion is then fragmented through unimolecular or collisionally induced dissociation to yield product or fragment ions from the precursor species. These ions can be rationalized to a structure. A MS/MS spectrum is provided for 3-glucose-28-glucose medicagenic acid (Fig.3.10). The spectrum shows ion peaks corresponding to the loss of two hexoses, the deprotonated aglycone, and a characteristic fragment ion of medicagenic acid.

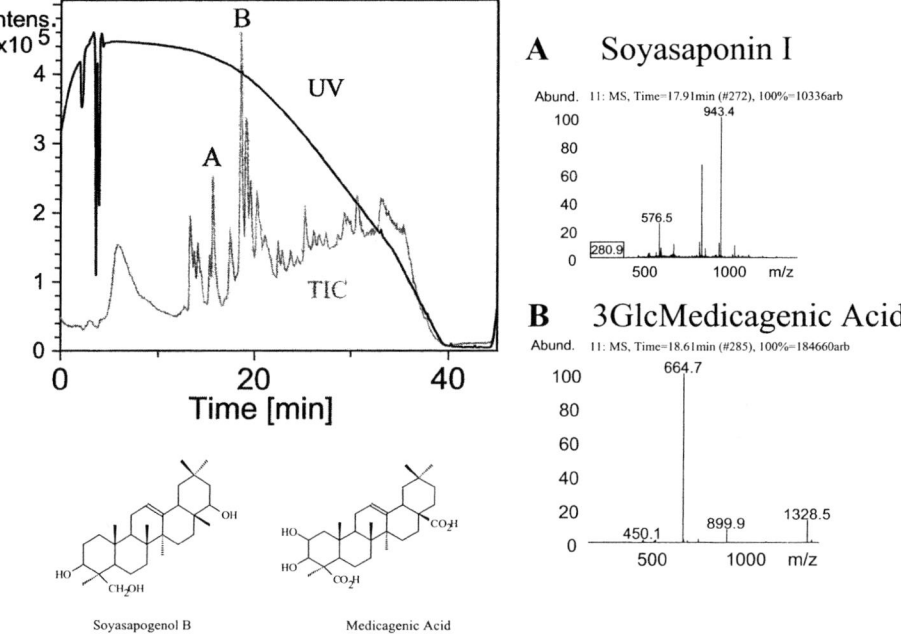

Figure 3.7: HPLC/PDA/MS data for an alfalfa root saponin. Comparison of the UV chromatogram and the total ion chromatogram (TIC) from the mass data illustrates the increased sensitivity of mass selective detection for saponins that possess only weak chromophores. Aglycome structures and mass spectra of two common saponins found in alfalfa and *M. truncatula* are provided in A) 3-rhamnose-galactose-glucuronic acid-soyasapogenol B (common name soysaponin I) and B) 3-glucose-medicagenic acid. The increased selectivity of MS is achieved through molecular weight and fragment information.

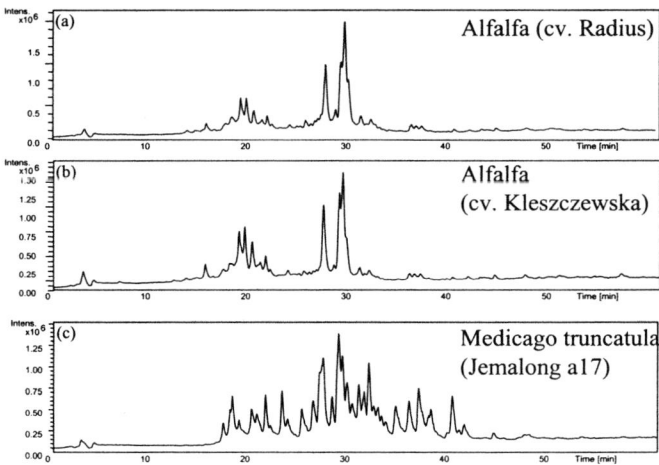

Figure 3.8: Comparative saponin profiles for two cultivars of alfalfa and one cultivar of *M. truncatula* obtained by reverse-phase HPLC/PDA/MS using electrospray ionization and an ion trap mass spectrometer. The profiles illustrate the increased complexity of saponins in *M. truncatula* and offer a richer source for bio-prospecting of natural products.

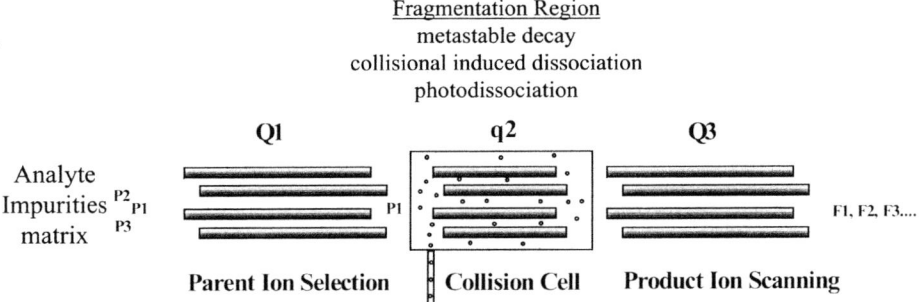

Figure 3.9: Conceptual view of tandem mass spectrometry with a tandem-in-space triple quadrupole mass analyzer. The first mass analyzer (Q1) selects the precursor ion of interest by allowing only it to pass, while discriminating against all others. The precursor ion is then fragmented, usually by energetic collisions, in the second quadrupole (q2) that is operated in transmissive mode allowing all fragment ions to be collimated and passed into the third quadrupole (Q3). Q3 performs mass analysis on the product ions that compose the tandem mass spectra and are rationalized to a structure.

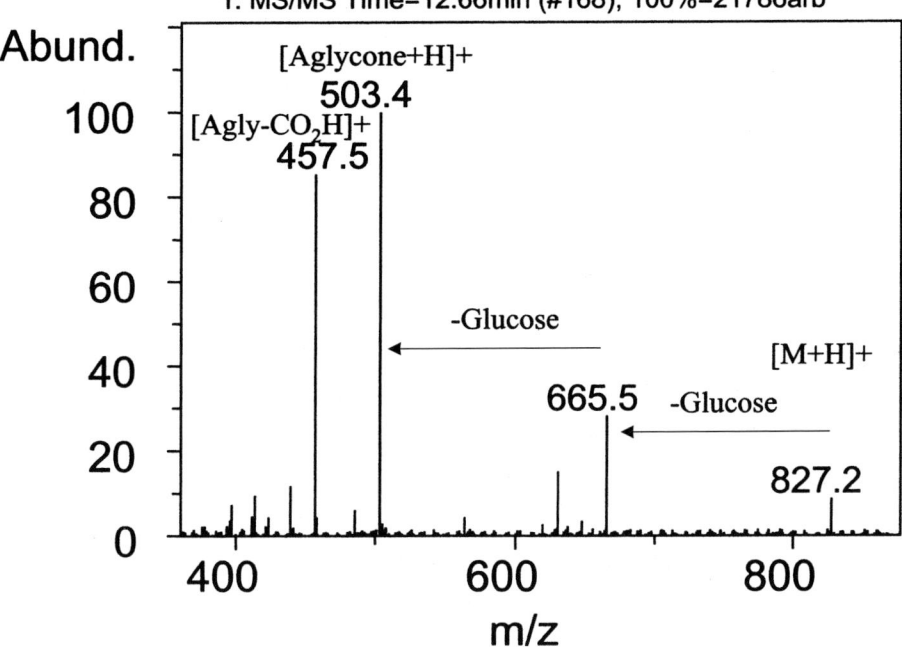

Figure 3.10: HPLC/MS/MS tandem mass spectrum of 3-Glc-28-Glc medicagenic acid obtained using an ion-trap mass analyzer. The spectrum illustrates the successive loss of two hexoses, the deprotonated aglycone, and a characteristic fragment ion associated with medicagenic acid saponins.

Multiple mass analyzers exist that can perform tandem mass spectrometry. Some use a tandem-in-space configuration, such as the triple quadrupole mass analyzers illustrated (Fig.3.9). Others use a tandem-in-time configuration and include instruments such as ion-traps (ITMS) and Fourier transform ion cyclotron resonance mass spectrometry (FTICRMS or FTMS). A triple quadrupole mass spectrometer can only perform the tandem process once for an isolated precursor ion (*e.g.*, MS/MS), but trapping or tandem-in-time instruments can perform repetitive tandem mass spectrometry (MS^n), thus adding n^{th} degrees of structural characterization and elucidation. When an ion-trap is combined with HPLC and photodiode array detection, the net result is a profiling tool that is a powerful tool for both metabolite profiling and metabolite identification.

We have used HPLC/PDA/MS/MS for the profiling and identification of saponins in legumes. Initial studies were performed with alfalfa because many of its saponins have been identified. Alfalfa served as a learning set for the HPLC/PDA/MS/MS characterization of saponins. We then applied this approach to the identification of approximately 20 previously unreported saponins in *M. truncatula*.[61] The identified saponins represent the end products of the triterpenoid biosynthetic pathway. The identified saponins, known entry points into the triterpene pathway and putative enzymes mined from the EST databases are being used to construct a biosynthetic pathway for these important natural products.[62]

The primary limitation of an HPLC/MS approach when compared to GC/MS is its lower separation efficiency. The separation efficiencies of GC/MS capillary columns can be in excess of 100,000, while those of HPLC are on the order of 10,000 to 40,000; thus, HPLC has a lower ability to separate complex mixtures. Because HPLC is generally utilized to separate compounds not possible via GC/MS, it provides complementary rather than competitive data.

High Resolution Accurate Mass Measurements

Metabolic profiling can also be performed by utilizing high resolution and accurate mass measurements. These analyses are generally employed in a batch mode and rely on instrument resolution to differentiate components in a mixture. Resolution is a mass spectrometric instrumental performance parameter defined as the peak mass divided by its peak width at half height (M/ΔM). Increasing resolution makes possible the separation and differentiation of species progressively closer in mass. Increased resolution is also directly related to mass accuracy. Higher resolutions result in narrower peaks and greater precision in assigning mass-to-charge ratios.

Resolutions in the range of 10,000 to 20,000 are achievable with modern time-of-flight mass spectrometry (TOFMS). Fourier transform ion cyclotron resonance mass spectrometry (FTICRMS) is more costly but capable of resolutions exceeding 100,000. Resolutions exceeding 10,000 can provide low to sub parts-per-million mass accuracies. One ppm is equivalent to a mass accuracy of 0.001 for a molecular weight of 1,000 Da.

Most molecules have unique "exact or accurate masses" because all elements except carbon have non-integer values. For example, the mass of the most abundant isotope of C is 12.00000 Da, H is 1.007825 Da, N is 14.00307 Da, and oxygen is 15.99491 Da.[63] The nominal masses for glutamine and lysine are the same (146 Da), but the accurate mass of Gln ($C_5H_{10}N_2O_3$) is 146.06912 Da, and the accurate mass of Lys ($C_6H_{14}N_2O_2$) is 146.10551 Da. The resolution or resolving power required to differentiate these two species would be (M/ΔM) 146/0.03639 or 4012; thus, these

METABOLOMICS

two species can be differentiated at 10,000 resolving power and 1 ppm mass accuracy.

We have used accurate mass measurements obtained by matrix-assisted laser desorption ionization time-of-flight mass spectrometry (MALDI-TOFMS) to differentiate and profile saponins from *M. truncatula* roots. An example is provided (Fig.3.11) showing the MALDI-TOFMS spectra of a solid-phase extract of *M. truncatula* root tissue. In this spectrum, we can identify multiple saponins.

Saponin	Emperical Formula	Theoretical Accurate M/Z	Experimental Accurate M/Z	Error (ppm)
3Glc-Med.A.	C36H56O11Na	687.37201	687.3718	-0.338
3Glc28Glc-Med.A.	C42H66O16Na	849.42483	849.4240	-0.945
3GlcAGalRha-SoyB	C48H78O18Na	965.50856	965.5017	-7.172
3AraGlcAra28Glc-Med.A.	C52H84O22Na	1083.53517	1083.5454	7.715
3GlcGlc28XylRhaAra-Med.A.	C52H82O23Na	1097.51443	1097.5188	3.922
3GlcGlc28XylRhaAra-Med.A	C58H92O28Na	1259.56725	1259.5708	3.126

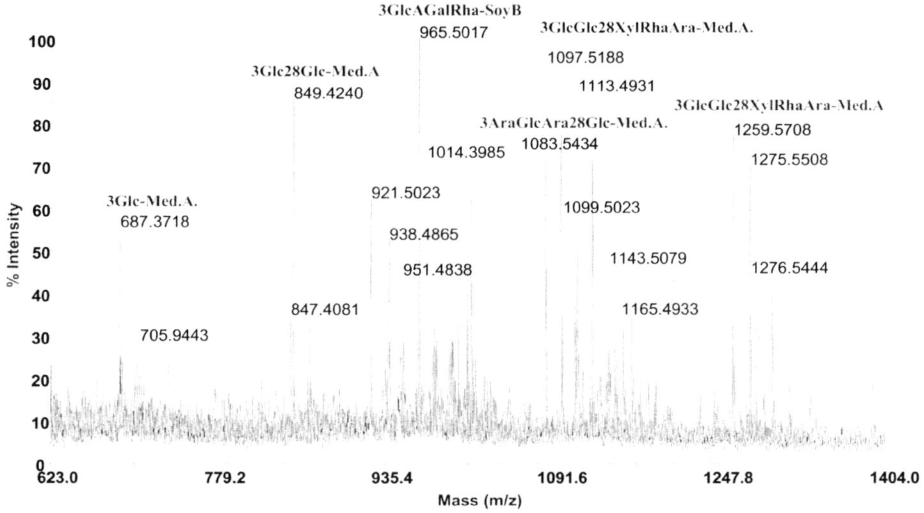

Figure 3.11: High resolution (*i.e.*, greater than10,000) MALDI-TOFMS analysis of an alfalfa saponin extract. Accurate mass measurements are obtained by internal calibration and used to rapidly identify multiple saponins.

The major limitation of high resolution accurate mass profiling is its inability to differentiate isomeric species with the same empirical formula. An example of isomers would be glucose ($C_6H_{12}O_6$) and galactose ($C_6H_{12}O_6$). In GC/MS and LC/MS methods, the isomers generally have different elution times that allow for differentiation even though the observed masses are similar. Another limitation of accurate mass measurement performed in a batch mode includes its susceptibility to changes in matrix composition. For example, variations in salt levels lead to variations in the observed cation adducts, *i.e.*, $[M+H]^+$, $[M+Na]^+$, and $[M+K]^+$ of the same metabolite and complicate the quantification and interpretation of the data. Variations in other matrix components can result in competitive ionization or ion suppression.

Capillary Electrophoresis

The past ten years have seen countless demonstrations of the potential of capillary electrophoresis (CE) in many areas of biotechnology.[64-71] Capillary electrophoresis offers higher separation efficiencies than attainable in HPLC and uses simple instrumentation. The fundamental difference between traditional slab format electrophoresis and CE is that the separation is achieved inside small fused-silica capillaries, typically 25-100 µm, as first demonstrated by Jorgenson and co-workers.[72] CE can be performed in several different modes. The most common is capillary zone electrophoresis (CZE), which is similar to traditional paper electrophoresis where charged species migrate in a continuous electrolyte based on their net charge and size. Many different detection strategies have been explored in CE.[73] UV detection is the most common; however, the narrow internal diameter of the separation capillary results in lower sensitivity relative to HPLC with UV detection. An extremely sensitive detection scheme is laser-induced fluorescence (LIF) following derivatization with a suitable fluorophore allowing sub-femtomole levels to be quantified. Indirect detection schemes have also been devised that allow the detection of species not amenable to traditional detection techniques due to the absence of specific chemical properties such as the lack of a chromophore.

Metabolic Profiling of Carbohydrates by Capillary Electrophoresis

Capillary electrophoresis has proven a valuable tool in the analysis of carbohydrates.[74,75] Initial studies on the utility of CE in metabolic profiling appear promising. We have explored the use of CE for carbohydrate profiling of *M. truncatula* tissue extracts. Carbohydrates present a significant challenge in detection, since most sugars do not posses a strong chromophore or fluorophore. To overcome this problem, samples are derivatized with 4-aminobenzonitrile (4-ABN) by using reductive amination. This forms a stable product, and the reaction occurs rapidly upon heating.[76] Once derivatized, the extracts are analyzed in less than 20 minutes. Carbohydrate profiles of various *M. truncatula* tissue extracts obtained using CZE are shown in Fig.3.12a. Differences in carbohydrate accumulation for various tissue regions are evident. The concentration detection limits are in the low ppm range for monosaccharides. If carbohydrates need to be analyzed at lower concentrations, 9-aminopyrene-1,4,6-trisulfonate (APTS) can be used to label them at trace levels.[77] The fluorescence properties of APTS match the emission wavelengths generated by the argon ion laser, which is most commonly used in LIF-CE. LIF-CE allows parts-per-billion carbohydrate profiling. A major limitation of all derivatization chemistry is the lack of a universal reagent that derivatizes all reducing and non-reducing sugars equally.

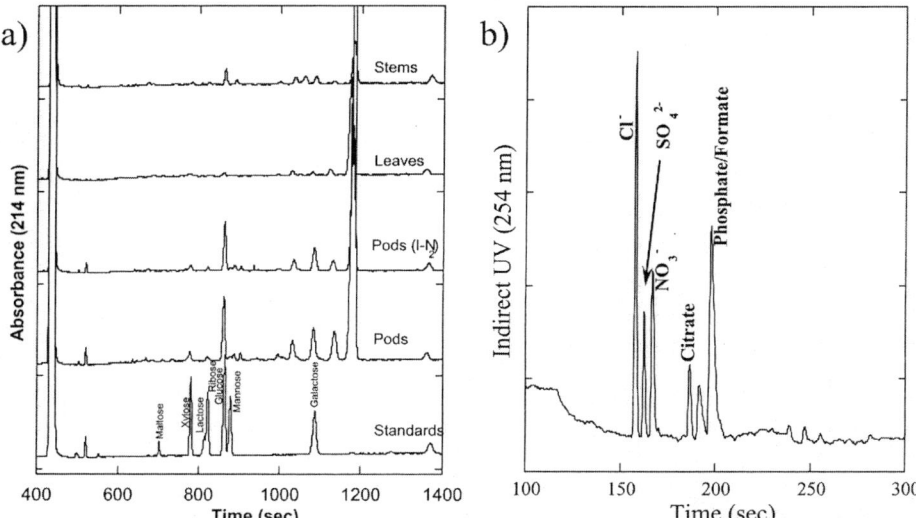

Figure 3.12: Metabolic profiling by capillary electrophoresis. (a) Comparative carbohydrate profiles of *M. truncatula* tissue obtained using 4-aminobenzonitrile derivatization, capillary electrophoresis with a 150 mM borate buffer, pH = 9, and on-column UV detection at 214 nm. (b) Anion profile from *M. truncatula* using capillary electrophoresis and indirect UV detection. The separation buffer was 5 mM K_2CrO_4, 1% Waters OFM-Anion BT, pH 8.0.

An alternative to derivatizing carbohydrates is the use of indirect photometric detection. In this method, a detectable co-ion in the electrolyte is added to the buffer system generating a steady state absorbance signal in the detector. As the analyte ions migrate in front of the detector window, they displace the detectable co-ion and cause a decrease or negative response in the detector signal. This method provides universal detection of all anions or cations. Since most carbohydrates are not ionized at neutral pH, high pH electrolytes (pH~12) are required to induce partial ionization. Soga and Heiger demonstrated the utility of indirect detection for the routine analysis of carbohydrates.[78] The lower sensitivity of indirect detection yields detection limits in 20-100 ppm range for most simple sugars that is suitable for profiling highly abundant carbohydrate species, *e.g.*, glucose, fructose, and sucrose, found in most plant tissue.

Metabolic Profiling of Anionic Species by Capillary Electrophoresis

CE possesses great potential for profiling small ionic species. The utilization of CE for ion analysis has proven to be competitive with HPLC in the areas of inorganic ions[79] and organic acids.[80] An anion profile from an aqueous extract of *M. truncatula* tissue obtained by using CE with indirect detection is illustrated (Fig.3.12b). Several of the anions are detectable even at a 1:100 dilution of the extract. The analytical sensitivities in ion analysis are quite good with detection limits in the sub- to low-ppm range for most ions. Even lower concentrations of ions can be observed by applying sample stacking injection techniques.[81,82] Drawbacks to ion analysis by CE include a limited dynamic range and potential interference by highly concentrated ions in the sample.

BIOINFORMATICS AND STATISTICAL PROCESSING OF METABOLOME DATA

Metabolite levels are expected to vary in response to genetic perturbation, biotic elicitation, and environmental stimuli. Changes may be dramatic or subtle. Subtle changes will necessitate statistical means to determine if changes are significant. Statistics begins with experimental design, including replicate sampling, controls, replicate analyses, and application of statistical tests. The data can be used to calculate values such as a mean and standard deviation and general statistical tests such as the Student t-test can be performed to eliminate erroneous data. F-ratios can then be used to determine whether or not a change is significant at a given confidence level.

It is also important to develop unbiased means of processing and visualizing large amounts of multivariant or metabolomic data.[83,84] What we desire are computer applications that tell us whether or not samples are statistically similar or different and what the exact differences/similarities are. Ideally, this would be performed in an automated manner. For example, we would like to compare the GC/MS, LC/MS, or CE profiles of a sample set in an automated mode and to be directed to the peak(s) that are statistically different. Identification of specific metabolites corresponding to the statistically different peaks would suggest function of the altered gene or response of the biological system. If identification is not desired, then the differences could be used as a unique or unbiased means of differentiation, phenotyping, genotyping, or relationship determinations.

A single GC/MS metabolite profile can potentially yield 300 to 500 distinct components. This provides a wealth of information to be interpreted and leads to challenges in data processing. To simplify the task, many researchers have incorporated data mining techniques to reduce the complexity of the data set and to visualize the data. There are several approaches to statistically processing and visualizing data, including principal component analysis (PCA), hierarchical cluster analysis (HCA), self-organizational maps (SOM), and neural networks. Many of these tools can be used with metabolite profiling as exploratory data analysis tools.

Principal Component Analysis (PCA)

Principal component analysis is one of the oldest and most widely used multivariate techniques.[85] The concept behind PCA is to describe the variability in a set of multivariate data in terms of a set of uncorrelated variables, each of which is a particular linear combination of the original variables. More simply stated, PCA is an attempt to explain a set of complex data in terms of a smaller number of dimensions than one originally starts with. The mathematical approach used in PCA involves determining the eigenvalues and eigenvectors of a square symmetric matrix with sums of squares and cross products. The eigenvector associated with the largest eigenvalue has the same direction as the first principal component. The eigenvector associated with the second largest eigenvalue determines the direction of the second principal component. The sum of the eigenvalues equals the trace of the square matrix, and the maximum number of eigenvectors equals the number of rows (or columns) of this matrix. Plotting the two largest eigenvalues or principal components provides a rapid means of visualizing similarities or differences. A PCA plot of replicate GC/MS analyses of different *M. truncatula* tissues is shown (Fig. 3.13). The plot illustrates good correlation in the replicate measurements, and it is obvious that root, leaf, and stem tissue differentiate. This approach will be used to determine similarity, differences, and relationships in a large number of t-DNA activation tagged mutants and various *M. truncatula* ecotypes.

Self-Organizing Maps (SOMs)

Self-organizing maps (SOMs) are gaining popularity due to their enhanced ability to differentiate and visualize data relative to PCA.[86,87] SOMs are unsupervised neural learning algorithms in which weighted vectors of neurons are initially set at random and then reiteratively processed until they best represent the input data. Recently, SOMs have been applied to the correlation of GC/MS data to compare the morphology of eighty-eight species of ants.[88]

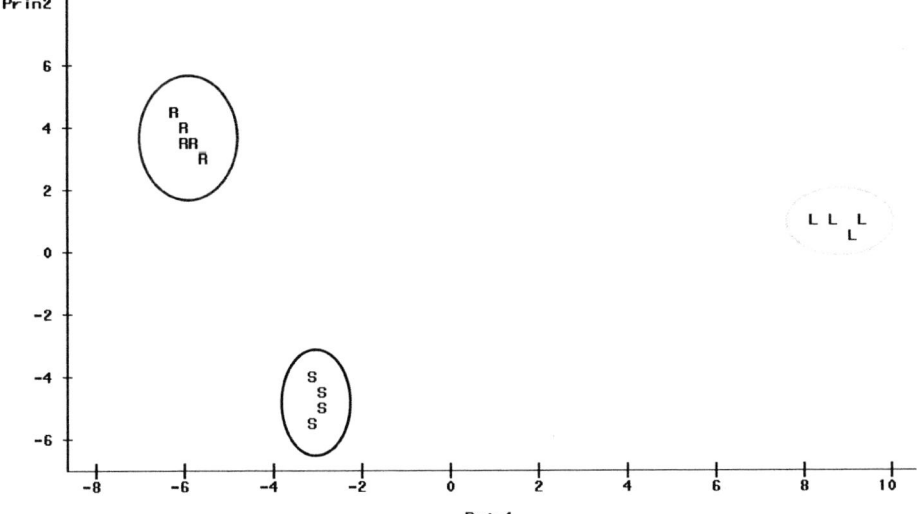

Figure 3.13: Principal component analysis of repetitive GC/MS profiles of *M. truncatula* root (R), stem (S) and leaves (L). The first and second principal component of each GC/MS analysis were calculated and plotted. The relative distance between points is a measure of similarity or difference. The clustering shows good reproducibility within the independent tissues but clear differentiation of tissues. The results also show that roots and stems are more similar to each other than to leaves.

Hierarchical Cluster Analysis (HCA)

Hierarchical cluster analysis (HCA) also provides a method of determining relationships among different data sets.[89] HCA involves the progressive pair-wise comparison of large data sets to determine relationships or relative similarity/homogeneity. The HCA output is usually visualized as a phylogenetic-type tree, or dendrogram, in which branch lengths are proportional to correlation coefficients. More closely related data sets appear closer together in the phylogenetic tree. The output provides an easy visualization of the interrelationships and relationship distances within sample sets.

The above bioinformatic tools provide methods of determining differences or similarities in datasets. The next step is to incorporate metabolomic data with other expression information, including mRNA and proteins, to infer gene function. To accomplish this, metabolomic data sets must be integrated and correlated in a global manner with genetic and enzymatic data, pathways assembled into systems, and literature references incorporated as learning tools to annotate existing data to yield in silico biological information.[90] Approaches and tools are now available for modeling metabolic systems and are vital in understanding metabolism.[91] The challenge of the future is to integrate these approaches and obtain complete integrated functional genomic systems to understand and visualize biology.[92,93]

SUMMARY

The sciences of functional genomics and metabolomics are maturing. Many of the approaches discussed above are still in developmental stages and have yet to reach their full potential. Most of the individual profiling components are still being operated independently, and they need to be integrated in the future. Current bioinformatic tools for processing, visualization, and integration of large metabolomic datasets are limited. These tools need to be refined and automated to allow for the processing of data on an "omic" scale. New bioinformatic tools are also needed for the integration of data from multiple levels of expression to provide an in-depth view of gene expression and function. Fortunately, we expect substantial progress in this area in the near future. Unfortunately, we believe there are still limitations in the ability to profile the entire metalobome. Sequential extraction followed by parallel analyses enhances our view, however, individual techniques still only provide fractional views of the entire metabolome. These views are sufficient to differentiate cell states or tissues, but do not always provide information on less abundant molecules that play important biochemical roles, such as in signaling. Superficial views are substantial limitations to transcriptomics, proteomics, and metabolomics. As functional genomics matures, we need to continue broadening our field of vision.

ACKNOWLEDGEMENTS

We thank Drs. Richard A. Dixon, Gregory D. May, and Barbara Wolf-Sumner for critical review of the manuscript. The work in the authors' laboratories were financially supported by The Samuel Roberts Noble Foundation.

REFERENCES

1. BARKER, D.G., BIANCHI, S., BLONDON, F., DATTÉE, Y., DUC, G., ESSAD, S., FLAMENT, P., GALLUSCI, P., GÉNIER, G., PIERRE, G., MUEL, X., TOURNEUR, J., DÉNARIÉ, J., HUGUET, T., *Medicago Truncatula*, a model plant for studying the molecular genetics of the *Rhizobium*-Legume symbiosis, *Plant Mol. Biol. Rep.*, 1990, **8**, 40-49.
2. COOK, D.R., VANDENBOSCII, K., DE BRUIJN, F.J., HUGUET, T., Model legumes get the nod, *Plant Cell*, 1997, 275-281.
3. COOK, D.R., *Medicago truncatula* - a model in the making!, *Curr. Opin. Plant Biol.*, 1999, **2**, 301-304.
4. TRIEU, A.T., BURLEIGH, S.H., KARDAILSKY, I.V., MALDONADO-MENDOZA, I.E., VERSAW, W.K., BLAYLOCK, L.A., SHIN, H., CHIOU, T.-J., KATAGI, H., DEWBRE, G.R., WEIGEL, D., HARRISON, M.J., Transformation of *Medicago truncatula* via infiltration of seedlings or flowering plants with *Agrobacterium*, *Plant J.*, 2000, **22**, 531-541.
5. BELL, C.J., DIXON, R.A., FARMER, A.D., FLORES, P., INMAN, J., GONZALES, R.A., HARRISON, M.J., PAIVA, N.L., SCOTT, A.D., WELLER, J.W., MAY, G.D., The *Medicago* genome initiative: A model legume database, *Nucl. Acids Res.*, 2001, **29**, 1-4.
6. VENTER, J.C., The human genome, *Science*, 2001, **291**, 1304-1351.
7. KING, R.D., KARWATH, A., CLARE, A., DEHASPE, L., Accurate prediction of protein functional class from sequence in *Mycobacterium tuberculosis* and *Escherichia coli* genomes using data mining, *Yeast*, 2000, **17**, 283-293.
8. HIETER, P., BOGUSKI, M., Functional genomics: It's all how you read it, *Science*, 1997, **278**, 601-602.
9. IDEKER, T., THORSSON, V., RANISH, J.A., CHRISTMAS, R., BUHLER, J., ENG, J.K., BUMGARNER, R., GOODLETT, D.R., AEBERSOLD, R., HOOD, L., Integrated genomic and proteomics analyses of a systematically perturbed metabolic network, *Science*, 2001, **292**, 929-934.
10. BLACKSTOCK, W.P., WEIR, M.P., Proteomics: quantitative and physical mapping of cellular protiens, *Trends in Biotech.*, 1999, **17**, 121-127.
11. THIELLEMENT, H., BAHRMAN, N., DAMERVAL, C., PLOMION, C., ROSSINGNOL, M., SANTONI, V., DE VIENNE, D., ZIVY, M., Proteomics for genetic and physiological studies in plants, *Electrophoresis*, 1999, **20**, 2013-2026.
12. FIEHN, O., KOPKA, J., DORMANN, P., ALTMANN, T., TRETHEWEY, R.N., WILLMITZER, L., Metabolite profiling for plant functional genomics, *Nature Biotechnol.*, 2000, **18**, 1157-1161.
13. TRETHEWEY, R.N., KROTZKY, A.J., WILLMITZER, L., Metabolic profiling: A Rosetta stone for genomics?, *Curr. Opin. Plant Biol.*, 1999, **2**, 83-85.
14. TRETHEWY, R.N., Gene discovery via metabolic profiling, *Current Opin. Biotechnol.*, 2001, **12**, 135-138.
15. SCHENA, M., SHALON, D., DAVIS, R.W., BROWN, P.O., Quantitative monitoring of gene expression patterns with a complementary DNA microarray, *Science*, 1995, **270**, 467-469.

16. VELCULESCU, V.E., ZHANG, L., VOGELSTEIN, B., KINZLET, K.W., Serial analysis of gene expression, *Science,* 1995, **270**, 484-487.
17. O'FARRELL, P.H., High resolution two-dimensional electrophoresis of proteins, *J. Biol. Chem.,* 1975, **250**, 4007-4021.
18. KLOSE, J., Protein mapping by combined isoelectric focusing and electrophoresis of mouse tissues. A novel approach to testing for induced point mutations in mammals, *Humangenetik,* 1975, **26**, 231-243.
19. KLOSE, J., KOBALZ, U., Two-dimensional electrophoresis of proteins: An updated protocol and implications for a functional analysis of the genome, *Electrophoresis,* 1995, **16**, 1034-1059.
20. JUNGBLUT, P.R., ZIMMY-ARNDT, U., ZEINDL-EBERHART, E., STULIK, J., KOUPILOVA, K., PLEISSNER, K.P., OTTO, A., MULLER, E.C., SOKOLOWSKA-KOHLER, W., GRABHER, G., STOFFLER, G., Proteomics in human disease: Cancer, heart and infectious disease, *Electrophoresis,* 1999, **20**, 2100-2110.
21. VAN WIJK, K.J., Challenges and prospects of plant proteomics, *Plant Physiol.,* 2001, **126**, 501-508.
22. MORRIS, A.C., DJORDEVIC, M.A., Proteome analysis of cultivar-specific interactions between *Rhizobium leguminosarum* biovar *trifolii* and subterranean clover cultivar Woogenenellup, *Electrophoresis,* 2001, **22**, 586-598.
23. PAPPIN, D.J.C., HOJRUP, P., BLEASBY, A.J., Rapid identification of proteins by peptide-mass fingerprinting, *Curr. Biol.,* 1993, **3**, 327-332.
24. ARNOTT, D., O'CONNELL, K.L., KING, K.L., STULTS, J.T., An integrated approach to proteome analysis: Identification of proteins associated with cardiac hypertrophy, *Anal. Biochem.,* 1998, **258**, 1-18.
25. YATES III, J.R., Mass spectrometry and the age of the proteome, *J. Am. Soc. Mass Spectrom.,* 1998, **33**, 1-19.
26. GODOVAC-ZIMMERMANN, J., BROWN, L.R., Perspectives for mass spectrometry and functional proteomics, *Mass Spec. Rev.,* 2001, **20**, 1-57.
27. CONRADS, T.P., ANDERSON, G.A., VEENSTRA, T.D., PASA-TOLIC, L., SMITH, R.D., Utility of accurate mass tags for proteome-wide protein identification, *Anal. Chem.,* 2000, **72**, 3349-3354.
28. CLAUSER, K.R., BAKER, P., BURLINGAME, A.L., Role of accurate mass measurement (±10 ppm) in protein identification strategies employing MS or MS/MS and database searching, *Anal. Chem.,* 1999, **71**, 2871-2882.
29. GYGI, S.P., RIST, B., GERBER, S.A., TURECEK, F., GELB, M.H., AEBERSOLD, R., Quantitative analysis of complex protein mixtures using isotope-coded affinity tags, *Nature Biotechnol.,* 1999, **17**, 994-999.
30. GLASSBROOK, N., BEECHER, C., RYALS, J., Metabolic profiling on the right path, *Nature Biotechnol.,* 2000, **18**, 1142-1143.
31. ROESSNER, U., LUEDEMANN, A., BRUST, D., FIEHN, O., LINKE, T., WILLMITZER, L., FERNIE, A.R., Metabolic profiling allows comprehensive phenotyping of genetically or environmentally modified plant systems, *Plant Cell,* 2001, **13**, 11-29.

32. RAAMSDONK, L.M., TEUSINK, B., BROADHURST, D., ZHANG, N., HAYES, A., WALSH, M.C., BERDEN, J.A., BRINDLE, K.M., KELL, D.B., ROWLAND, J.J., WESTERHOFF, H.V., VAN DAM, K., OLIVER, S.G., A functional genomics strategy that uses metabolome data to reveal the phenotype of silent mutations, *Nature Biotechnol.,* 2001, **19**, 45-50.
33. GYGI, S.P., ROCHON, Y., FRANZA, B.R., AEBERSOLD, R., Correlation between protein and mRNA abundance in yeast, *Mol. Cell. Biol.,* 1999, **19**, 1720-1730.
34. ROESSNER, U., WAGNER, C., KOPKA, J., TRETHEWEY, R.N., WILLMITZER, L., Simultaneous analysis of metabolites in potato tuber by gas chromatography-mass spectrometry, *Plant J.,* 2000, **23**, 131-142.
35. SMITH, R.D., LOO, J.A., OGORZALEK-LOO, R.R., BUSMAN, M., UDSETH, H.R., Principles and practice of electrospray ionization-mass spectrometry for large polypeptides and proteins, *Mass Spec. Rev.,* 1991, **31**, 472-485.
36. SIUZDAK, G., Mass Spectrometry for Biotechnology, Academic Press, San Diego, 1996, 161 p.
37. MCCLOSKEY, J.A., Methods in Enzymology. Vol. 193: Mass Spectrometry, Academic Press, San Diego, 1990, 960 p.
38. WATSON, J.T., 3rd Edition, Introduction to Mass Spectrometry, Lippincott Williams & Wilkins, New York, 1997, 512 p.
39. KATONA, Z.F., SASS, P., MOLNÁR-PERL, I., Simultaneous determination of sugars, sugar alcohols, acids, and amino acids in apricots by gas chromatography-mass spectrometry, *J. Chromatogr.,* 1999, **847**, 91-102.
40. ADAMS, M.A., CHEN, Z., LANDMAN, P., COLMER, T., Simultaneous determination by capillary gas chromatography of organic acids, sugars, and sugar alcohols in plant tissue extracts as their trimethylsilyl derivatives, *Anal. Biochem.,* 1999, **266**, 77-84.
41. MCLAFFERTY, F.W., ZHANG, M.Y., STAUFFER, D.B., LOH, S.Y., Comparison of algorithms and databases for matching unknown mass spectra, *J. Am. Soc. Mass Spectrom.,* 1998, **9**, 92-95.
42. MCLAFFERTY, F.W., STAUFFER, D.A., LOH, S.Y., WESDEMIOTIS, C., Unknown identification using reference mass spectra. Quality evaluation of databases, *J. Am. Soc. Mass Spectrom.,* 1999, **10**, 1229-1240.
43. POOL, W.G., LEEUW, J.W., VAN DE GRAAF, B.J., Automated extraction of pure mass spectra from gas chromatographic/mass spectrometric data, *J. Mass Spectrom.,* 1997, **32**, 438-443.
44. HALKET, J.M., PRZYBOROWSKA, A., STEIN, S.E., MALLARD, W.G., DOWN, S., CHALMERS, R.A., Deconvolution gas chromatography/mass spectrometry of urinary organic acids--potential for patern recognition and automated identification of metabolic disorders, *Rapid Commun. Mass Spectrom.,* 1999, **13**, 279-284.
45. HERRON, N.R., DONNELLY, J.R., SOVOCOOL, G.W., Software-based mass spectal enhancement to remove interferences from spectra of unknowns, *J. Am. Soc. Mass Spectrom.,* 1996, **7**, 598-604.

46. DAGAN, S., Comparison of gas chromatography-pulsed flame photometric detection-mass spectrometry, automated mass spectral deconvolution and identification system and gas chromatography-tandem mass spectrometry as tools for trace level detection and identification, *J. Chromatogr., A.,* 2000, **868**, 229-247.
47. MARCH, R.E., TODD, J.F., Practical Aspects of Ion Trap Mass Spectrometry. Vol. II, Ion Trap Instrumentation, CRC Press, Boca Raton, 1995, 320 p.
48. MARCH, R.E., TODD, J.F., Practical Aspects of Ion Trap Mass Spectrometry. Vol I, Fundamentals of Ion Trap Mass Spectrometry, CRC Press, Boca Raton, 1995, 430 p.
49. JENNINGS, K.R., MS/MS instrumentation. In: Applications of Modern Mass Spectrometry in Plant Science Research (R. P. Newton and T. J. Watson, eds.), Oxford Science Publications, Oxford, 1996, pp. 25-43
50. COSTELLO, C.E., Application of tandem mass spectral approach to structural determination of saponins. In: Advances in Experimental Medicine and Biology (G. R. Waller and K. Yamaski, eds.), 405, Plenum Publishing, New York. 1996, pp. 317-329
51. FRASER, P.D., ELISABETE, M., PINTO, S., HOLLOWAY, D.E., BRAMLEY, P.M., Application of high-performance liquid chromatography with photodiode array detection to the metabolic profiling of plant isoprenoids, *Plant J.,* 2000, **24**, 551-558.
52. SILVA, O., GOMES, E.T., WOLFENDER, J.L., MARSTON, A., HOSTETTMANN, K., Application of high performance liquid chromatography coupled with ultraviolet spectroscopy and electrospray mass spectrometry to the characterisation of ellagitannins from *Terminalia macroptera* roots, *Pharm. Res.,* 2000, **17**, 1396-1401.
53. WALLER, G.R., JURZYSTA, M., THORNE, R.L.Z., Allelopathic activity of root saponins from alfalfa (*Medicago sativa* L.) on weeds and wheat, *Bot Bull. Academy. Sin,* 1993, **34**, 1-11.
54. OLESZEK, W., Alfalfa saponins: structure biological activity and chemotaxonomy. In: Saponins Used in Food and Agriculture; Advances in Experimental Medicine and Biology, (G. R. Waller and K. Yamasaki, eds.), Plenum Press, New York. 1996, pp. 155-170
55. TAVA, A., ODOARDI, M., Saponins from *Medicago* SPP: chemical characterization and biological activity against insects. In: Saponins Used in Food and Agriculture; Advances in Experimental Medicine and Biology, (G. R. Waller and K. Yamasaki, eds.), Vol. 405, Plenum Press, New York. 1997, pp. 97-109
56. NAGATA, T., TSUSHIDA, T., HAMAYA, E., ENOKI, N., MANABE, S., NISHINO, C., Camellidins: Antifungal saponins isolated from *Camellia japonica*, *Agric. Bio. Chem.,* 1985, **49**, 1181-1186.
57. WALLER, G.R., YAMASAKI, K., eds., Saponins Used in Food and Agriculture; Advances in Experimental Medicine and Biology, Vol. 405, Plenum Press, New York. 1996, 441 p.
58. JURZYSTA, M., NOWACKI, E., Saponins of the genus *Medicago*, *Acta Agro.,* 1979, **32**, 13.

59. MALINOW, M.P., MAC NULTRY, W.P., HOUGHTON, D.C., KESSLER, S., STENZEL, P., GOODNIGHT, S.H., JR., BARDANA, E.J., JR., PALOTAY, S.L., MAC LAUGHLIN, P., LIVINGSTON, A.L., Lack of toxicity of alfalfa saponins in *Cynomolgus macaues*, *J. Med. Prima.*, 1982, **11**, 106-118.
60. HARIDAS, V., HIGUCHI, M., JAYATILAKE, G.S., BAILEY, D., MUJOO, K., BLAKE, M.E., ARNTZEN, C.J., GUTTERMAN, J.U., Avicins: Triterpenoid saponins from *Acacia victoriae* (Bentham) induce apoptosis by mitochondrial perturbation, *Proc. Natl. Acad. Sci., USA*, 2001, **98**, 5821-5826.
61. HUHMAN, D.V., SUMNER, L.W., Metabolioc profiling of saponins in *Medicago sativa* and *Medicago truncatula* using HPLC coupled to an electrospray ion-trap mass spectometer, *Phytochemistry*, in press.
62. SUZUKI, S., ACHNINE, L., HUHMAN, D., SUMNER, L.W., DIXON, R.A., A functional genomics approach to the triterpene saponin biosynthetic pathway in *Medicago truncatula,* Phytochemical Society of North America, 2001, Oklahoma City, OK
63. LIDE, D.R., 73rd ed., Handbook of Chemistry and Physics, (D. R. Lide, ed.), CRC Press, Boca Raton, 1992, 11-28.
64. WEHR, T., RODRIGUEZ-DIAZ, R., LIU, C.-M., Capillary electrophoresis of proteins, 1997, **37**, 237-361.
65. ISSAQ, H.J., A decade of capillary electrophoresis, *Electrophoresis*, 2000, **21**, 1921-1939.
66. MANABE, T., Capillary electrophoresis of proteins for proteomic studies, *Electrophoresis*, 1999, **20**, 3116-3121.
67. LURIE, I.S., Capillary electrophoresis for drug analysis, *Proc. SPIE-Int. Soc. Opt. Eng.*, 1999, **3576**, 125-135.
68. LAGU, A.L., Applications of capillary electrophoresis in biotechnology, *Electrophoresis*, 1999, **20**, 3145-3155.
69. CHEN, S.H., CHEN, Y.-H., Pharmacokinetic applications of capillary electrophoresis, *Electrophoresis*, 1999, **20**, 3259-3268.
70. BOONE, C.M., WATERVAL, J.C.M., LINGEMANN, H., ENSING, K., UNDERBERG, W.J.M., Capillary electrophoresis as a versatile tool for the bioanalysis of drugs - a review, *J. Pharm. Biomed. Anal.*, 1999, **20**, 831-863.
71. ALTRIA, K.D., Overview of capillary electrophoresis and capillary electrochromatograhy, *J. Chromatogr., A.*, 1999, **856**, 443-463.
72. JORGENSON, J.W., LUKACS, K.D., Zone electrophoresis in open tubular glass capillaries, *Anal. Chem.*, 1981, **53**, 1298-1302.
73. SWINNEY, K., BORNHOP, D.J., Detection in capillary electrophoresis, *Electrophoresis*, 2000, **21**, 1239-1250.
74. PAULUS, A., KLOWCKOW-BECK, A., Analysis of carbohydrates by capillary electrophoresis. In: Chromatographia: CE Series, (K. D. Altria, ed.), Vol. 3, Vieweg & Sohn, Wiesbanden, Germany, 1999, pp. 93-170.
75. EL RASSI, Z., Recent developments in capillary electrophoresis and capillary electrochromatography of carbohydrate species, *Electrophoresis*, 1999, **20**, 3134-3144.

76. SCHWAIGER, H., OEFNER, P.J., HUBER, C., GRILL, E., BONN, G.K., Capillary zone electrophoresis and micellar electrokinetic chromatography of 4-aminobenzonitrile carbohydrate, *Electrophoresis*, 1994, **15**, 941-952.
77. EVANGELISTA, R.A., LIU, M.-S., CHEN, F.-T.A., Characterization of 9-aminopyrene-1,4,6-trisulfonate-derivatize sugars by capillary electrophoresis with laser-induced fluorescence detection, *Anal. Chem.*, 1995, **67**, 2239-2245.
78. SOGA, T., HEIGER, D.N., Simultaneous determination of monosaccharides in glycoproteins by capillary electrophoresis, *Anal. Chem.*, 1998, **261**, 73-83.
79. BUCHBERGER, W., Inorganic ions. In: Handbook of Capillary Electrophoresis Applications (H. Shintani and J. Polonsky, eds.), Blackie Academic & Professional, London. 1997, pp. 531-549
80. STOVER, F.S., Organic acids and organic ions. In: Handbook of Capillary Electrophoresis Applications (H. Shintani and J. Polonsky, eds.), Blackie Academic & Professional, London. 1997, pp. 550-567
81. QUIRINO, J.P., TERABE, S., Sample stacking of fast-moving anions in capillary zone electrophoresis, *J. Chromatogr., A.*, 1999, **850**, 339-344.
82. YANG, Y., KANG, J., LU, H., OU, Q., LIU, F., Determination of trace level anions in snow samples by capillary electrophoresis with sample stacking, *J. Chromatogr., A.*, 1999, **834**, 287-291.
83. EVERITT, B.S., DER, G., 2nd Ed., A Handbook of Statistical Analysis Using SAS, Chapman & Hall/CRC, Boca Raton, 2001, 376 p.
84. TABACHNICK, B.G., FIDELL, L.S., 4th Ed., Using Multivariate Statistics, Allyn & Bacon, Boston, 2000, 966 p.
85. HOTELLIN, H., Analysis of a complex of statistical variables into principal components, *J. Educ. Physchol.*, 1933, **24**, 417-441.
86. KOHONEN, T., 3rd Ed., Self-Organizing Maps, Springer, Berlin, 2000, 528 p.
87. TÖRÖNEN, P., KOLEHMAINEN, M., WONG, G., CASTRÉN, E., Analysis of gene expression data using self-organizing maps, *FEBS Lett.*, 1999, **451**, 142-146.
88. NIKIFOROW, A., SCHLICK-STEINER, B., STEINER, F., KALB, R., MISTRIK, R., Classification of GC-MS data of epicuticular hydrocarbon from T*etramorium* ants by self-organizing maps for morphological determinations, Proceedings of the 49th ASMS Conference on Mass Spectrometry and Allied topics, 2001, Chicago, IL
89. EISEN, M.B., SPELLMAN, P.T., BROWN, P.O., BOTSTEIN, D., Cluster analysis and display of genome-wide expression patterns, *Proc. Natl. Acad. Sci., USA,* 1998, **95**, 14863-14868.
90. PALSSON, B., The challenges of in silico biology, *Nature Biotechnol.*, 2000, **18**, 1147-1150.
91. MENDES, P., KELL, D., Non-linear optimization of biochemical pathways: applications to metabolic engineering and parameter estimation, *Bioinformatics*, 1998, **14**, 869-883.
92. MENDES, P., Modeling large biological systems from functional genomic data: parameter estimation. In: Foundations of Systems Biology (H. Kitano, ed.), MIT Press, Cambridge, MA. 2001, pp. In press
93. VOIT, E.O., RADIVOYEVITHC, T., Biochemical systems analysis of genome-wide expression data, *Bioinformatics*, 2000, **16**, 1023-1037.

Chapter Four

METABOLITE PROFILING: FROM METABOLIC ENGINEERING TO FUNCTIONAL GENOMICS

Richard N. Trethewey

Metanomics GmbH & Co KG
Tegeler Weg 33
10589 Berlin
Germany

e-mail: richard.trethewey@metanomics.de

Introduction.. 64
Metabolic Engineering in Plants.. 64
    Metabolic Engineering of Starch Biosynthesis in Potato Tubers............ 65
    Production of PHB in *Arabidopsis*.. 68
    Maximizing Production of PHB in *Arabidopsis*...................................... 70
Metabolite Profiling.. 71
    The Case for More Extensive Analysis to Support Metabolic Engineering.. 71
    Basic Approach to Metabolite Profiling.. 71
Metabolite Profiling of Plants... 72
    Application to Study of Starch Metabolism in Transgenic Potato Tubers... 72
    Application to Analysis of *Arabidopsis* Mutants...................................... 73
    Application to Study of *Arabidopsis* Lines Producing PHB.................... 74
Functional Genomics.. 75
Summary.. 76

INTRODUCTION

Metabolite profiling, in a range of forms, is emerging as a powerful tool for functional genomics.[1-3] The need for metabolite profiling has been clearly demonstrated by a range of metabolic engineering approaches in plants. In this chapter, examples of the metabolic engineering of existing plant metabolism and the introduction of novel pathways are reviewed, with emphasis on the limitations of these approaches. The examples covered led directly to the development of GC/MS based metabolite profiling protocols, and the value of such profiling when applied to the original metabolic engineering projects is highlighted. However, the full power of metabolite profiling may be realized in the area of functional genomic research, and the opportunity for metabolite profiling to play an important role in this emerging discipline is reviewed.

METABOLIC ENGINEERING IN PLANTS

The last few decades have seen many significant advances in the field of molecular plant physiology.[4,5] In particular, progress has been driven by the rapidly expanding ability to clone and manipulate the activities of particular genes in transgenic plants.[6] Over the last 12 years, this has opened up a range of experimentation in physiology and metabolism mediated either through the direct manipulation of endogenous enzyme activities or through the overexpression of heterologous enzyme activities. In the latter case, there have been studies on the overexpression of enzymes that have the same function as an endogenous enzyme but have altered regulatory properties,[7-9] and there have been examples where novel enzymes[10,11] or pathways[12] have been introduced into plant tissues. The ability to modulate the activities of enzymes with exquisite precision through the use of specific promoters, targeting sequence or the careful selection of transgenic lines, has led to a renaissance in the study of the relatively old discipline of plant metabolism.[5]

In general terms, the wide range of recent studies using the technologies of molecular plant physiology has been characterized by mixed success. Success is, of course, subjective but judged through the prism of metabolic engineering the majority of approaches to flux manipulation have failed, with only a few examples where the desired alteration in flux was actually achieved. Even in the cases where metabolism has been successfully manipulated as intended, there are normally unforeseen secondary consequences that arise.

To illustrate the problems that are faced in contemporary metabolic engineering it is worth considering two fields that have recently received much attention: starch biosynthesis in potato tubers and the production of bioplastics in *Arabidopsis*.

Metabolic Engineering of Starch Biosynthesis in Potato Tubers

Yield from crop plants has been successively improved through the action of long term breeding programs and the refinement of methods in agriculture. In the case of potato, the harvest index (the ratio of the dry weight of harvestable organs to the total plant dry weight) has increased from 0.09 in wild species to 0.81 in modern cultivars.[13] Such improvements, although dramatic, have been the result of very long term incremental progress in breeding programs. The relatively new ability to undertake direct metabolic engineering offers at least the prospect of achieving more rapid advancement in enhancing both the harvest index and the absolute yields from crop plants.[14]

Metabolism in potato tubers has been extensively studied by conventional approaches, and the pathway for the conversion of sucrose to starch is relatively well defined (Figure 4.1).[5,15] Sucrose is transported from the leaves via the phloem and enters into the parenchyma cells of the potato tuber via symplastic connections.[16,17] It is then subsequently cleaved by sucrose synthase to give rise to UDPglucose and fructose. The UDPglucose is converted to glucose 1-phosphate via the action of UDPglucose pyrophosphorylase, while the fructose is phosphorylated by fructokinase and gives rise to fructose 6-phosphate. The hexose phosphates are interconverted in the cytosol via the action of phosphoglucomutase and phosphoglucoisomerase.[15] It has recently been possible to demonstrate that glucose 6-phosphate is taken up into the amyloplast and serves as the precursor for starch biosynthesis.[18,19] Starch biosynthesis is often described as a simple linear pathway involving phosphoglucomutase, ADPglucose pyrophosphorylase, starch synthases, and branching enzymes. This view overlooks the complexity associated with the fact that there are a range of isoforms of starch synthases and branching enzymes, and the combination of these together with the action of certain degradative enzymes actually gives rise to the complex structure of insoluble starch granules.[20,21]

The starch biosynthetic pathway has been the subject of a wide range of studies whereby particular enzymes have been cloned and genetically manipulated in transgenic potato tubers. One central target has been enhancing starch yield via the manipulation of the cytosolic metabolism of sucrose. Elegant biochemical studies have demonstrated that sucrose synthase, UDPglucose pyrophosphorylase, phosphoglucomutase, and phosphoglucosisomerase are all close to equilibrium during the developmental stage of starch deposition.[22-24] This means that the ratio between sucrose and hexose phosphates in the cytosol is effectively fixed and close to equilibrium. Therefore, one approach that was taken to the metabolic engineering of this pathway was to introduce an invertase from yeast into the cytosol of the potato tubers, with the intention of facilitating an irreversible cleavage of sucrose.[10,25] This was anticipated to have two important effects: the lowering of the sucrose levels could promote the transport of sucrose from the source leaves through the phloem and into the tuber, and the elevation of the cytosolic hexose phosphate levels could promote the uptake of this precursor into the amyloplast and, thus, enhance the rate of starch synthesis.

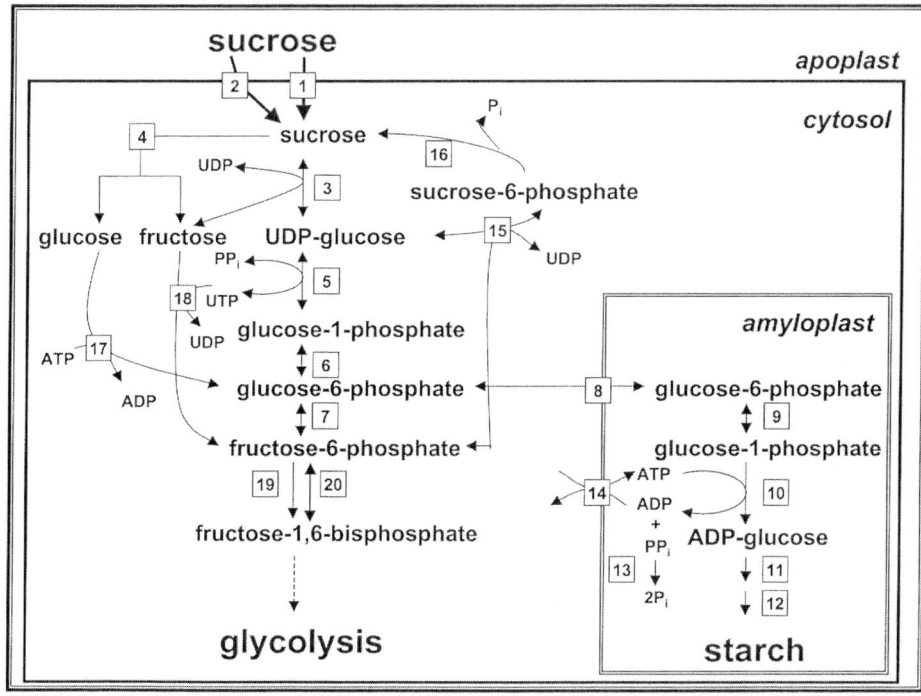

Figure 4.1: Primary carbon metabolism in potato tubers.
Primary flow of carbon between sucrose, starch, and glycolysis in potato tubers. The numbers denote the principle enzymes or steps: (1) sucrose transporter; (2) plasmodesmatal movement; (3) sucrose synthase; (4) invertase; (5) UDPglucose pyrophosphorylase; (6) cytosolic phosphoglucomutase; (7) cytosolic phosphoglucose isomerase; (8) glucose 6-phopshate transporter; (9) plastidial phosphoglucomutase; (10) ADPglucose pyrophosphorylase; (11) starch synthases; (12) starch branching enzymes; (13) inorganic pyrophosphatase; (14) amyloplast adenylate transporter; (15) sucrose phosphate synthase; (16) sucrose phosphate phosphatase; (17) hexokinase; (18) fructokinase; (19) phosphofructokinase; (20) pyrophosphate: fructose-6-phosphate 1-phosphotransferase. Pi (inorganic phosphate), PPi (pyrophosphate).

The introduction of the invertase from yeast alone was not sufficient. Invertase cleaves sucrose to release the two component sugars, glucose and fructose. While fructose can be readily metabolized by fructokinase in potato tubers, there is insufficient hexokinase activity in developing potato tubers to bring the glucose into intermediary metabolism. Therefore, it was necessary to introduce a second transgene, a bacterial glucokinase, in order to ensure that the hexoses became available for subsequent metabolism.[25]

So much for the theory, in practice this approach led to a significant reduction in the starch yield rather than the expected increase.[25] Measurements of the intermediates revealed that at the first approximation the approach was working as intended: there was in excess of a 90% reduction in the sucrose content, while the hexose phosphate levels increased by a factor of 7. However, detailed measurements of both starch levels in developing tubers and density measurements on tubers following large scale greenhouse trials confirmed that there was in fact a 30% reduction in the yield of starch.

These surprising results provoked the question why the genetic manipulation of sucrose metabolism produced the desired results of shifting the ratio of sucrose to hexose phosphates, but did not lead to an enhanced accumulation of starch. After a detailed analysis of these lines, a complex picture of metabolism through individual determinations of metabolite contents became clear. The carbon was being diverted from starch biosynthesis and towards glycolysis. Direct determination of carbon dioxide production from the tubers revealed an increase in flux of up to five fold: a quantitative assessment of carbon flows indicated that the increased release of carbon dioxide was equivalent to the amount "missing" from starch. Further analysis of metabolism related to the tricarboxylic acid cycle revealed that there were also substantial increases in the levels of certain organic acids and some rather complex changes. On the whole there was also an increase in amino acid levels.

The conclusions from measurements of the levels of metabolic intermediates were confirmed by direct determination of fluxes by using tracer amounts of radiolabel supplied either directly into tubers still growing on the plant, or by supplying them to excised tuber discs.[26] These experiments confirmed that there was an elevated glycolytic flux in the transgenic lines. However, they also revealed a further complexity in that they provided evidence for a massive cycling of sucrose between cleavage by the heterologous invertase and re-synthesis via sucrose phosphate synthase. The latter enzyme was shown to be activated in the transgenic tubers, presumably as a direct result of a well described regulatory mechanism that is triggered by the elevated hexose phosphate levels in the transgenic tubers.[27] This is a clear indication of how the engineering of metabolism can lead to changes in intermediate levels, which in turn themselves exert effects on the behavior of the system through further regulatory valves. Of particular interest is that this type of mechanism does not involve any changes at the genetic, transcriptional, or translational level: indicating that classical approaches of genomics (sequence, expression arrays, proteomics) would be blind to determining such behavior in this metabolic system.

Further analysis of the central enzymes of primary metabolism in these lines indicated that maximum catalytic activity of several key enzymes of the glycolytic sequence and tricarboxylic acid cycle was induced.[25] It was concluded that this induction, particularly in phosphofructokinase, triose phosphate isomerase, pyruvate kinase, and citrate synthase was the primary cause of the change in partitioning between starch and glycolysis. However, the question as to the signal and mechanism whereby this increase in maximum catalytic activity occurs remains unsolved. Analysis of the steady state levels of mRNAs did not reveal any changes, indicating that the induction was not occurring at the level of gene expression. More fundamental knowledge about the nature of the regulation of primary metabolism in plants will be required before this question can be adequately answered.

Production of PHB in Arabidopsis

The previous example related to manipulation of primary plant metabolism through creating a bypass within the existing system. There are also a range of studies that have looked at the production of novel substances within plants. One particularly relevant example is the production of a polyhydroxyalkanoate (PHA), polyhydroxybutyrate (PHB), in *Arabidopsis thaliana*.[28]

PHA is a class of biodegradable plastic, first identified in *Bacillus megaterum* as long ago as 1926. Subsequently, a wide range of different types of PHA have been identified in around 90 bacterial genera, and some 91 different hydroxyalkanoic acids have been found to be constituents of PHA polymers.[29] If such plastics were to be produced in plants, this would be a renewable resource and theoretically could have a cost comparable to that of non-biodegradable plastics produced from oil.[28] Given the large and rising demand for plastic, the possibility of economic production of biodegradable plastic is an attractive one that has led a large number of groups and companies to work in this area in the last decade.

In order to produce PHAs in plants it is necessary to introduce the biosynthetic enzymes from bacteria. PHB represents the best characterized and simplest form of PHA, and the synthetic pathway (Figure 4.2) has been extensively studied in *Ralstonia eutropha*.[30,31] Starting from acetyl-CoA, a β-ketothiolase is required in order to form acetoacetyl-CoA. This is then reduced by a NADPH-dependent acetoacetyl-CoA reductase, which gives rise to 3-hydroxybutyryl-CoA. The latter intermediate is the substrate for the polymerization reaction catalyzed by polyhydroxybutyrate synthase.[30] In *Ralstonia eutropha*, the thiolase, reductase, and synthase genes make up an operon.[31]

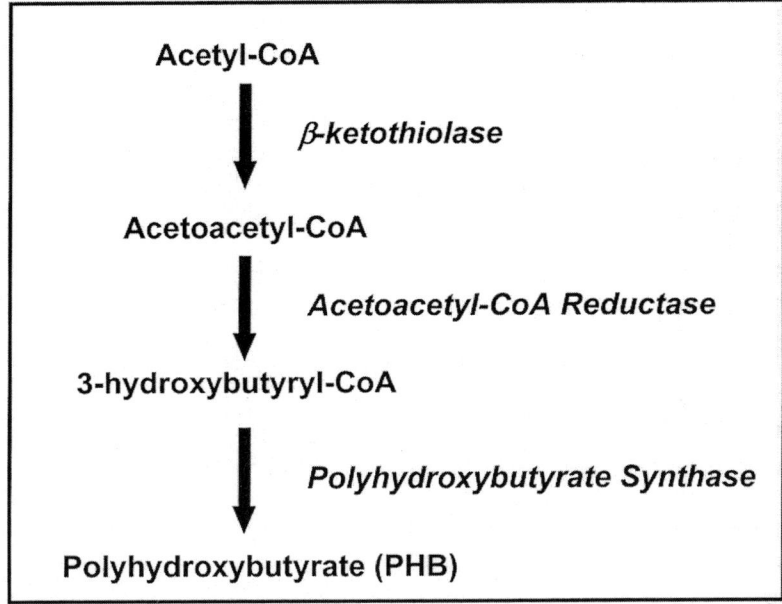

Figure 4.2: Synthesis of polyhydroxybutyrate (PHB).
The three step pathway of polyhydroxybutyrate synthesis (PHB) as found in *Ralstonia eutropha*.

There is a range of genes available from different sources for metabolic engineering of plants. The approach taken in the pioneering work of the group of Somerville was to overexpress the genes from *Ralstonia eutropha* in the cytosol of *Arabidopsis thaliana* in order to demonstrate initially whether PHB could be produced.[32] Poirier *et al.*[32] transformed *Arabidopsis* with the reductase and synthase individually and then crossed lines homozygous for each transgene in order to obtain lines that expressed both genes. This approach relied upon the presence of endogenous activity in the cytosol that could produce acetoacetyl-CoA from acetyl-CoA. Screening of the crossed transgenic lines led to the identification of progeny that produced PHB, albeit at relatively low levels (20-100 μg/g FW). Granules of PHB could be seen in the cytosol, nucleus, and vacuole following microscopic examination of the leaves. Despite the relatively low levels of PHB found in these lines, the growth and development of the plants were strongly affected. In particular, expression of the reductase alone led to a reduction in biomass production of up to 45% over the first 22 days. Interestingly, expression of the synthase alone did not lead to any negative effects, but the combination of both transgenes led to more severe adverse affects than those following single expression of the reductase. The authors proposed that the toxic effects were due either to the presence of the PHB granules, which might exert some type of adverse effect, or are a consequence of the depletion of acetyl CoA.[32]

The same group then attempted to determine if expression of the transgenes in the plastid would lead to reduced pleiotropic effects.[33] If it was acetyl CoA depletion that was the primary cause of the reduced growth, then the authors proposed that the plastid might be a better host for PHB production because there should be more acetyl CoA available there due to it being the site of fatty acid production. The group first generated three single transgenic lines expressing the thiolase, reductase, and synthase from *Ralstonia eutropha* directed to the plastid by using the targeting sequence from the pea ribulose-bisphosphate carboxylase enzyme. The three enzymes were then assembled into a pathway by using sexual crossing to generate the desired genotype. This time much higher levels of PHB were produced, with levels of 20-700 µg/g FW of PHB being found in leaves from 20- to 30-day old plants, and astonishingly up to 10 mg/g FW (14% of the dry weight) found in the leaves of senescent plants. PHB granules could be distinguished in the chloroplasts by electron microscopy.[33] Importantly, plants producing PHB in the plastid showed wild type growth and fertility.[33]

Maximizing Production of PHB in Arabidopsis

The economic viability of transgenic plants for PHB production depends directly on the levels of PHB produced in the plants. A second group set out to explore the upper limit of PHB production in transgenic *Arabidopsis* plants.[12] In the approaches described previously, the three genes for PHB production were transformed individually into plants and were later combined into a single line via crossing. The lines taken for the crossing were selected according to the expression level of the individual genes. This approach holds the risk that the levels of the three different activities may not be adequately balanced in the combined transgenic line, and in addition the crossing of lines with similar genetic elements may give rise to undesirable and unpredictable silencing effects. Therefore, the alternative, but bold, approach was taken to prepare a "quadruple construct" containing all three genes for plastidial PHB production, and the selectable marker gene, in one plasmid. This allowed the introduction of the entire pathway in a single transformation event. In order to avoid the problems of selecting the lines with the correct expression level and to ignore the potential problem of silencing, some 87 primary transformants were screening directly by using gas chromatography – mass spectrometry (GC-MS) for the accumulation of PHB. With this direct approach, the authors identified lines that produced PHB levels up to 4% of their fresh weight, some 4 times more than the maximum figures recorded in the earlier studies using crossed lines.[12]

METABOLITE PROFILING

Although this jump in PHB production represents an important step towards generating economically viable levels of PHB in plants, in this study, a strong negative correlation between PHB accumulation and plant growth was found. Thus, it appears that there is a level of PHB production beyond which there are other factors in metabolism that then give rise to negative consequences for the overall metabolism and physiology of the plants.

METABOLITE PROFILING

The Case for More Extensive Analysis to Support Metabolic Engineering

The above examples, one where endogenous pathways are bypassed and the other where a novel pathway is introduced, illustrate the central challenges facing metabolic engineers when working with plants. These can be summarized as follows:
1) Too little is understood currently for there to be a high success rate with a rational design of engineering strategies.
2) Manipulation of one enzymatic step in a system can have wide reaching consequences because of the interplay between metabolite levels and a wide range of regulatory circuits. These circuits can operate at the level of transcription, translation, post-translational modification, or through allosteric and competitive influences on the kinetic properties of enzymes.
3) Measurements of gene expression or protein abundance, which are often favored in metabolic engineering, do not necessarily provide a route to understand what is occurring within metabolic systems.
4) While it is now relatively easy to generate a transgenic plant, altered with specificity at a target step, the more significant technical challenge is to obtain quality data on the levels of a wide range of metabolites and on the nature of the fluxes in the metabolic system. The availability of such data is often the step that limits progress in metabolic engineering.

Basic Approach to Metabolite Profiling

Given the challenges facing metabolic engineering, a direct approach to improving the situation is to seek and establish methods that rapidly provide data on a range of metabolites *i.e.*, metabolite profiling. These methods must provide a compromise between the breadth of the metabolites that can be specifically measured and the quality of the measurement (quantification, sensitivity, and reliability). These requirements inevitably lead to the conclusion that mass spectrometry, coupled to a separation principle (*e.g.*, liquid chromatography, gas chromatography, or capillary electrophoresis) is the most appropriate approach.[1-3] Metabolite profiling that uses mass spectrometry, as such, is not a particularly novel technology, having been used for nearly two decades in medical applications such as the screening of blood or urine samples.[34-37] However there are relatively few publications, perhaps 10 per year in the recent past, on the development and application of metabolite profiling.

METABOLITE PROFILING OF PLANTS

The first reported application of a metabolite profiling approach to plants was a study by Sauter et al.[38] working at BASF AG on herbicide development. This group treated barley leaves with a range of herbicides whose modes of action were only partially known. Following treatment, a GC/MS analysis was performed in order to obtain a complex profile of the major components of the barley leaves. The changes in the profiles following herbicide treatment allowed the modes of action of the herbicides to be categorized, and facilitated conclusions on the area of metabolism affected.

Despite the power of this approach, it took another decade before GC/MS profiling of plants was developed further and applied by Roessner et al. to transgenic plants.[39] These authors modified the method used by Sauter et al.[38] by introducing some extra steps in the derivatization procedures in order to reduce the isomer complexity of the chromatograms, and they optimized the method for potato tubers. This allowed the qualitative and quantitative determination of more than 150 tuber metabolites including sugars, sugar alcohols, dimeric and trimeric saccharides, amines, amino acids, and organic acids. The robustness of the method was proven through a range of control studies that determined both the variability of the procedure and the overall recovery of 25 defined metabolites, representing different chemical classes. Reassuringly, the biological variability in wild type developing potato tubers was found to be an order of magnitude higher than the variability due to the analytical methods. The majority of metabolites were shown to have standard deviations among different tubers in the 10-25% range. Importantly, the error associated with metabolite measurements with profiling technologies are similar to or lower than those found in microarray expression analysis: the average coefficient of variation for a 2500-element *Arabidopsis* array has been reported to be 37%.[40]

Application to Study of Starch Metabolism in Transgenic Potato Tubers

Having established and validated the method, Roessner et al.[39] applied it to the transgenic potato lines that overexpressed a yeast invertase specifically in the tubers. As described, these lines are characterized by a reduced starch accumulation and an altered partitioning of carbon flux away from starch accumulation and towards glycolysis. The central conclusions of the conventional studies described earlier, particularly the alteration in partitioning, could be deduced following a single set of GC-MS analyses taking less than one week to conduct from extraction to data analysis.[39] In addition to a speedy repetition of the results of previous studies, because the GC-MS analysis provides data on a wide range of metabolites in a non-biased way, and is potentially open to the identification of completely novel peaks, additional unexpected changes in the metabolism of the transgenic potato tubers were identified. For example, there was a strong accumulation of maltose and shikimate, and a massive reduction in inositol content in the invertase expressing potato tubers.[39] In general terms, the changes in metabolism in these lines were pleiotropic, and such changes only become visible by using technologies that offer a broad approach. It is unlikely that a metabolic engineer working on starch metabolism in potato tubers would think it necessary to measure shikimate or inositol. However, once information is available on these changes, then new connections and approaches are fostered. These results underline the fact that a systems approach ultimately will be necessary to support metabolic engineering.

Roessner et al.[41] extended their studies to analyze four genotypes characterized by modifications in sucrose metabolism. Given the complex profiles, the authors adopted data mining tools, such as hierarchical cluster analysis and principle component analysis, in order to assign clusters to the individual genotypes. Then by using extraction analysis, the most important metabolites defining the formation of the clusters could be identified. An example of the importance of this approach is seen in the comparison between the transgenic lines overexpressing invertase and glucokinase and a line that overexpresses a sucrose phosphorylase from *Pseudomonas saccharophila*.[42] This latter enzyme cleaves sucrose into glucose 1-phosphate and fructose, thus avoiding the formation of free glucose as occurs through the action of invertase. The metabolite profiles of the two lines separated out into two distinct clusters. The primary reason for this was that several compounds, for example maltose, trehalose, and isomaltose, increased sharply in the invertase lines, presumably as a direct consequence of the elevated glucose production in these lines. Conversely, in the lines that overexpressed the sucrose phosphorylase, a novel and unknown peak (PT00) appeared, and remarkably this had a magnitude that was greater than sucrose in the transgenic lines. The authors did not elucidate the chemical structure of this peak but speculated that it may be the result of a side activity of sucrose phosphorylase.[41] Again, this illustrates the benefit of using broad and open systems in characterizing the results of metabolic engineering studies. The authors concluded that metabolite profiling coupled with advanced data analysis is a potent approach to the phenotyping of plant genotypes.

Application to Analysis of Arabidopsis Mutants

The basic approach used by Roessner et al.[39,41] has also been developed by Fiehn et al. for *Arabidopsis*.[43] Using GC-MS analysis, these authors distinguished some 326 distinct compounds in *Arabidopsis* leaves, of which roughly half could be assigned a chemical structure. They applied the GC-MS method to the analysis of four different *Arabidopsis* genotypes: the ecotype Columbia (Col-2) and the *dgd1* mutant (reduced digalactosyldiacylglycerol accumulation[44-46]) in the Col-2 genetic background, as well as the ecotype C24 and the *sdd1-1* mutant in the C24 genetic background (reduced stomatal density[47]). Given the complexity of the data sets, these authors also explored the use of different data analysis tools and found the most utility in principle component analysis. It was surprising that the "metabolic phenotypes" of the two ecotypes were found to be more divergent, as assessed via principle component analysis, than the mutants were from their respective parental ecotypes. A statistical assessment of the data set revealed 41 significant changes in the *sdd1-1* mutant; two of the most significant were in unknown hydrophilic substances. The *dgd1* mutation gave rise to an even more pleiotropic pattern: a total of 153 significant changes were found in comparison to the parental Columbia ecotype.[43]

Application to Study of Arabidopsis Lines Producing PHB

Another example of the ability of metabolite profiling to reveal unexpected changes has come from the application of GC/MS analysis to the study of the PHB-producing *Arabidopsis* lines described earlier. Bohmert *et al.*[12] generated lines that contained up to 4% of the leaf fresh weight as PHB, but these were characterized by growth retardation. Surprisingly, metabolite profiling of the *Arabidopsis* leaves revealed no appreciable changes in either the composition or the quantity of any of the fatty acids analyzed in the transgenic lines versus wild types. However, substantial pleiotropic changes were observed upon analysis of metabolites present in the polar phase of the extraction. Interestingly, in most cases, the alterations correlated to the amount of PHB accumulation in the transgenic lines. Of particular note is that decreased amounts of fumarate and isocitrate were observed, which might indicate a reduction in TCA cycle activity. Such a reduction may result in a depletion of acetyl-CoA pools, which could explain the retardation of growth. However, an increase was also found in the levels of osmoprotectants such as proline and mannitol, and also in the levels of glucose, fructose, and sucrose with respect to the wild type. These data may be indicative of strong osmotic stress, however, such an interpretation may be an oversimplification since the osmoprotectant *myo*-inositol and its derivatives decreased in these lines. It is also important to keep in mind that the PHB granules themselves might also disturb the structure and physiology of the chloroplasts through unknown mechanisms.

This example further serves to illustrate the argument that the deeper one looks the more one finds. A metabolic engineer working exclusively on PHB production may not have set out to measure the metabolites that were found to have altered in the GC/MS analysis. However, ignorance of such changes could seriously restrict the speed of development of such projects, and if the objective is to generate a commercial product, then knowledge about potential multiple changes and an evaluation of their consequence has to occur.

FUNCTIONAL GENOMICS

The impressive success of genome sequencing projects has given rise to a wealth of information on DNA sequences present within plants,[48] and in the case of *Arabidopsis*, the whole genome sequence is now available.[49] Current estimates of the numbers of genes in *Arabidopsis* are around the 27,000 mark, although as these estimates are based upon gene finding algorithms that are less than mature, the exact number of genes may still vary significantly from this figure.

The determination of the existence of genes is, of course, just the first step in the genomics revolution. More profound and sophisticated is the challenge to understand the functions of the genes. The assignment of function must occur at the level of the determination of the activity associated with a particular gene product. It is also necessary, however, to take a more systems approach and define the function associated with a gene product in the context of the performance of a complex organism. This latter objective is a particularly complex challenge, but it is only through an understanding of the role of gene products in the context of a whole organism that we can begin to understand biology.

The primary technologies of functional genomics have, until now, been expression arrays operating upon different technology platforms and proteomics.[50-58] Both approaches, although powerful and important, have the inherent disadvantage that they do not necessarily directly determine the function of a gene. For example, expression arrays provide information on the mRNAs that alter in steady state abundance following a particular developmental transition or as a consequence of an environmental signal or stress.[40,59] However, a change in steady state mRNA abundance does not necessarily mean that the corresponding gene actually has a function within the biological process in question. By analogy, it is known and accepted that changes in signal strength on a Northern blot do not necessarily indicate that a gene plays a role in a particular function or phenotype. Results from expression arrays or Northern blots inevitably lead to further experimental work before an unambiguous assignment of functionality can occur. Similarly, at the proteomic level, if a particular protein is found to be more abundant, it cannot directly be concluded that there will be more activity within the structural, metabolic, or signaling network associated with this protein.

It is likely that the greatest success will be achieved by functional genomics programs that adopt a wide range of different technologies and utilize them in accordance with the particular question under evaluation. Analysis at the biochemical level will certainly come to play an enhanced and particularly important role in plant genomics.[1-3] In addition, given the considerable commercial interest in plant biotechnology, it is likely that many of the future products will be produced from metabolic engineering, and, thus, a metabolic approach to functional genomics can be expected to be particularly fruitful. The number of plant functional genomics initiatives that take a balanced approach are currently low. One important and pioneering exception is an initiative targeted towards the generation of genomic information on the growth, development and environmental interactions of the model legume, *Medicago truncatula*.[60] This project aims to integrate data from a large scale EST sequencing program, gene expression profiling, metabolic profiling, and proteomics. In addition, populations of promoter trapped and activation tagged lines are being generated to allow the direct experimental testing of hypotheses.

The initial results from the application of metabolite profiling approaches to mutant and transgenic plants, as described above, indicate that the complexity of changes in plant metabolism should not be underestimated.[39,41,43] This supports the point made at the start of this section that the assignment of gene function should not be limited to the simple determination of the nature of the activity of a gene product. The process of gene discovery must also include the acquisition of information on the behavior of a particular gene within the complex regulatory networks present in an organism such as a higher plant. Metabolite profiling has now been demonstrated to provide general insight into the behavior of complex metabolic and physiological networks, and will certainly play an indispensable role in the characterization of gene function.

SUMMARY

In this chapter, some of the progress that has been made by using the modern tools of molecular plant physiology in the study of plant metabolism has been reviewed with respect to examples of projects targeted towards the metabolic engineering of starch and bioplastic production in plants. These studies reveal that the current limitation is no longer the generation of transgenic lines specifically altered in the desired enzymatic steps, but rather the speed of analysis of the metabolic networks in these lines. Where such analysis has occurred it has invariably generated surprises, indicating that we do not have enough knowledge of plant metabolism to accurately generate rational metabolic engineering strategies. A drive to increase the speed and width of metabolic analysis through the development of metabolite profiling procedures based upon GC/MS has enhanced the metabolic snapshot that can be quickly taken, and in all applications so far reported has led to substantial surprises. The power of metabolite profiling combined with advanced data mining approaches as a phenotyping tool recently has been demonstrated, and this raises prospects of an exciting future for accelerating progress in assembling the large jigsaw puzzle of plant functional genomics.

ACKNOWLEDGEMENTS

The author thanks Lothar Willmitzer, Karen Bohmert, Alisdair Fernie, Joachim Kopka, Oliver Fiehn and Ute Roessner at the Max Planck Institute für

Molekulare Pflanzenphysiologie in Golm, Germany for important and stimulating discussions.

REFERENCES

1. TRETHEWEY, R.N., KROTZKY, A.J., WILLMITZER, L., Metabolic profiling: a rosetta stone for genomics, *Cur. Opin. Plant Biol.*, 1999, **2**, 83-85.
2. TRETHEWEY, R.N., Gene discovery via metabolic profiling, *Cur. Opin. Biotechnol.*, 2001, **12**, 135-138.
3. GLASSBROOK, A., RYALS, J., A systematic approach to biochemical profiling, *Cur. Opin. Plant Biol.*, 2001, **4**, 186-90.
4. SIEDOW, J., STITT, M., Plant metabolism: where are all those pathways leading us?, *Cur. Opin. Plant Biol.*, 1998, **1**, 197-200.
5. FERNIE, A.R., WILLMITZER, L., TRETHEWEY, R.N., Sucrose to starch: a transition in molecular plant physiology, *Trends Plant Sci.*, 2002 (in press).
6. STITT, M., SONNEWALD, U., Regulation of metabolism in transgenic plants, *Annu. Rev. Plant Physiol. Plant Mol. Biol.*, 1995, **46**, 341-368.
7. STARK, D.M., TIMMERMANN, K.P., BARRY. G.F., PREISS, J., KISHORE, G.M., Regulation of the amount of starch in plant tissues by ADP glucose pyrophosphorylase, *Science*, 1992, **258**, 287-292.
8. SWEETLOVE, L.J., BURRELL, M.M., AP REES, T., Starch metabolism in tubers of transgenic potato (*Solanum tuberosum*) with increased ADPglucose pyrophosphorylase, *Biochem. J.*, 1996, **320**, 493-498.
9. HESSE, H., KREFT, O., MAIMANN, S., ZEH, M., WILLMITZER, L., HÖFGEN, R., Approaches towards understanding methionine biosynthesis in higher plants, *Amino Acids*, 2001, **20**, 281-289.
10. SONNEWALD, U., HAJIRAEZAEI, M.-R., KOSSMANN, J., HEYER, A., TRETHEWEY, R.N., WILLMITZER, L., Expression of a yeast invertase in the apoplast of potato tubers increases tuber size, *Nature Biotech.*, 1997, **15**, 794-797.
11. FARRÉ, E.M., BACHMANN, A., WILLMITZER, L., TRETHEWEY, R.N., Acceleration of potato tuber sprouting by the expression of a bacterial pyrophosphatase, *Nature Biotech.*, 2001, **19**, 268-72.
12. BOHMERT, K., BALBO, I., KOPKA, J., MITTENDORF, V., NAWRATH, C., POIRIER, Y., TISCHENDORF, G., TRETHEWEY, R.N., WILLMITZER, L., Transgenic *Arabidopsis* plants can accumulate polyhydroxybutyrate to up to 4% of their fresh weight, *Planta,* 2000, **211**, 841-845.
13. INOUE, H., TANAKA, A., Comparison of source and sink potentials between wild and cultivated potatoes, *J. Sci. Soil Management Japan,* 1978, **49**, 321-327.
14. DUNWELL, J.M., Transgenic approaches to crop improvement, *J. Exp. Bot.*, 2000, **51**, 487-496.
15. AP REES, T., MORRELL, S., Carbohydrate metabolism in developing potatoes, *Am. Potato J.*, 1990, **67**, 835-847.
16. OPARKA, K.J. PRIOR, D., C-14 sucrose efflux from the perimedulla of growing potato tubers, *Plant Cell Environ.*, 1987, **10**, 667-675.

17. OPARKA, K.J., WRIGHT, K.M., Osmotic regulation of starch synthesis in potato-tubers, *Planta*, 1988, **174**, 123-126.
18. KAMMERER, B., FISHCER, K., HILPERT, B., SCHUBERT, S., GUTENSOHN, M., WEBER, A., FLUGGE, U.-I., Molecular characterization of a carbon transporter in plastids from heterotrophic tissues: the glucose 6-phosphate phosphate antiporter, *Plant Cell*, 1998, **10**, 105-117.
19. TAUBERGER, E., FERNIE, A.R., EMMERMANN, M., KOSSMANN, J., WILLMITZER, L., TRETHEWEY, R.N., Antisense inhibition of plastidial phosphoglucomutase provides compelling evidence that potato tuber amyloplasts import carbon from the cytosol in the form of glucose-6-phosphate, *Plant J.*, 2000, **23**, 43-53.
20. SMITH, A.M., DENYER, K., MARTIN, C., The synthesis of the starch granule, *Annu. Rev. Plant. Physiol. Plant Mol. Biol.*, **48**, 1997, 65-87.
21. KOSSMANN, J., LLOYD, J., Understanding and influencing starch biochemistry, *Crit. Rev. Plant Sci.*, **19**, 2000, 171-226.
22. GEIGENBERGER, P., STITT, M., Sucrose synthase catalyses a readily reversible reaction *in vivo* in developing potato tubers and other plant systems, *Planta*, 1991, **189**, 329-339.
23. ZRENNER, R., SALANOUBAT, M., SONNEWALD, U., WILLMITZER, L., Evidence of the crucial role of sucrose synthase for sink strength using transgenic potato plants (*Solanum tuberosum* L.), *Plant J.*, 1995, **7**, 97-107.
24. ZRENNER, R., WILLMITZER, L., SONNEWALD, U., Analysis of the expression of potato uridinediphosphate-glucose pyrophosphorylase and its inhibition by antisense RNA, *Planta*, 1993, **190**, 247-252.
25. TRETHEWEY, R.N., GEIGENBERGER, P., RIEDEL, K, HAJIREZAEI, M.R., SONNEWALD, U., STITT, M., RIESMEIER, J.W., WILLMITZER, L., Combined expression of glucokinase and invertase in potato tubers leads to a dramatic reduction in starch accumulation and a stimulation of glycolysis, *Plant J.*, 1998, **15**, 109-118.
26. TRETHEWEY, R.N., REISMEIER, J.W., WILLMITZER, L., STITT, M., GEIGENBERGER, P., Tuber specific expression of a yeast invertase and a bacterial glucokinase in potato leads to an activation of sucrose phosphate synthase and the creation of a futile cycle, *Planta*, 1999, **208**, 227-238.
27. REIMHOLZ, R., GEIGER, M., HAAKE, V., DEITING, U., KRAUSE, K.P., SONNEWALD, U., STITT, M., Sucrose phosphate synthase is regulated by metabolites and protein phosphorylation in potato tubers, in a manner analogous to the enzyme in leaves, *Planta*, 1994, **192**, 480-488.
28. POIRIER, Y., Production of polyesters in transgenic plants, *Adv. Biochem. Eng. Biotechnol.*, 2001, **71**, 209-40.
29. STEINBÜCHEL, A., VALENTIN, H.E., Diversity of bacterial polyhydroxyalkanoic acids, *FEMS Microbiol. Lett.*, 1995, **128**, 219-228.
30. STEINBÜCHEL, A., SCHLEGEL, H.G., Physiology and molecular genetics of poly(β-hydroxy-alkanoic acid) synthesis in *Acaligenes eutrophus*, *Mol. Microbiol.*, 1991, **5**, 535-542.

31. STEINBÜCHEL, A., HUSTEDE, E., LIEBERGESELL, M., PIEPER, U., TIMM, A., VALENTIN, H., Molecular basis for biosynthesis and accumulation of polyhydroxyalkanoic acid in bacteria, *FEMS Micribiol. Rev.*, 1992, **103**, 217-230.
32. POIRIER, Y., DENNIS, D.E., KLOMPARENS, K., SOMMERVILLE, C., Polyhydroxybutyrate, a biodegradable thermoplastic, produced in transgenic plants, *Science*, 1992, **256**, 520-523.
33. NAWRATH, C., POIRIER, Y., SOMERVILLE C., Targeting of the polyhydroxybutyrate biosynthetic pathway to the plastids of *Arabidopsis thaliana* results in high levels of polymer accumulation, *Proc. Natl. Acad. Sci. USA*, 1994, **91**, 12760-12764.
34. MATSUMOTO, I., KAHURA, T., A new chemical diagnostic method for inborn errors of metabolism by mass spectrometry – rapid, practical, and simultaneous urinary metabolites analaysis, *Mass Spec. Reviews*, 1996, **15**, 43-57.
35. NING, C., KUHARA, T., INOUE, Y., ZHANG, C.H., MATSUMOTO, M., SHINKA, T., FURUMOTO, T., YOKOTA, K., MATSUMOTO, I., Gas chromatographic mass spectrometric metabolic profiling of patients with fatal infantile mitochondrial myopathy with De Toni-Fanconi-Debre syndrome, *Acta Paediatrica Japonica*, 1996, **38**, 661-666.
36. RASHED, M.S., BUCKNALL, M.P., LITTLE, D., AWAD, A., JACOB, M., ALAMOUDI, M., ALWATTAR, M., OZAND, P.T., Screening blood spots for inborn errors of metabolism by electrospray tandem mass spectrometry with a microplate batch process and a computer algorithm for automated flagging of abnormal profiles, *Clin. Chem.*, 1997, **43**, 1129-1141.
37. GOPAUL, S.V., FARRELL, K., ABBOTT, F.S., Gas chromatography/negative ion chemical ionization mass spectrometry and liquid chromatography /electrospray ionization tandem mass spectrometry quantitative profiling of N-acetylcysteine conjugates of valproic acid in urine: application in drug metabolism studies in humans, *J. Mass Spectrom.*, 2000, **35**, 698-704.
38. SAUTER, H., LAUER, M., FRITSH, H., Metabolic profiling of plants a new diagnostic technique. In: American Chemical Society Symposium Series (D.R. Baker, J.G. Fenyes and W.K. Moberg, eds.), American Chemical Society, Washington DC, USA; 1991, 443, pp. 288-299
39. ROESSNER, U., WAGNER, C., KOPKA, J., TRETHEWEY, R.N., WILLMITZER, L., Simultaneous analysis of metabolites in potato tuber by gas chromatography – mass spectrometry, *Plant J.*, 2000, **23**, 131-142.
40. RICHMOND, T., SOMERVILLE, S., Chasing the dream: plant EST microarrays, *Cur. Opin. Plant Biol.*, 2000, **3**, 108-116.
41. ROESSNER, U., LUEDEMANN, A., BRUST, D., FIEHN, O., LINKE, T., WILLMITZER, L., FERNIE, A.R., Metabolic profiling allows comprehensive phenotyping of genetically or environmentally modified plant systems, *Plant Cell* 2001, **13**, 11-29.
42. TRETHEWEY, R.N., FERNIE, A.R., BACHMANN, A., FLEISCHER-NOTTER, H., GEIGENBERGER, P, WILLMITZER, L., Expression of a bacterial sucrose phosphorylase in potato tubers results in a glucose-independent induction of glycolysis, *Plant Cell Environ.*, 2001, **24**, 357-365.

43. FIEHN, O., KOPKA, J., DÖRMANN, P., ALTMANN, T., TRETHEWEY, R.N., WILLMITZER, L., Metabolic profiling for plant functional genomics, *Nature Biotech.*, 2000, **18**, 1157-1161.
44. DÖRMANN, P., HOFFMANN-BENNING, S., BALBO, I., BENNING, C., Isolation and characterization of an *Arabidopsis* mutant deficient in the thylakoid lipid digalactosyl diacylglycerol, *Plant Cell*, 1995, **7**, 1801-1810.
45. HÄRTEL, H., LOKSTEIN, H., DÖRMANN, P., TRETHEWEY, R.N., BENNING, C., Photosynthetic light utilization and xantophyll cycle activity in the galactolipid deficient dgd1 mutant of *Arabidopsis thaliana*, *Plant Physiol. Biochem.*, 1998, **36**, 407-417.
46. DÖRMANN, P., BALBO, I., BENNING, C., *Arabidopsis* galactolipid biosynthesis and lipid trafficking mediated by DGD1, *Science*, 1999, **284**, 2181-2184.
47. BERGER, D., ALTMANN; T., A subtilisin-like serine protease involved in the regulation of stomatal density and distribution in *Arabidopsis thaliana*, *Gene Dev.*, 2000, **14**, 1119-1131.
48. http://www.ncbi.nlm.nih.gov/Genbank/genbankstats.html
49. http://www.ncbi.nlm.nih.gov/cgi-bin/Entrez/map_search?chr=arabid.inf
50. TERRYN, N., ROUZÉ, P., VAN MONTAGU, M., Plant genomics, *FEBS Lett.*, 1999, **452**, 3-6.
51. SOMERVILLE, C., SOMERVILLE, S., Plant functional genomics, *Science* 1999, **285**, 380-383.
52. FIELDS, S., KOHARA, Y., LOCKHART, D.J., Functional genomics, *Proc. Natl. Acad. Sci. USA*, 1999, **96**, 8825-8826.
53. FIEHN, O., KLOSKA, S., ALTMANN, T., Integrated studies on plant biology using multiparallel techniques, *Cur. Opin. Biotech.*, 2001, **12**, 82-86.
54. YATES, J.R., Mass spectrometry – from genomics to proteomics, *Trends in Genet.*, 2000, **16**, 5-8.
55. CELIS, J.E., KRUHOFFER, M., GROMOVA, I., FREDERIKSEN, C., OSTERGAARD, M., THYKJAER, T., GROMOV, P., YU, J.S., PALSDOTTIR, H., MAGNUSSON, N., ORNTOFT, T.F., Gene expression profiling: monitoring transcription and translation products using DNA microarrays and proteomics, *FEBS Lett.*, 2000, **480**, 2-16.
56. THIELLEMENT, H., BAHRMAN, N., DAMERVAL, C., PLOMION, C., ROSSIGNOL, M., SANTONI, V., DE VIENNE, D., ZIVY, M., Proteomics for genetic and physiological studies in plants, *Electrophoresis*, 1999, **20**, 2013-2026.
57. JACOBS, D.I., VAN DER HEIJDEN, R., VERPOORTE, R., Proteomics in plant biotechnology and secondary metabolism research, *Phytochem. Anal.*, 2000, **11**, 277-287.
58. VAN WIJK; K.J., Challenges and prospects of plant proteomics, *Plant Physiol.*, 2001, **126**, 501-508.
59. CUSHMANN, J.C., BOHNERT, H.J., Genomic approaches to plant stress tolerance, *Cur. Opin Plant Biol.*, 2000, **3**, 117-124.
60. BELL, C.J., DIXON, R.A., FARMER, A.D., FLORES, R., INMAN, J., GONZALES, R.A., HARRISON, M.J., PAIVA, N.L., SCOTT, A.D., WELLER, J.W., MAY, G.D., The *Medicago* genome initiative: a model legume database, *Nuc. Acids Res.*, 2001, **29**, 114-117.

Chapter Five

TRITERPENOID SAPONIN BIOSYNTHESIS IN PLANTS

A.E. Osbourn* and K. Haralampidis

*Sainsbury Laboratory,
John Innes Centre,
Colney Lane, Norwich NR4 7UH, UK.*

* *Author for correspondence, email: annie.osbourn@bbsrc.ac.uk*

Introduction	82
Cyclization of 2,3-Oxidosqualene – The First Committed Step in Triterpenoid Biosynthesis	82
Triterpenoid Saponin Biosynthesis in Monocots	84
Avenacins – Antimicrobial Phytoprotectants in Oat	84
Characterization of β-Amyrin Synthase from *Avena strigosa* – A Novel Oxidosqualene Cyclase4	85
sad1 Mutants of *A. strigosa* are Specifically Defective in AsbAS1	88
Future Prospects	88
Summary	89

INTRODUCTION

Plants synthesize a diverse array of secondary metabolites.[1] Since the ability to synthesize particular classes of secondary metabolites is restricted to certain plant groups, these compounds are clearly not essential for survival. However, evidence is accumulating to indicate that they confer selective advantages by protecting against pests, pathogens, and stress.[1,2] Some of these molecules may also have subtle physiological roles in plants that are as yet uncharacterized. In addition to their natural roles in plants, secondary metabolites also represent a vast resource of complex molecules that are valued and exploited by man for pharmacological and other uses.[1]

Saponins are glycosylated secondary metabolites that are widely distributed in the Plant Kingdom.[3,4] They are a diverse and chemically complex family of compounds that can be divided into three major groups depending on the structure of the aglycone, which may be a steroid, a steroidal alkaloid, or a triterpenoid. These molecules have been proposed to contribute to plant defense.[3-6] Saponins are also exploited as drugs and medicines and for a variety of other purposes.[4] Despite the considerable commercial interest in this important group of natural products, little is known about their biosynthesis. This is due in part to the complexity of the molecules, and also to the lack of pathway intermediates for biochemical studies.

A more detailed understanding of the biochemical pathways and enzymes involved in saponin biosynthesis will facilitate the development of plants with altered saponin content. In some cases, enhanced levels of saponins or the synthesis of novel saponins may be desirable (for example, for drug production[4] or improved disease resistance[3,5,6]), while for other plants, reduction in the content of undesirable saponins would be beneficial (for example, for legume saponins that are associated with antifeedant properties in animal feed[7]). This chapter is concerned with recent progress that has been made in the characterization of the enzymes and genes involved in the synthesis of these complex molecules and focuses on triterpenoid saponins.

CYCLIZATION OF 2,3-OXIDOSQUALENE – THE FIRST COMMITTED STEP IN TRITERPENOID BIOSYNTHESIS

Triterpenoid saponins are synthesized via the isoprenoid pathway.[4] The first committed step in triterpenoid saponin biosynthesis involves the cyclization of 2,3-oxidosqualene to one of a number of different potential products (Fig. 5.1).[4,8] Most plant triterpenoid saponins are derived from oleanane or dammarane skeletons although lupanes are also common.[4] This cyclization event forms a branchpoint with the sterol biosynthetic pathway in which 2,3-oxidosqualene is cyclized to cycloartenol in plants, or to lanosterol in animals and fungi.

TRITERPENOID SAPONIN BIOSYNTHESIS 83

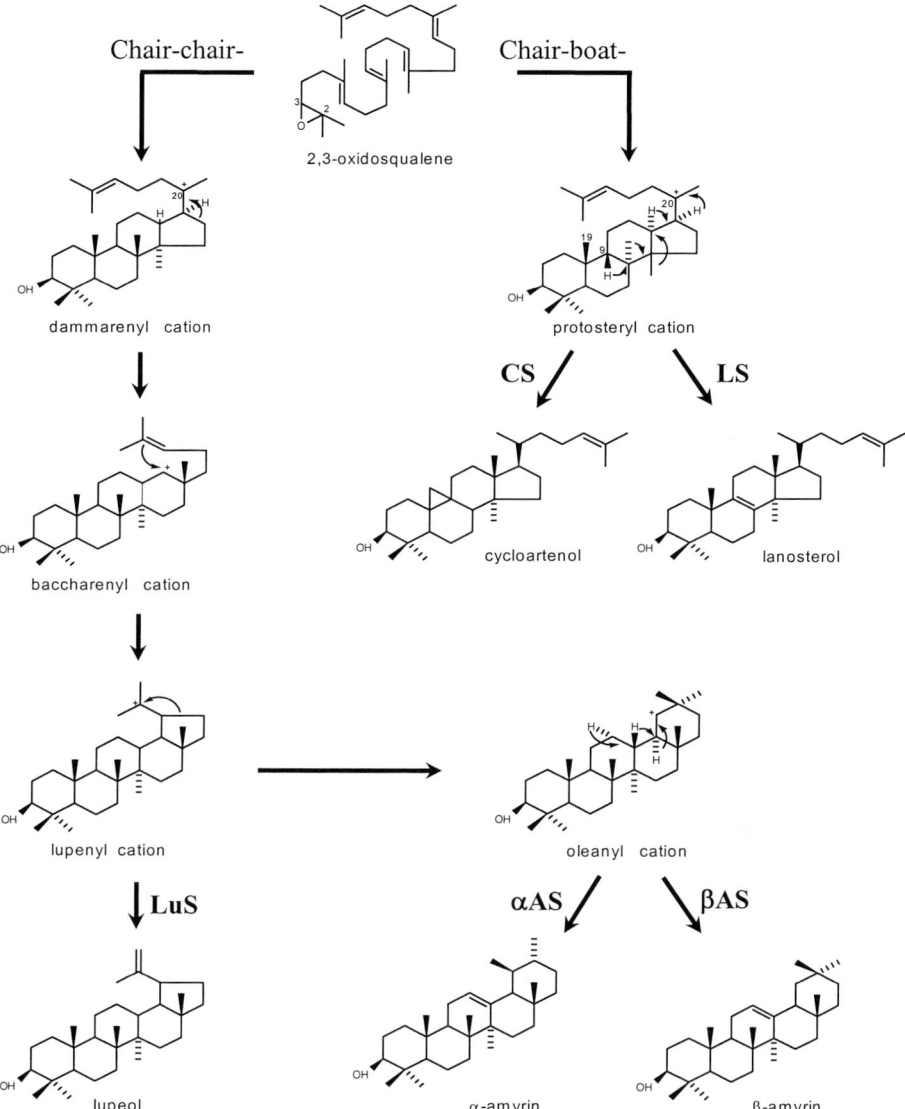

Fig. 5.1: Cyclization of 2,3-oxidosqualene to sterols and triterpenoids. The 2,3-oxidosqualene cyclase enzymes that catalyse the formation of the different products are indicated: LS, lanosterol synthase; CS, cycloartenol synthase; LuS, lupeol synthase; βAS, β-amyrin synthase; αAS, α-amyrin synthase.

Considerable advances have recently been made in the area of 2,3-oxidosqualene cyclization, and a number of genes encoding the oxidosqualene cyclase (OSC) enzymes that give rise to the diverse array of plant triterpenoid skeletons have been cloned. These include α-/β-amyrin synthase enzymes from *Panax ginseng*,[9,10] *Pisum sativum*,[11] and *Glycyrrhiza glabra*,[12] lupeol synthases from *Olea europaea* and *Taraxacum officinale*,[13] and multifunctional triterpene synthases from *Arabidopsis thaliana*.[14-16] These triterpene synthases all share sequence similarity with OSCs required for sterol biosynthesis but form discrete subgroups within the OSC superfamily that reflect the nature of their products.[8-16] It is clear that while some triterpenoid OSCs are highly specific in the products that they generate, others are multifunctional and form a number of different products, at least when expressed in heterologous expression systems.[9-11,13,14,16,17] Oleanane, lupane, and dammarane skeletons are the most common triterpenoid structures associated with saponin biosynthesis, but plants are likely to produce over 80 different oxidosqualene cyclase products.[18] Given the multifunctional properties of some triterpene synthases[14-16] and the fact that single amino acid changes can alter the nature of the product,[19,20] it seems likely that this structural diversity may be generated by a core set of "flexible" enzymes.

OSCs represent attractive tools for investigating the regulation of synthesis and the physiological role of triterpenoids, and potentially for manipulation of sterol and triterpenoid content.[21-24] Several lines of evidence indicate that manipulation of OSCs is likely to alter the flux through the isoprenoid pathway leading to the sterol and triterpenoid pathways.[21,22] OSCs have also been implicated as key regulatory steps in the synthesis of sterols and triterpenes in *Gypsophila paniculata*, *Saponaria officinalis*,[24] and *Tabernaemontana divaricata*.[22,23] Also, chemicals that inhibit lanosterol and cycloartenol synthases but that are not effective against β-amyrin synthases cause blocking of sterol biosynthesis and accumulation of β-amyrin at the expense of cycloartenol and methylene-24-cycloartenol.[25-28]

TRITERPENOID SAPONIN BIOSYNTHESIS IN MONOCOTS

Avenacins – Antimicrobial Phytoprotectants in Oat

Work in our laboratory has focussed on avenacins, a family of antifungal triterpenoid saponins that accumulate in the roots of oat (*Avena* spp.) (Fig. 5.2).[29,30] These secondary metabolites accumulate in the root epidermis[31,32] and have been implicated as chemical defenses against attack by soil fungi.[30,31] We have isolated a collection of sodium azide-generated saponin-deficient (*sad*) mutants of diploid oat that are defective in avenacin biosynthesis, and we have demonstrated that these mutants are impaired in their resistance to fungal pathogens,[6] providing good

TRITERPENOID SAPONIN BIOSYNTHESIS

evidence to indicate that these molecules do indeed play a role in plant defense. *sad* Mutants show enhanced susceptibility to the root pathogen *Gaeumannomyces graminis* var. *tritici*, which causes "take-all" disease of cereals, and also to other fungal pathogens such as *Fusarium* spp. Saponin deficiency and disease susceptibility were inseparable in segregating F_2 progeny, indicating that saponin deficiency is likely to be the cause of enhanced disease susceptibility.[6] Interestingly, other cereals and grasses do not appear to synthesize avenacins and are generally deficient in antifungal saponins.[3,4,30,32] Thus, the isolation of genes for avenacin biosynthesis may offer potential for the development of improved disease resistance in cultivated cereals.[6] Our efforts have, therefore, focussed on molecular genetic and biochemical dissection of this secondary metabolite pathway.

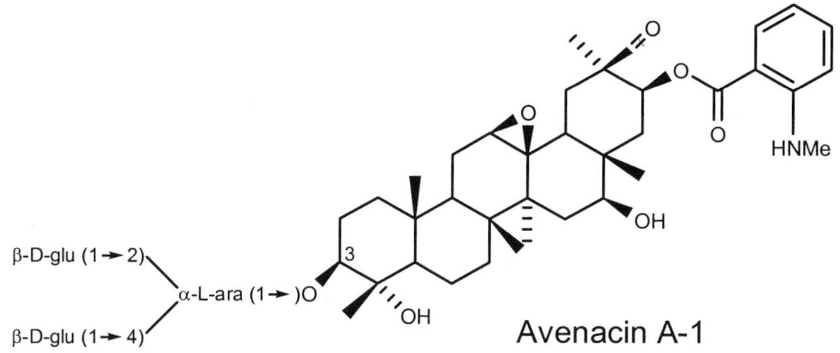

Fig. 5.2: The oat root triterpenoid saponin avenacin A-1.

Characterization of β-Amyrin Synthase from Avena strigosa – A Novel Oxidosqualene Cyclase

Avenacins are synthesized via β-amyrin.[4,33,34] These saponins are found primarily in young oat roots and do not occur in the foliar parts of the plant.[31] Incorporation of radioactivity from R [2-^{14}C]MVA into β-amyrin and avenacins occurs primarily in the root tips, and β-amyrin synthase activity is also highest in this region,[33] indicating that this is the site of synthesis. The subsequent conversion of β-amyrin into antifungal avenacins has not been biochemically characterized but is predicted to be a multi-step process involving cytochrome P450-dependent monooxygenases, glycosyltransferases, and other enzymes.[6,35]

Expressed sequence tag (EST) analysis of cDNAs from specific plant tissues has proved to be a valuable tool for the identification of genes for secondary metabolite biosynthesis.[36] We have used this approach to identify two distinct sequences predicted to encode OSCs from cDNA libraries from roots of diploid oat (*Avena strigosa*).[35] One of these sequences is highly homologous to cycloartenol synthases from other plants (e.g., amino acid sequence identities of 87% with *Oryza sativa* CS (AF169966) and 75% with *Arabidopsis thaliana* CS (U02555)), and represents the *A. strigosa* CS gene *AsCS1*. The second shares 55% amino acid identity with *AsCS1*, and was shown by expression in yeast (*Saccharomyces cerevisiae*) to encode β-amyrin synthase (AsbAS1).[35] *AsbAS1* is present as a single copy gene in the *A. strigosa* genome. The gene is expressed strongly in the epidermal cell layer of the root tips with little or no detectable transcript in the leaves, flowers, and shoots,[35] consistent with the organ-specific accumulation of the saponins[31,32] and also with the biochemical information indicating that the root tips are the site of synthesis.[33]

The deduced amino acid sequence of *AsbAS1* contains the conserved DCTAE motif implicated in substrate binding in OSCs,[37] and also four conserved QW motifs that are characteristic for the OSC superfamily.[38] Remarkably, AsbAS1 is clearly distinct from the other cloned βAS enzymes that have been characterized to date from other plant species, and is more closely related to lanosterol synthases from animals and fungi than to triterpenoid synthases or cycloartenol synthases from plants (Fig. 5.3).[35] There are substantial mechanistic differences in the processes of

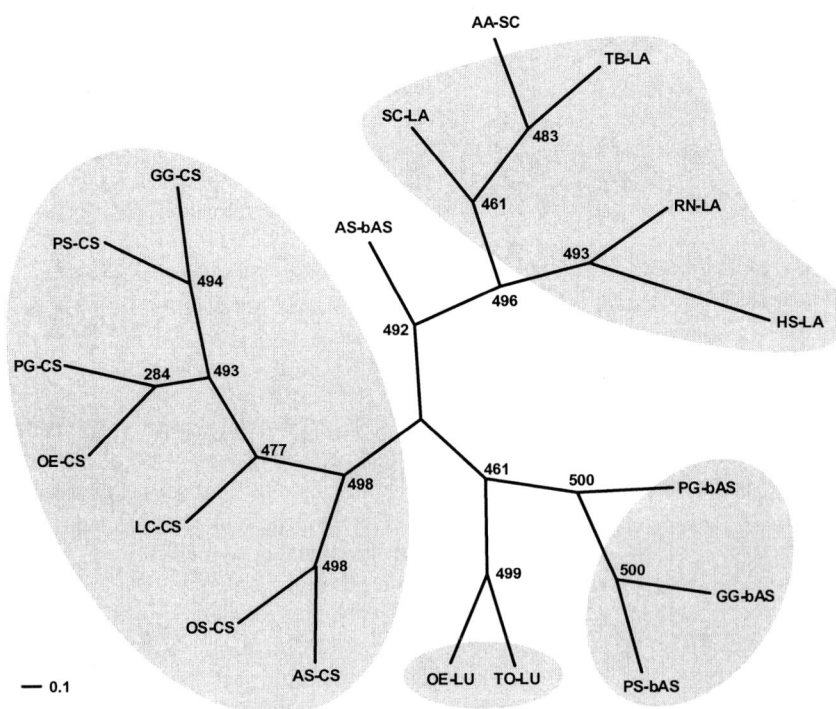

Fig. 5.3: Phenogram showing the relatedness between deduced amino acid sequences of members of the OSC superfamily. Clockwise from the top: AS-bAS, *A. strigosa* AsbAS1 (AJ311789); SC-LA, *Saccharomyces cerevisiae* lanosterol synthase (U04841); AA-SC, *Alicyclobacillus acidocaldarius* squalene-hopene cyclase (AB007002); TB-LA, *Trypanosoma brucei* lanosterol synthase (AF226705); RN-LA, *Rattus norvegicus* lanosterol synthase (U31352); HS-LA, *Homo sapiens* lanosterol synthase (U22526); β-amyrin synthases: PG-bAS, *Panax ginseng* (AB009030); GG-bAS, *Glycyrrhiza glabra* (AB037203); PS-bAS, *Pisum sativum* (AB034802); lupeol synthases: TO-LU, *Taraxacum officinale* (AB025345); OE-LU, *Olea europaea* (AB025343); cycloartenol synthases: AS-CS, *A, strigosa* AsCS1 (AJ311790); OS-CS, *Oryza sativa* (AF169966); LC-CS, *Luffa cylindrica* (AB033334); OE-CS, *Olea europaea* (AB025344); PG-CS, *Panax ginseng* (AB009029); PS-CS, *Pisum sativum* (D89619); GG-CS, *Glycyrrhiza glabra* (AB025968). The phylogenetic tree was constructed by using the UPGMA method as implemented in the "Neighbor" program of the PHYLIP package (Version 3.5c).[48] Amino acid distances were calculated using the Dayhoff PAM matrix method of the "Protdist" program of PHYLIP. The numbers indicate the numbers of bootstrap replications (out of 500) in which the given branching was observed. The protein parsimony method (the "Protpars" program of PHYLIP) produced trees with essentially identical topologies.

cyclization of 2,3-oxidosqualene to sterols and triterpenoids. Although these processes are catalyzed by related subgroups of enzymes, cyclization to yield sterols proceeds with the substrate in the "chair-boat-chair" conformation, while triterpenoid synthesis involves cyclization of the "chair-chair-chair" conformation of 2,3-oxidosqualene.[39] The amino acid residues that are required for sterol and triterpenoid determination in oxidosqualene cyclases have not yet been fully resolved, but the closer relatedness of AsbAS1 to lanosterol synthases rather than to triterpenoid synthases raises intriguing questions about the enzymology and evolution of this superfamily of enzymes.[35]

sad1 Mutants of A. strigosa Are Specifically Defective in AsbAS1

Genetic analysis indicates that two of the 10 *sad* mutants of *A. strigosa* that we isolated represent different mutant alleles at the *Sad1* locus.[6] These mutants accumulate radiolabelled 2,3-oxidosqualene but not β-amyrin when the roots are fed with ^{14}C-labelled precursor mevalonic acid, suggesting that the triterpenoid pathway is blocked between 2,3-oxidosqualene and β-amyrin.[34] The roots of these mutants also lack detectable β-amyrin synthase activity, but, like the wild type and the other mutants, are unimpaired in cycloartenol synthase (CS) activity and sterol biosynthesis.[34] The transcript levels for *AsbAS1* are substantially reduced in roots of *sad1* mutants, while *AsCS1* transcript levels are unaffected,[35] suggesting that the *sad1* mutants are either mutated in the *AsbAS1* gene itself or in a gene involved in its regulation.

DNA sequence analysis of the *AsbAS1* genes of both *sad1* mutants has resolved this by identifying single point mutations in the *AsbAS1* gene in each mutant that would be predicted to give rise to premature termination of translation.[35] The reduced *AsbAS1* transcript levels in the *sad1* mutants may, therefore, be due to nonsense-mediated mRNA decay.[40-44] Genotyping of F_2 populations that segregate for the saponin-deficient phenotype indicated that there was no recombination between *AsbAS1* and *Sad1*. Taken together, the evidence that *Sad1* encodes AsbAS1 is compelling.[35] To our knowledge the *sad1* mutants are the first βAS mutants to be isolated for any plant. Although it has been suggested that β-amyrin may act as a structural component of plant membranes,[45,46] the *sad1* mutants were not obviously affected in root morphology, growth, tillering, flowering time, or seed production.[6,35] Thus, while AsbAS1 is essential for saponin biosynthesis and disease resistance, it is not required for normal plant growth and development.

FUTURE PROSPECTS

There have been a number of exciting recent advances in the molecular characterization of the OSC enzymes that catalyze the first committed step in the synthesis of triterpenoid plant secondary metabolites. Further advances in this area will give a comprehensive insight into the mechanistic enzymology and evolution of this fascinating superfamily of enzymes, and should enable the development of new enzymes that generate novel products by "accelerated evolution". The oat enzyme AsbAS1 is the first triterpene synthase to be characterized from monocots.[35] This enzyme is clearly distinct from other β-amyrin synthases that have been cloned from other plants, and defines a new class of triterpene synthases within the OSC superfamily. Orthologs of *AsbAS1* are absent from modern cereals,[35] raising the possibility that this gene could be exploited to enhance disease resistance in crop plants. The availability of cloned genes for plant triterpene synthases will now enable the regulation of triterpenoid biosynthesis to be investigated in more detail, and the effects of altering the levels of these enzymes on sterol and triterpenoid synthesis to be assessed via the generation of transgenic plants.

The synthesis of saponins from the cyclization product of 2,3-oxidosqualene involves a series of further modifications, including a variety of oxidation and substitution events and glycosylation.[4,8] Little is known about the enzymes and genes involved in the elaboration of the triterpenoid skeleton, although genetic and biochemical analysis of saponin-deficient mutants of plants is likely to accelerate the dissection of these processes. Progress has been made in the characterization of saponin glucosyltransferases (primarily for steroidal and steroidal alkaloid saponins) (reviewed in 8), and the first of these enzymes has recently been cloned from potato.[47] Since glycosylation at the C-3 hydroxyl position confers amphipathic properties on the molecule and is normally critical for biological activity,[2-5] this is clearly an important area in which to invest effort in the future.

SUMMARY

Saponins are an important group of glycosylated plant secondary metabolites. Their natural role is likely to be in the protection of plants against attack by pests and pathogens.[3-6,35] They are also exploited commercially as drugs and medicines, and for a variety of other purposes.[4] Saponins are synthesized via the isoprenoid pathway from 2,3-oxidosqualene,[4] which is then cyclized to sterol or triterpenoid products by different 2,3-oxidosqualene cyclases. We have recently cloned and characterized the novel oxidosqualene cyclase *AsbAS1*, which catalyzes the synthesis of the triterpenoid saponin precursor β-amyrin in oat.[35] This enzyme is unusual in that it is more closely related to lanosterol synthases from animals and fungi than to other oxidosqualene cyclases (triterpenoid synthases or cycloartenol synthases) from plants. *AsbAS1* is required for the synthesis of triterpenoid avenacin saponins and for resistance to a variety of pathogens, indicating that avenacins contribute to disease resistance in oat. *AsbAS1* and other as yet uncharacterized genes required for saponin biosynthesis have potential for the development of plants with altered saponin content through metabolite engineering. In some cases, enhanced levels of saponins or the synthesis of novel saponins may be beneficial (for drug production or improved disease resistance, for example), while in others the objective may be to reduce the content of undesirable saponins (such as those associated with antinutritional effects in legumes).

ACKNOWLEDGMENTS

The Sainsbury Laboratory is supported by the Gatsby Charitable Foundation. We acknowledge G. Bryan, K. Papadopoulou, X. Qi, S. Bakht, Rachel Melton, who all contributed to the work that has been summarized in this manuscript and DuPont Agricultural Products for DNA sequence analysis and financial support.

REFERENCES

1. WINK, M., Functions of Plant Secondary Metabolites and their Exploitation in Biotechnology, Sheffield Academic Press, 1999, 362 p.
2. TSCHESCHE, R., Advances in the chemistry of antibiotic substances from higher plants. In: Pharmacognosy and Phytochemistry, (H. Wagner and L. Hörhammer, eds,), Springer-Verlag, Berlin. 1971, pp. 274-289.
3. PRICE, K.R., JOHNSON, I.T., FENWICK, G.R., The chemistry and biological significance of saponins in food and feedingstuffs, *CRC Crit. Rev. Food Sci. Nutr.,* 1987, **26,** 27-133.
4. HOSTETTMANN, K.A., MARSTON, A., Saponins. Chemistry and Pharmacology of Natural Products, Cambridge University Press, 1995.
5. MORRISSEY, J.P., OSBOURN, A.E., Fungal resistance to plant antibiotics as a mechanism of pathogenesis, *Microbiol. Mol. Biol. Revs.*, 1999, **63,** 708-724.
6. PAPADOPOULOU, K., MELTON, R.E., LEGGETT, M., DANIELS, M.J., OSBOURN, A.E., Compromised disease resistance in saponin-deficient plants, *Proc. Natl. Acad. Sci., USA*, 1999, **96,** 12923-12928.
7. FENWICK, G.R., PRICE, K.R., TSUKAMOTA., C, OKUBO, K., Saponins. In: Toxic Substances in Crop Plants (J.P. D'Mello, C.M. Duffus, J.H. Duffus, eds,), The Royal Society of Chemistry, Cambridge. 1992, pp. 284-327.
8. HARALAMPIDIS, K., TROJANOWSKA, M., OSBOURN, A.E., Biosynthesis of triterpenoid saponins in plants, *Adv. Biochem. Eng.*, in press.
9. KUSHIRO, T., SHIBUYA, M., EBIZUKA, Y., β-Amyrin synthase. Cloning of oxidosqualene cyclase that catalyzes the formation of the most popular triterpene among higher plants, *Eur. J. Biochem.*, 1998, **256,** 238-244.
10. KUSHIRO, T., SHIBUYA, M., EBIZUKA, Y., Molecular cloning of oxidosqualene cyclase cDNA from *Panax ginseng*: The isogene that encodes β-amyrin synthase. In: Towards Natural Medicine Research in the 21st Century, (H. Ageta, N. Aimi, Y. Ebizuka, T. Fujita and G. Honda, eds,), Elsevier Science, Amsterdam. 1998, pp. 421-427.
11. MORITA, M., SHIBUYA, M., KUSHIRO, T., MASUDA, K., EBIZUKA, Y., Molecular cloning and functional expression of triterpene synthases from pea (*Pisum sativum*), *Eur. J. Biochem.*, 2000, **267,** 3453-3460.
12. HAYASHI, H., HUANG, P.Y., KIRAKOSYAN, A., INOUE, K., HIRAOKA, N., IKESHIRO, Y., KUSHIRO, T., SHIBUYA, M., EBIZUKA, Y., Cloning and characterization of a cDNA encoding beta-amyrin synthase involved in glycyrrhizin and soyasaponin biosynthesis in licorice, *Biol. Pharm. Bull.*, 2001, **24,** 912-916.

13. SHIBUYA, M., ZHANG, H., ENDO, A., SHISHIKURA, K., KUSHIRO, T., EBIZUKA, Y., Two branches of the lupeol synthase gene in the molecular evolution of plant oxidosqualene cyclases, *Eur. J. Biochem.*, 1999, **266**, 302-307.
14. HERRERA, J.B.R., BARTEL, B., WILSON, W.K., MATSUDA, S.P.T., Cloning and characterization of the *Arabidopsis thaliana* lupeol synthase gene, *Phytochemistry*, 1998, **49**, 1905-1911.
15. HUSSELSTEIN-MULLER, T., SCHALLER, H., BENVENISTE, P., Molecular cloning and expression in yeast of 2,3-oxidosqualene-triterpenoid cyclases from *Arabidopsis thaliana*, *Plant Mol. Biol.*, 2001, **45**, 75-92.
16. KUSHIRO, T., SHIBUYA, M., MASUDA, K., EBIZUKA, Y., A novel multifunctional triterpene synthase from *Arabidopsis thaliana*, *Tetrahedron Letts.*, 2000, **41**, 7705-7710.
17. SEGURA, M.J.R., MEYER, M.M., MATSUDA, S.P.T., *Arabidopsis thaliana* LUP1 converts oxidosqualene to multiple triterpene alcohols and a triterpene diol, *Org. Letts.*, 2000, **2**, 2257-2259.
18. MATSUDA, S.P.T., On the diversity of oxidosqualene cyclases. In: Biochemical Principles and Mechanisms of Biosynthesis and Degradation of Polymers (A. Steinbüchel, ed,), Wiley-VCH, Weinheim. 1998, pp. 300-307.
19. KUSHIRO, T., SHIBUYA, M., EBIZUKA, Y., Chimeric triterpene synthase. A possible model for multifunctional triterpene synthase, *J. Am. Chem. Soc.*, 1999, **121**, 1208-1216.
20. KUSHIRO, T., SHIBUYA, M., MASUDA, K., EBIZUKA, Y., Mutational studies of triterpene synthases: Engineering lupeol synthase into β-amyrin synthase, *J. Am. Chem. Soc.*, 2000, **122**, 6816-6824.
21. BAISTED, D.J., Sterol and triterpene synthesis in the developing and germinating pea seed, *Biochem. J.* 1971, **124**, 375-383.
22. THRELFALL, D.R., WHITEHEAD, I.M., Redirection of terpenoid biosynthesis in elicitor-treated plant cell suspension cultures. In: Plant Lipid Biochemistry (P.J. Quinn and J.L. Harwood, eds,), Portland Press, London. 1990, pp. 344-346.
23. VAN DER DEIJDEN, R., THRELFALL, D.R., VERPOORTE, R., WHITEHEAD, I.M., Regulation and enzymology of pentacyclic triterpenoid phytoalexin biosynthesis in cell suspension cultures of *Tabernaemontana divaricata*, *Phytochemistry*, 1989, **28**, 2981-2988.
24. HENRY, M., RAHIER, A., TATON, M., Effect of gypsogenin 3,O-glucuronide pretreatment of *Gypsophila paniculata* and *Saponaria officinalis* cell suspension cultures on the activities of microsomal 2,3-oxidosqualene cycloartenol and amyrin cyclases, *Phytochemistry*, 1992, **31**, 3855-3859.
25. TATON, M., BENVENISTE, P., RAHIER, A., Inhibition of 2,3-oxidosqualene cyclases, *Biochemistry*, 1992, **31**, 7892-7898.
26. TATON, M., BENVENISTE, P., RAHIER, A., N-[(1,5,9)-trimethyl-decyl]-4α,10-dimethyl-8-aza-trans-decal-3β-ol; A novel potent inhibitor of 2,3-oxidosqualene cycloartenol and lanosterol cyclases, *Biochem. Biophys. Res. Comm.*, 1986, **138**, 764-770.

27. CATTEL, L., CERUTI, M., 2,3-Oxidosqualene cyclase and squalene epoxidase: Enzymology, mechanism and inhibitors. In: Physiology and Biochemistry of Sterols (G.W. Patterson and W.D. Nes, eds,), American Oil Chemists' Society, Champaign. 1992, pp. 50-82.
28. TATON, M., CERUTI, M., CATTEL, L., RAHIER, A., Inhibition of higher plant 2,3-oxidosqualene cyclases by nitrogen-containing oxidosqualene analogues, *Phytochemistry*, 1996, **43,** 75-81.
29. CROMBIE, L., CROMBIE, W.M.L., WHITING, D.A., Structures of the oat root resistance factors to take-all disease, avenacins A-1, A-2, B-1 and B-2 and their companion substances, *J. Chem. Soc., Perkins,* 1985, **I,** 1917-1922.
30. CROMBIE, W.M.L., CROMBIE, L., Distribution of avenacins A-1, A-2, B-1 and B-2 in oat roots: Their fungicidal activity towards 'take-all' fungus, *Phytochemistry*, 1986, **25,** 2069-2073.
31. TURNER, E.M., The nature of resistance of oats to the take-all fungus, *J. Exp. Bot.*, 1953, **4,** 264-271.
32. OSBOURN, A.E., CLARKE, B.R., LUNNESS, P., SCOTT, P.R., DANIELS, M.J., An oat species lacking avenacin is susceptible to infection by *Gaeumannomyces graminis* var. *tritici, Physiol. Mol. Plant Pathol.*, 1994, **45,** 457-467.
33. TROJANOWSKA, M.R., OSBOURN, A.E., DANIELS, M.J., THRELFALL, D.R., Biosynthesis of avenacins and phytosterols in roots of *Avena sativa* cv. Image, *Phytochemistry*, 2000, **54,** 153-164.
34. TROJANOWSKA, M.R., OSBOURN, A.E., DANIELS, M.J., THRELFALL, D.R., Investigation of avenacin-deficient mutants of *Avena strigosa*, *Phytochemistry*, 2001, **56,** 121-129.
35. HARALAMPIDIS, K., BRYAN, G., Qi, X., PAPADOPOULOU, K., BAKHT, S., MELTON, R, OSBOURN, A.E., A new class of oxidosqualene cyclases directs synthesis of antimicrobial phytoprotectants in monocots, *Proc. Natl. Acad. Sci., USA*, 2001, **98,** 13431-13436.
36. OHLROGGE, J., BENNING, C., Unraveling plant metabolism by EST analysis, *Curr. Opin. Plant Biology*, 2000, **3,** 224-228.
37. ABE, I., PRESTWICH, G.D., Identification of the active site of vertebrate oxidosqualene cyclase, *Lipids*, 1995, **30,** 231-234.
38. PORALLA, K., HEWELT, A., PRESTWICH, G.D., ABE, I., REIPEN, I., SPRENGER, G., A specific amino acid repeat in squalene and oxidosqualene cyclases, *TIBS*, 1994, **19,** 157-158.
39. ABE, I., ROHMER, M., PRESTWICH, G.D., Enzymatic cyclization of squalene and oxidosqualene to sterols and triterpenes, *Chem. Rev*, 1993, **93,** 2189-2206.
40. HILLEREN, P., PARKER, R., Mechanisms of mRNA surveillance in eukaryotes. *Annu. Rev. Genet,* 1999, **33,** 229-260.
41. JOFUKU, K.D., SCHIPPER, R.D., GOLDBERG, R.B., A frameshift mutation prevents Kunitz trypsin-inhibitor messenger-RNA accumulation in soybean embryos, *Plant Cell*, 1989, **1,** 427-435.
42. VOELKER, T.A., MORENO, J., CHRISPEELS, M.J., Expression analysis of a pseudogene in transgenic tobacco - a frameshift mutation prevents messenger-RNA accumulation, *Plant Cell*, 1990, **2,** 255-261.

43. VAN HOOF, A., GREEN, P.J., Premature nonsense codons decrease the stability of phytohemagglutinin mRNA in a position-dependent manner, *Plant J.*, 1996, **10,** 415-424.
44. PETRACEK, M.E., NUYGEN, T., THOMPSON, W.F., DICKEY, L.F., Premature termination codons destabilize ferredoxin-1 mRNA when ferredoxin-1 is translated, *Plant J.*, 2000, **21,** 563-569.
45. NES, W.D., HEFTMANN, E., A comparison of triterpenoids with steroids as membrane- components, *J. Nat. Prod.*, 1981, **44,** 377-400.
46. FANG, T-Y., BAISTED, D.J., 2,3-Oxidosqualene cyclase and cycloartenol-S-adenosylmethionine methyltransferase activitites *in vivo* in the cotyledon and axis tissues of germinating pea seeds, *Biochem. J.*, 1975, **150,** 323-328.
47. MOEHS, C.P., ALLEN, P.V., FRIEDMAN, M., BELKNAP, W.R., Cloning and expression of solanidine UDP-glucose glycosyltransferase from potato, *Plant J.*, 11: 227-236.
48. FELSENSTEIN, J., Inferring phylogenies from protein sequences by parsimony, distance, and likelihood methods, *Methods Enzymol.*, 1996, **266,** 418-427.

Chapter Six

A MUTATIONAL APPROACH TO DISSECTION OF FLAVONOID BIOSYNTHESIS IN *ARABIDOPSIS*

Brenda Winkel-Shirley

Department of Biology
Virginia Tech
Blacksburg, VA, USA 24061-0406

e-mail: winkel@vt.edu

Introduction 96
Flavonoid Mutants in *Arabidopsis* 98
The Mechanism of Transport of Flavonoids into the Vacuole 101
Flavonoids and Polar Auxin Transport 101
Intracellular Organization of The Flavonoid Pathway as a Membrane
    Associated Multienzyme Complex 103
Use of Recombinant scFv Antibodies to Disrupt Flavonoid
    Metabolism in Transgenic Plants 105
Summary 106

INTRODUCTION

Flavonoids are an important class of plant secondary metabolites that are derived from phenylalanine via the general phenylpropanoid pathway and malonyl CoA, which is generated from citrate (Fig. 6.1).[1,2] These compounds play a diverse array of essential roles in the interaction of plants with the environment, from providing pigments for protection against UV radiation and the recruitment of pollinators, to communicating with other plants and microbes, including those involved in nitrogen fixation.[3-5] Flavonoids also appear to function in regulating polar auxin transport, thus helping mediate plant growth in response to light and gravity,[6] and in some species they are essential for male fertility.[7] Despite these varied roles, flavonoids are generally dispensable under greenhouse conditions, and mutations in the biosynthetic pathway result in easily-identifiable phenotypes such as altered flower or seed color. As a result, genetic approaches have proven to be useful for studying the regulation of this pathway and identifying its diverse physiological functions, particularly in maize, petunia, snapdragon, and, more recently, *Arabidopsis*. Each of these species offers unique attributes for these studies and together have facilitated the accumulation of a significant body of knowledge in the area of flavonoid metabolism. This chapter discusses the use of mutational approaches in *Arabidopsis* for the identification of regulatory factors controlling flavonoid gene expression, and for understanding the mechanism of transport of flavonoids into the vacuole, the role of flavonoids in auxin transport, and the intracellular organization of the pathway as a membrane-associated multienzyme complex. The use of the flavonoid pathway to test the feasibility of a new method of mutagenesis, involving expression of recombinant scFv antibodies in transgenic plants, is also discussed. Together, these studies provide a good example of how classical genetic approaches can be combined with contemporary technologies to dissect the biochemistry, cell biology, and physiological functions of plant metabolism.

Figure 6.1: Major branch pathways of flavonoid biosynthesis in *Arabidopsis*. Branch pathways, enzymes, and end products present in other plants but not *Arabidopsis* are shown in light gray. Abbreviations: cinnamate-4-hydroxylase (C4H), chalcone isomerase (CHI), chalcone synthase (CHS), 4-coumarate:CoA-ligase (4CL), dihydroflavonol 4-reductase (DFR), flavanone 3-hydroxylase (F3H), flavonoid 3' or 3'5' hydroxylase (F3'H, F3'5'H), leucoanthocyanidin dioxygenase (LDOX), leucoanthocyanidin reductase (LCR), *O*-methyltransferase (OMT), phenylalanine ammonia-lyase (PAL), rhamnosyl transferase (RT), and UDP flavonoid glucosyl transferase (UFGT).

MUTATIONAL APPROACH -- FLAVONOID BIOSYNTHESIS

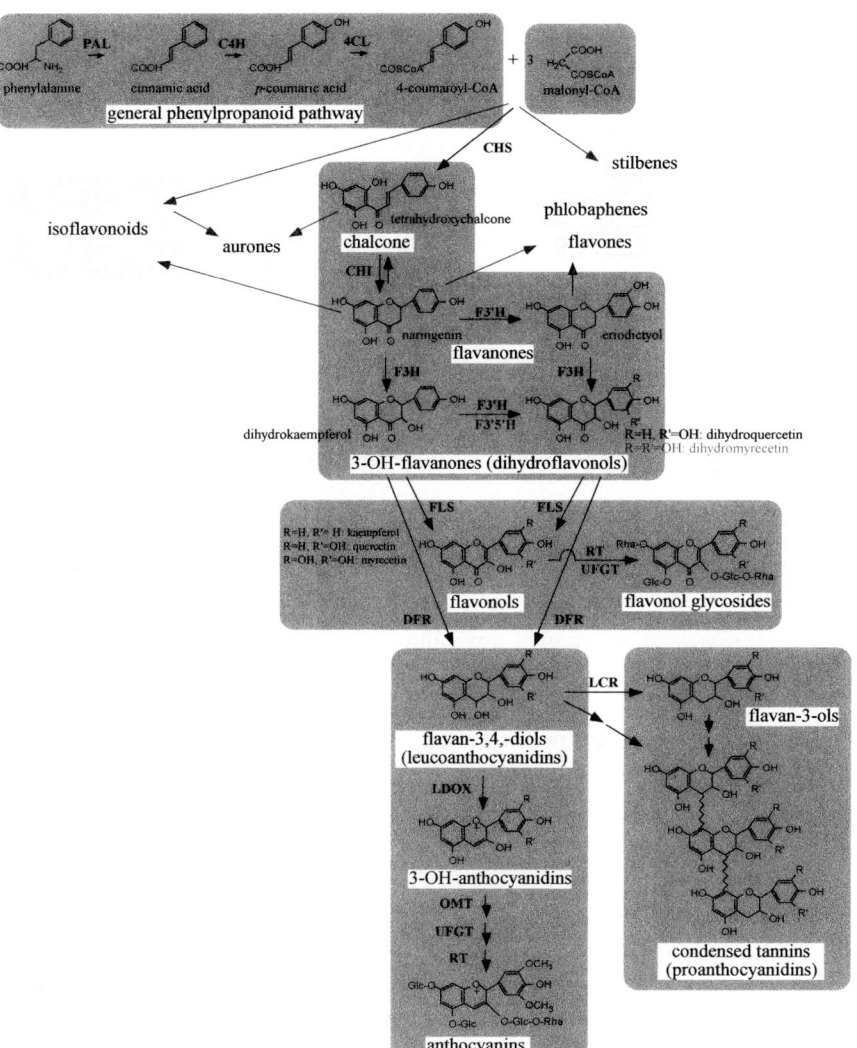

FLAVONOID MUTANTS IN ARABIDOPSIS

Genes required for the production of flavonoids in *Arabidopsis* have been identified largely on the basis of changes in the accumulation of a polymerized form of flavonoids, known as proanthocyanidins or condensed tannins, in the seed coat. The *transparent testa* (*tt*) phenotype, characterized by a distinct yellow or pale brown seed color, is the most common result of mutations in the proanthocyanidin pathway (Fig. 6.2A). The first *Arabidopsis* flavonoid mutants were described in 1971, identified in screens of mutagenized populations that were carried out by Dorothea Bürger at the University of Göttingen.[8] Isolation of additional *tt* mutants and the first detailed functional characterizations were performed by Maarten Koornneef's group at the Agricultural University at Wageningen in the 1980's.[9-11] A number of additional flavonoid mutants have been isolated since that time, several in screens for reduced seed dormancy that uncovered mutants with more subtle effects on pigmentation (Table 6.1). There are currently 19 *tt* mutants, plus several others with somewhat different phenotypes, that are known to affect flavonoid biosynthesis in *Arabidopsis*.

Genes encoding all of the enzymes required for synthesis of the basic flavonoid skeleton have been isolated in *Arabidopsis* and correlated with genetic loci identified by *tt* mutants (Table 6.1). These include genes for chalcone synthase (CHS, *tt4*), chalcone isomerase (CHI, *tt5*), flavanone 3-hydroxylase (F3H, *tt6*), flavonoid 3'-hydroxylase (F3'H, *tt7*), dihydroflavonol reductase (DFR, *tt3*), and leucoanthocyanidin dioxygenase (LDOX, *tt18*), all of which are single-copy in this species[12-17](Atsushi Tanaka, personal communication). Mutations in these genes affect not only seed coat pigmentation, but also the accumulation of colored anthocyanins in developing seedlings (Fig. 6.2B). Erich Grotewold's goup has recently shown that maize genes can complement the *tt3, tt4,* and *tt5* mutants, indicating that flavonoid enzymes have been functionally conserved over large evolutionary distances.[18] Interestingly, the enzyme that catalyzes the branch step leading to flavonols is encoded by multiple genes in *Arabidopsis*. Six sequences with high homology to flavonol synthase (FLS) are present in the *Arabidopsis* genome, although two appear to be pseudogenes (Owens, Bandara, and Winkel-Shirley, unpublished data). No mutant locus for FLS has been identified based on the *tt* phenotype or in any other screen of *Arabidopsis* mutants, presumably because flavonols do not contribute significantly to seed coat pigmentation in this species and/or because the genes are functionally redundant. Bernd Weisshaar's group has, however, identified an *En-1* insertional mutant for FLS1 by using a reverse genetic screen.[16] Analysis of the end products accumulating in the various *tt* mutants has confirmed that the basic scheme for the flavonoid pathway is maintained in *Arabidopsis*, resulting in the synthesis of three major classes: the anthocyanins, flavonols, and proanthocyanidins (Fig. 6.1). Characterization of flavonoid mRNA, protein, and end product levels in these lines has also uncovered evidence for feedback effects within the flavonoid pathway and for crosstalk among phenylpropanoid branch pathways, particularly sinapate ester biosynthesis, which is upregulated in *tt4* and downregulated in *tt5*[19,20] (Barthet and Winkel-Shirley, unpublished data). The mechanisms underlying these interactions remain to be elucidated.

The majority of *Arabidopsis* flavonoid structural genes have been isolated based on homology to genes from other plant species. Recently, however, *Arabidopsis* contributed to the isolation of a novel gene, named *BANYULS*, which encodes a DFR-like protein and results in the accumulation of red pigments in the seed coat when disrupted.[21] The BAN protein has been suggested to have leucoanthocyanidin reductase (LCR) activity and may, thus, be important in condensed tannin biosynthesis. Loss of function is hypothesized to shift flux into anthocyanin biosynthesis, resulting in the overaccumulation of anthocyanins in the seed coat, rather than a *tt* phenotype. This gene is of particular interest in light of efforts to engineer the synthesis of condensed tannins for nutritional enhancement of crop species and to minimize pasture bloat in forage crops.[22]

Although the enzymatic features of central flavonoid metabolism are highly conserved among plant species, the regulatory machinery appears to be somewhat dissimilar. This may reflect variations in the physiological functions of flavonoids among plants and a requirement for diverse expression patterns. Genetic approaches to identifying these genes have been essential, as homology-based approaches are limited by the presence of sequences such as basis helix-loop-helix (bHLH) and myb domains in many otherwise-unrelated regulatory factors. A number of the *tt* loci have been useful in this regard. For example, the *TRANSPARENT TESTA GLABRA1 (TTG1)* gene, which is required for expression of the "late" flavonoid genes, *DFR* and *LDOX,* was isolated by positional cloning and found to encode a protein containing four WD-40 repeats.[23] This protein has homology with AN11, a regulator of flavonoid biosynthesis in petunia, but also has some important differences that suggest that these proteins may regulate different transcription factors. Interestingly, mutations at the *TTG1* locus have pleiotropic effects, disrupting not only flavonoid accumulation, but also trichome development, root epidermal cell patterning and the production of mucilage by seeds.[10,24] The connection with trichome development has been reported in only one other plant species, the related crucifer *Matthiola incana*.[25] A second locus with a similar phenotype, *TTG2,* was more recently identified in *Arabidopsis* and found to encode a protein belonging to the WRKY family of transcription factors.[26] Two other transcriptional regulators have been cloned in *Arabidopsis*, *TT2* and *TT8,* which are also required for expression of "late" genes.[27,28] The *TT2* and *TT8* genes were isolated by T-DNA tagging in Loïc Lepiniec's group and found to encode a myb domain protein and a bHLH protein, respectively. One other myb regulator, *PAP1,* has also been identified in *Arabidopsis* and appears to be a general regulator of

Figure 6.2: Flavonoid pigmentation in *Arabidopsis*. (A) Comparison of the color of wild-type (brown), *tt4* (yellow), and *tt7* (pale brown) seeds, shown here in black and white. The altered pigmentation of the mutants is due to a reduction or absence of proanthocyanidins in the seed coat, revealing the yellow color of the underlying cotyledons. (B) Anthocyanin pigments are visible as a reddish color in the hypocotyl and around the edge of cotyledons (dark areas in this photgraph) in wild-type seedlings grown in the presence of 2M sucrose. This coloration reaches a peak on day 4 of germination. The levels of flavonoid gene mRNA's are maximal around day 3, while the pattern of enzyme expression more closely parallels the accumulation of pigments. [15,20,50,56,57]

phenylpropanoid metabolism, including the flavonoid pathway.[29] The nature of the relationships among these genes, including whether any of the gene products physically interact with one another, remains to be determined. At the other end of the signal transduction spectrum, information on regulation of CHS in response to blue and UV light has come from studies of cryptochrome- and phytochrome-deficient mutants in *Arabidopsis*.[30] A more complete understanding of how regulation of flavonoid gene expression differs in *Arabidopsis* and other plant species may provide insights into the evolution of flavonoid metabolism and the recruitment of its endproducts for specific biological functions.

THE MECHANISM OF TRANSPORT OF FLAVONOIDS INTO THE VACUOLE

For many years, the issue of how flavonoids were transported from the site of synthesis in the cytosol to one of their major destinations, the vacuole, remained a mystery. Studies in maize and petunia have recently tied glutathione *S*-transferases (GST) to the vacuolar uptake of flavonoids, suggesting that flavonoids are conjugated to glutathione and then transported via a glutathione-conjugate pump, or that unconjugated flavonoids are transported by a cotransport mechanism with reduced GSH.[31,32] However, an *Arabidopsis* EST with 50% amino acid identity to the petunia GST, AN9, was not able to complement the vacuolar uptake deficiency in an *an9* mutant, suggesting that a different more distantly-related GST performs this function.[33] However, it is also possible that a different mechanism is used in *Arabidopsis*. Debeaujon *et al.* have described the isolation and characterization of a member of the MATE family of transporters, identified by the *TT12* locus, that appears to be involved in sequestration of flavonoids in vacuoles of the seed coat endothelium.[34] It remains to be determined whether different plants use different transport mechanisms, or whether each contains multiple mechanisms, perhaps for the transport of different flavonoid end products.

FLAVONOIDS AND POLAR AUXIN TRANSPORT

Another area in which *Arabidopsis* genetics is proving to be useful is the study of the proposed role of flavonoids in regulating auxin transport. Several years ago, Jacobs and Rubery published the first evidence that flavonoids could specifically compete with naphthylphthalamic acid for binding to the auxin efflux carrier in etiolated zucchini hypocotyls.[35] This finding was somewhat controversial, however, and no additional evidence for the connection between flavonoids and auxin transport was reported for some time. Brown *et al.* have now used *Arabidopsis* flavonoid mutants to generate new evidence in support of a role for flavonoids in the regulation of polar auxin movement.[6] These experiments took

Table 6.1: *Arabidopsis* flavonoid biosynthetic loci.

Locus	Gene product	Tissue specificity	Citation(s)
Structural genes			
tt3	DFR	whole plant	13
tt4	CHS	whole plant	12,49
tt5	CHI	whole plant	13
tt6	F3H	whole plant	50
tt7	F3'H	whole plant	11,14,17
tt12	MATE transporter	seed coat endothelium	34
tt15	enzyme	NR	[51] Loïc Lepiniec, PC
tt18	LDOX	NR	Atsushi Tanaka, PC
ban, ast	LCR	seed-specific	21,52
Regulatory genes			
tt1	regulator	seed coat	[53] Bernd Weisshaar, PC
tt2	regulator (myb); required for DFR and BAN expression	seed coat	28,53
tt8	regulator (bHLH/myc); required for DFR and BAN expression	seed coat	27,53
tt16	regulator	NR	Loïc Lepiniec, PC
ttg1	regulator (WD40), required for DFR, LDOX, and BAN expression	whole plant	10,23,27,53
ttg2	regulator (WRKY)	seed coat	26
icx1	regulator - not yet cloned	epidermal cells	54
anl2	developmental regulator (homeobox)	subepidermal cells	55
Uncharacterized genes			
tt9	unknown	seed coat	53
tt10	unknown	seed coat	53
tt11	unknown	NR	24
tt13	unknown	NR	24
tt14	unknown	NR	24
tt17	unknown	NR	Jitendra P. Khurana, PC

Table 6.1 Abbreviations: BAN, BANYULS; bHLH, basic helix-loop-helix; CHS, chalcone synthase; CHI, chalcone isomerase; DFR, dihydroflavonol reductase; F3H, flavonol 3-hydroxylase; F3'H, flavonoid 3'-hydroxylase; FLS, flavonol synthase; *icx*, increased chalcone synthase expression; LDOX, leucoanthocyanidin dioxygenase; LCR, leucoanthocyanidin reductase; MATE, multidrug and toxic compound extrusion; NR, not yet reported; *tt, transparent testa; ttg, transparent testa glabrous;* the WD40 and WRKY transcription factors are named for conserved amino acid sequences within these proteins. PC = personal communication.

advantage of the availability of a *tt4* null mutant, which is defective in the first enzyme of flavonoid biosynthesis and, therefore, does not synthesize any of these compounds. The *tt4* plants were found to exhibit a number of phenotypes consistent with altered auxin transport, including reduced apical dominance, a shorter primary inflorescence, longer primary roots, and increased root branching. Moreover, it was shown that basipetal transport of [^{3}H]indole-3-acetic acid was elevated in the inflorescence of this mutant relative to wild type plants and that this could be reversed by application of the flavonoid intermediate, naringenin. It is intriguing that the first two enzymes of flavonoid biosynthesis are located at the opposite end of the cell from the efflux carrier in cortex cells of the root elongation zone in *Arabidopsis*, suggesting that localization of the biosynthetic machinery may be important in mediating this effect.[17] Additional work on flavonoid mutants in *Arabidopsis* and other species should help further define the mechanism(s) by which flavonoids affect the directional movement of auxin in plant tissues.

INTRACELLULAR ORGANIZATION OF THE FLAVONOID PATHWAY AS A MEMBRANE-ASSOCIATED MULTIENZYME COMPLEX

The organization of biosynthetic pathways as multienzyme complexes has important implications for efforts to engineer metabolism in plants and other organisms. Some 26 years ago, Helen Stafford first suggested that the enzymes of phenylpropanoid metabolism might be organized as one or more enzyme complexes in order to control competition for shared substrates, sequester reactive or toxic intermediates, and help determine the subcellular distribution of end products.[36] Subsequent work, largely in Geza Hrazdina's laboratory, provided experimental evidence for channeling in the general phenylpropanoid pathway and for the membrane association of several "soluble" enzymes of flavonoid biosynthesis in various plant species.[2] This work has now been extended in *Arabidopsis* and, again, has been aided by the availability of mutants that lack specific flavonoid enzymes. In the first experiments, two-hybrid analysis, coimmunoprecipitation, and affinity chromatography were used to demonstrate specific interactions among *Arabidopsis* CHS, CHI, F3H, and DFR.[37] In the affinity chromatography experiments, in which recombinant CHS or CHI was used to capture flavonoid enzymes from plant lysates, much more efficient recovery of the other enzymes was observed with extracts from the corresponding (*tt4* or *tt5*) mutant lines.[37] This suggested that endogenous CHS and CHI competed with the immobilized recombinant protein for binding of the other enzymes. More recently, it has been shown that CHS and CHI are colocalized at the endoplasmic reticulum in *Arabidopsis* root cells.[17] These experiments also uncovered a distinct asymmetric distribution of CHS and CHI in cortex cells of the root elongation zone, which is speculated to be important in controlling auxin transport, as discussed above. The availability of CHS and CHI mutants significantly aided in demonstrating the specificity of immunolabeling in these experiments. Moreover, the *tt7* mutant, which lacks the C-terminal cytosolic domain of the cytochrome P450, flavonoid 3'-hydroxylase, was useful for exploring the proposed role of this membrane protein as an anchor for the rest of the flavonoid enzyme complex. Interestingly, CHS and CHI are still associated with the ER in this mutant, however, the asymmetric labeling pattern was abolished, suggesting some role for this protein in the correct localization of other flavonoid enzymes. Current efforts center on testing the utility of GFP fusion proteins for studying changes in flavonoid enzyme localization in living plant cells in response to various external stimuli. Here, too, comparisons of the response in wild type and mutant lines should provide insights into the effects of protein interactions on the organization and function of the flavonoid enzyme system.

A long-term goal of this work is to understand the mechanisms underlying assembly of the flavonoid enzyme complex, with an eye to determining how this organization allows plant cells to control the types and amounts of flavonoids that are produced in different tissues. One aspect involves the identification of protein domains involved in the assembly of the flavonoid enzyme complex. Preliminary results suggest that surface plasmon resonance may provide a useful tool for functional analysis of such domains (Dana and Winkel-Shirley, unpublished results). Another important step in this direction has been the recent elucidation of the crystal structures of alfalfa CHS and CHI by Joe Noel's group at the Salk Institute.[38,39] It has been possible to model the *Arabidopsis* proteins by using the alfalfa structures as templates[40] (Dana and Winkel-Shirley, unpublished results). However, there are significant differences in the functions of these proteins in legumes and other angiosperms.[41] Moreover, *Arabidopsis* CHI has N- and C-terminal extensions relative to the alfalfa enzyme, the structure of which, therefore, cannot be predicted by homology modeling. Thus, there is an effort underway to determine the crystal structures of the *Arabidopsis* enzymes, as well as other enzymes of flavonoid biosynthesis. The ultimate goal of this work is to develop a three-dimensional model of the flavonoid enzyme complex.

USE OF RECOMBINANT scFv ANTIBODIES TO DISRUPT FLAVONOID METABOLISM IN TRANSGENIC PLANTS

Because the flavonoid pathway is not essential under greenhouse conditions, it provides a useful experimental model for evaluating new mutational approaches. As one example, we have recently used the *Arabidopsis* flavonoid pathway to test the feasibility of engineering metabolism by expressing recombinant single chain variable fragment (scFv) antibodies in transgenic plants (Santos and Winkel-Shirley, unpublished results). There are several published reports of the use of scFv antibodies to interfere effectively with protein function in plants, including efforts to modulate the activities of abscisic acid, gibberellin, and the light receptor, phytochrome, as well as engineering protection against a variety of viruses, a pathogenic bacterium, and possibly nematodes.[42,43] However, no examples of using this approach to modify metabolism have been described to date, despite recent success with similar attempts in mammalian cells.[44-47] In general, these approaches have been limited by the availability of monoclonal antibodies needed to provide access to the genes required for engineering antibody expression in plants. This limitation has recently been circumvented by the development of bacteriophage libraries expressing scFv's as fusions to the PIII coat protein, providing a potentially rich source of genes for engineering plants by using antibody technology.[48]

To test the possibility that recombinant antibodies can be used to modify plant metabolism, bacteriophage-expressing antibodies against *Arabidopsis* CHI were isolated from a human synthetic scFv library (Santos and Winkel-Shirley, unpublished results). Genes encoding scFv's specific for CHI were isolated from two phage clones and expressed in transgenic *Arabidopsis* plants under control of the double-enhanced 35S promoter. In one line expressing low levels of the scFv, a reduction in both measurable flavonol glycosides and visible pigmentation was observed. The scFv was shown to bind directly to CHI in these plants by a mobility shift assay. Surprisingly, other plants expressing much higher levels of the same transgene showed no phenotypes. The mobility shift assay suggested that the antibody in these plants existed as a high-molecular weight aggregate that did not bind to CHI. Interestingly, previous efforts to disrupt flavonoid metabolism by expressing anti-DFR scFv antibodies at high levels in petunia were unsuccessful.[42] These experiments indicate that, although expression levels may have to be carefully tuned in order to obtain the desired effect, recombinant antibody technology offers a promising new avenue for metabolic engineering. It may also offer a novel method for probing protein domains as an approach to defining regions important in functions, such as protein-protein interactions, that are distinct from well-established functions, such as the catalytic activities of enzymes.

SUMMARY

Significant progress has been made during the past fifteen years toward developing *Arabidopsis* as a model for studying flavonoid metabolism. The availability of defined mutants in single-copy genes encoding the enzymes of the central flavonoid pathway is certainly an advantage of this system compared to many other plant species. This feature has simplified a number of research efforts, including analyses of the biological activities of flavonoids, as in seed dormancy and the control of auxin transport, as well as investigations into the subcellular organization of the pathway. At the same time, it raises questions about the need for multiple FLS genes in *Arabidopsis* and the potential role of gene families in providing tissue-specific expression and/or endproduct specificity for important biologically active metabolites such as flavonols. The ability to clone genes by T-DNA tagging and positional cloning is further facilitating the identification of novel genes, such as the putative leucoanthocyanidin reductase encoded by *BAN* and transcription factors such as TT2 and TT8. These are helping to close the gaps in our understanding of both the biochemical pathway and how its expression is controlled in different species. The *Arabidopsis* flavonoid pathway is also providing a useful experimental model for efforts to understand fundamental properties of plant metabolism, including its subcellular organization. *Arabidopsis* is clearly proving to be a useful complement to maize, petunia, snapdragon, and other plant species for the ongoing dissection of the complexities of flavonoid biosynthesis and cellular metabolism in general.

ACKNOWLEDGEMENTS

The author would like to thank Loïc Lepiniec, Jitendra Khurana, Atsushi Tanaka, and Bernd Weisshaar for sharing data prior to publication. Work in the author's laboratory is supported by funding from the National Science Foundation (MCB-9808117) and the U.S. Department of Agriculture (2001-35318-11266).

REFERENCES

1. FATLAND, B., ANDERSON, M., NIKOLAU, B.J., WURTELE, E.S., Molecular biology of cytosolic acetyl-CoA generation, *Biochem. Soc. Trans.*, 2000, **28**, 593-595.
2. WINKEL-SHIRLEY, B., Evidence for enzyme complexes in the phenylpropanoid and flavonoid pathways, *Physiol. Plant.*, 1999, **107**, 142-149.
3. WINKEL-SHIRLEY, B., Flavonoid biosynthesis: a colorful model for genetics, biochemistry, cell biology and biotechnology, *Plant Physiol.*, 2001, **126**, 485-493.
4. KOES, R.E., QUATTROCCHIO, R., MOL, J.N.M., The flavonoid biosynthetic pathway in plants: function and evolution, *BioEssays*, 1994, **16**, 123-132.

5. MOL, J., GROTEWOLD, E., KOES, R., How genes paint flowers and seeds, *Trends Plant Sci.*, 1998, **3**, 212-217.
6. BROWN, D.E., RASHOTTE, A.M., MURPHY, A.S., NORMANLY, J., TAGUE, B.W., PEER, W.A., TAIZ, L., MUDAY, G.K., Flavonoids act as negative regulators of auxin transport *in vivo* in *Arabidopsis thaliana*, *Plant Physiol.*, 2001, **126**, 524-535.
7. SHIRLEY, B.W., Flavonoid biosynthesis: "new" functions for an "old" pathway, *Trends Plant Sci.*, 1996, **1**, 377-382.
8. BÜRGER, D., Die morphologischen mutanten des Göttinger Arabidopsis-sortiments, einschließlich der mutanten mit abweichender samenfarbe, *Arabid. Inf. Serv. (Frankfurt am Main)*, 1971, **8**, 36-42.
9. KOORNNEEF, M., Mutations affecting the testa color in *Arabidopsis*, *Arabid. Inf. Serv.*, 1990, **28**, 1-4.
10. KOORNNEEF, M., The complex syndrome of *ttg* mutants, *Arabid. Inf. Serv.*, 1981, **18**, 45-51.
11. KOORNNEEF, M., LUITEN, W., DE VLAMING, P., SCHRAM, A.W., A gene controlling flavonoid-3'-hydroxylation in Arabidopsis, *Arabid. Inf. Serv.*, 1982, **19**, 113-115.
12. FEINBAUM, R.L., AUSUBEL, F.M., Transcriptional regulation of the *Arabidopsis thaliana* chalcone synthase gene, *Mol. Cell. Biol.*, 1988, **8**, 1985-1992.
13. SHIRLEY, B.W., HANLEY, S., GOODMAN, H.M., Effects of ionizing radiation on a plant genome: analysis of two Arabidopsis *transparent testa* mutations, *Plant Cell*, 1992, **4**, 333-347.
14. SCHOENBOHM, C., MARTENS, S., EDER, C., FORKMANN, G., WEISSHAAR, B., Identification of the *Arabidopsis thaliana* flavonoid 3'-hydroxylase gene and functional expression of the encoded P450 enzyme, *Biol. Chem.*, 2000, **381**, 749-753.
15. PELLETIER, M.K., MURRELL, J., SHIRLEY, B.W., Arabidopsis flavonol synthase and leucoanthocyanidin dioxygenase: further evidence for distinct regulation of "early" and "late" flavonoid biosynthetic genes, *Plant Physiol.*, 1997, **113**, 1437-1445.
16. WISMAN, E., HARTMANN, U., SAGASSER, M., BAUMANN, E., PALME, K., HAHLBROCK, K., SAEDLER, H., WEISSHAAR, B., Knock-out mutants from an En-1 mutagenized *Arabidopsis thaliana* population generate phenylpropanoid biosynthesis phenotypes, *Proc. Natl. Acad. Sci. USA*, 1998, **95**, 12432-12437.
17. SASLOWSKY, D., WINKEL-SHIRLEY, B., Localization of flavonoid enzymes in *Arabidopsis* roots, *Plant J.*, 2001, **27**, 37-48.
18. DONG, X., BRAUN, E.L., GROTEWOLD, E., Functional conservation of plant secondary metabolic enzymes revealed by complementation of *Arabidopsis* flavonoid mutants with maize genes, *Plant Physiol.*, 2001, **127**, 46-57.
19. LI, J., OU-LEE, T.-M., RABA, R., AMUNDSON, R.G., LAST, R.L., Arabidopsis flavonoid mutants are hypersensitive to UV-B irradiation, *Plant Cell*, 1993, **5**, 171-179.
20. PELLETIER, M.K., BURBULIS, I.E., SHIRLEY, B.W., Disruption of specific flavonoid genes enhances the accumulation of flavonoid enzymes and endproducts in *Arabidopsis* seedlings, *Plant Mol. Biol.*, 1999, **40**, 45-54.
21. DEVIC, M., GUILLEMINOT, J., DEBEAUJON, I., BECHTOLD, N., BENSAUDE, E., KOORNNEEF, M., PELLETIER, G., DELSENY, M., The *BANYULS* gene encodes a DFR-like protein and is a marker of early seed coat development, *Plant J.*, 1999, **19**, 387-398.

22. MORRIS, P., ROBBINS, M.P., Manipulating condensed tannins in forage legumes. In: Biotechnology and the Improvement of Forage Legumes (B. D. McKersie and D. C. W. Brown, eds.), CAB International, Wallingford, CT. 1997, pp. 147-173
23. WALKER, A.R., DAVISON, P.A., BOLOGNESI-WINFIELD, A.C., JAMES, C.M., SRINIVASAN, N., BLUNDELL, T.L., ESCH, J.J., MARKS, M.D., GRAY, J.C., The *TRANSPARENT TESTA GLABRA1* locus, which regulates trichome differentiation and anthocyanin biosynthesis in Arabidopsis, encodes a WD40 repeat protein, *Plant Cell*, 1999, **11**, 1337-1350.
24. DEBEAUJON, I., LÉON-KLOOSTERZIEL, K.M., KOORNNEEF, M., Influence of the testa on seed dormancy, germination, and longevity in Arabidopsis, *Plant Physiol.*, 2000, **122**, 403-413.
25. KAPPERT, H., Die Genetik des *incana*-Characters und der Anthozyanbildung bei der Levkoje, *Der Züchter*, 1949, **19**, 289-297.
26. JOHNSON, C.S., SMYTH, D.R., The TTG2 gene of Arabidopsis encodes a WRKY family transcription factor that regulates trichome development and the production of pigment and mucilage in seed coats, 9th International Conference on Arabidopsis Research. 1998, 186.
27. NESI, N., DEBEAUJON, I., JOND, C., PELLETIER, G., CABOCHE, M., LEPINIEC, L., The *TT8* gene encodes a basic helix-loop-helix domain protein required for expression of *DFR* and *BAN* genes in Arabidopsis siliques, *Plant Cell*, 2000, **12**, 1863-1878.
28. NESI, N., JOND, C., DEBEAUJON, I., CABOCHE, M., LEPINIEC, L., The Arabidopsis TT2 gene encodes an R2R3 MYB domain protein that acts as a key determinant for accumulation in developing seed, *Plant Cell*, 2001, **13**, 2099-2114.
29. BOREVITZ, J.O., XIA, Y., BLOUNT, J., DIXON, R.A., LAMB, C., Activation tagging identifies a conserved MYB regulator of phenylpropanoid biosynthesis, *Plant Cell*, 2000, **12**, 2383-2393.
30. WADE, H.K., BIBIKOVA, T.N., VALENTINE, W.J., JENKINS, G.I., Interactions within a network of phytochrome, cryptochrome and UV-B phototransduction pathways regulate chalcone synthase gene expression in Arabidopsis leaf tissue, *Plant J.*, 2001, **25**, 675-685.
31. MUELLER, L., GOODMAN, C.D., SILADY, R.A., WALBOT, V., AN9, a petunia glutathione *S*-transferase required for anthocyanin sequestration is a flavonoid-binding protein, *Plant Physiol.*, 2000, **123**, 1561-1570.
32. MARRS, K.A., ALFENITO, M.R., LLOYD, A.M., WALBOT, V., A glutathione *S*-transferase involved in vacuolar transfer encoded by the maise gene *Bronze-2*, *Nature*, 1995, **375**, 397-400.
33. ALFENITO, M.R., SOUER, E., GOODMAN, C.D., BUELL, R., MOL, J., KOES, R., WALBOT, V., Functional complementation of anthocyanin sequestration in the vacuole by widely divergent glutathione *S*-transferases, *Plant Cell*, 1998, **10**, 1135-1149.
34. DEBEAUJON, I., PEETERS, A.J.M., LÉON-KLOOSTERZIEL, K.M., KOORNNEEF, M., The *TRANSPARENT TESTA12* gene of Arabidopsis encodes a multidrug secondary transporter-like protein required for flavonoid sequestration in vacuoles of the seed coat endothelium, *Plant Cell*, 2001, **13**, 853-871.
35. JACOBS, M., RUBERY, P.H., Naturally occurring auxin transport regulators, *Science*, 1988, **241**, 346-349.

36. STAFFORD, H.A., Possible multi-enzyme complexes regulating the formation of C_6-C_3 phenolic compounds and lignins in higher plants, *Rec. Adv. Phytochem.*, 1974, **8**, 53-79.
37. BURBULIS, I.E., WINKEL-SHIRLEY, B., Interactions among enzymes of the *Arabidopsis* flavonoid biosynthetic pathway, *Proc. Natl. Acad. Sci. USA*, 1999, **96**, 12929-12934.
38. FERRER, J.-L., JEZ, J.M., BOWMAN, M.E., DIXON, R.A., NOEL, J.P., Structure of chalcone synthase and the molecular basis of plant polyketide biosynthesis, *Nature Struc. Biol.*, 1999, **6**, 775-784.
39. JEZ, J.M., BOWMAN, M.E., DIXON, R.A., NOEL, J.P., Structure and mechanism of chalcone isomerase: an evolutionarily unique enzyme in plants, *Nature Struct. Biol.*, 2000, **7**, 786-791.
40. SASLOWSKY, D.E., DANA, C.D., WINKEL-SHIRLEY, B., An allelic series for the chalcone synthase locus in Arabidopsis, *Gene*, 2000, **255**, 127-138.
41. DIXON, R.A., STEELE, C.L., Flavonoids and isoflavonoids - a gold mine for metabolic engineering, *Trends Plant Sci.*, 1999, **4**, 394-400.
42. DE JAEGER, G., DE WILDE, C., EECKHOUT, D., FIERS, E., DEPICKER, A., The plantibody approach: expression of antibody genes in plants to modulate plant metabolism or to obtain pathogen resistance, *Plant Mol. Biol.*, 2000, **43**, 419-428.
43. XIAO, X.W., CHU, P.W.G., FRENKEL, M.J., TABE, L.M., SHUKLA, D.D., HANNA, P.J., HIGGINS, T.J.V., MÜLLER, W.J., WARD, C.W., Antibody-mediated improved resistance to ClYVV and PVY infections in transgenic tobacco plants expressing a single-chain variable region antibody, *Mol. Breed.*, 2000, **6**, 421-431.
44. WANG, W., ZHOU, J., XU, L., ZHEN, Y., Antineoplastic effect of intracellular expression of a single-chain antibody directed against type IV collagenase, *J. Environ. Pathol. Toxicol. Oncol*, 2000, **19**, 61-68.
45. LENER, M., HORN, I.R., CARDINALE, A., MESSINA, S., NIELSEN, U.B., RYBACK, S.M., HOOGENBOOM, H.R., CATTANEO, A., BIOCCA, S., Diverting a protein from its cellular location by intracellular antibodies, The case of p21 Ras., *Eur. J. Biochem.*, 2000, **267**, 1196-1205.
46. RAJPAL, A., TURI, T.G., Intracellular stability of anti-caspase-3 intrabodies determines efficacy in re-targeting antigen, *J. Biol. Chem.*, 2001, **2001**, 33139-33146.
47. ZHU, Q., ZENG, C., HUHALOV, A., YAO, J., TURI, T.G., DANLEY, D., HYNES, T., CONG, Y., DIMATTIA, D., KENNEDY, S., DAUMY, G., SCHAEFFER, E., MARASCO, W.A., HUSTON, J.S., Extended half-life and elevated steady-state level of a single-chain Fv intrabody are critical for specific intracellular retargeting of its antigen, caspase-7, *J. Immunol. Methods*, 1999, **231**, 207-222.
48. HOOGENBOOM, H.R., DE BRUÏNE, A.P., HUFTON, S.E., HOET, R.M., ARENDS, J.-W., ROOVERS, R.C., Antibody phage display technology and its applications, *Immunotechnol.*, 1998, **4**, 1-20.
49. BURBULIS, I.E., IACOBUCCI, M., SHIRLEY, B.W., A null mutation in the first enzyme of flavonoid biosynthesis does not affect male fertility in Arabidopsis, *Plant Cell*, 1996, **8**, 1013-1025.

50. PELLETIER, M.K., SHIRLEY, B.W., Analysis of flavanone 3-hydroxylase in Arabidopsis seedlings: Coordinate regulation with chalcone synthase and chalcone isomerase, *Plant Physiol.*, 1996, **111**, 339-345.
51. FOCKS, N., SAGASSER, M., WEISSHAAR, B., BENNING, C., Characterization of *tt15*, a novel *transparent testa* mutant of *Arabidopsis thaliana* (L.) Heynh, *Planta*, 1999, **208**, 352-357.
52. TANAKA, A., TANO, S., CHANTES, T., YOKOTA, Y., SHIKAZONO, N., WATANABE, H., A new *Arabidopsis* mutant induced by ion beams affects flavonoid synthesis with spotted pigmentation in testa, *Genes Genet. Syst.*, 1997, **72**, 141-148.
53. SHIRLEY, B.W., KUBASEK, W.L., STORZ, G., BRUGGEMANN, E., KOORNNEEF, M., AUSUBEL, F.M., GOODMAN, H.M., Analysis of *Arabidopsis* mutants deficient in flavonoid biosynthesis, *Plant J.*, 1995, **8**, 659-671.
54. JACKSON, J.A., FUGLEVAND, G., BROWN, B.A., SHAW, M.J., JENKINS, G.I., Isolation of *Arabidopsis* mutants altered in the light-regulation of chalcone synthase gene expression using a transgenic screening approach, *Plant J.*, 1995, **8**, 369-380.
55. KUBO, H., PEETERS, A.J., AARTS, M.G., PEREIRA, A., KOORNNEEF, M., ANTHOCYANINLESS2, a homeobox gene affecting anthocyanin distribution and root development in *Arabidopsis*, *Plant Cell*, 1999, **11**, 1217-1226.
56. KUBASEK, W.L., SHIRLEY, B.W., MCKILLOP, A., GOODMAN, H.M., BRIGGS, W., AUSUBEL, F.M., Regulation of flavonoid biosynthetic genes in germinating Arabidopsis seedlings, *Plant Cell*, 1992, **4**, 1229-1236.
57. CAIN, C.C., SASLOWSKY, D.E., WALKER, R.A., SHIRLEY, B.W., Expression of chalcone synthase and chalcone isomerase proteins in *Arabidopsis* seedlings, *Plant Mol. Biol.*, 1997, **35**, 377-381.

Chapter Seven

BIOPANNING BY ACTIVATION TAGGING

Yiji Xia,[*,a,b,c,1], Justin Borevitz,[a,1] Jack W. Blount,[b] Richard A. Dixon,[b] Chris Lamb[a,d]

[a]*Plant Biology Laboratory, Salk Institute for Biological Studies, 10010 North Torrey Pines Road, La Jolla, California 92037;*

[b]*Plant Biology Division, Samuel Roberts Noble Foundation, 2510 Sam Noble Parkway, Ardmore, Oklahoma 73401;*

[c]*Donald Danforth Plant Science Center, 975 N. Wason, St. Louis, Missouri 63132;*

[d]*John Innes Center, Colney Lane, Norwich NR4 7UH, UK;*

[1]*Joint first authors*

[*]*Author for correspondence, e-mail: yxia@danforthcenter.org*

Introduction...	112
The *pap1-D* Mutation Hyperactivates the Phenylpropanoid Pathway	114
PAP1 Encodes a MYB-Like Transcription Factor...	115
Over-Expression of *PAP2*, Another *MYB* Gene, Enhances Anthocyanin Accumulation...	118
Over-Expression of *PAP1* and *PAP2* Enhances Pigmentation in Tobacco.........	118
Potential of Activation Tagging...	118
Summary..	120

INTRODUCTION

T-DNA insertion mutagenesis is a powerful tool to dissect biological pathways and clone underlying plant genes. This approach has been especially successful in *Arabidopsis* because of the ease of generating a large number of independent T-DNA tagged lines.[1] However, since T-DNA tagging generally causes gene disruption and results in a recessive loss-of-function mutation, the mutant phenotype can usually only be observed in the individuals homozygous for the mutant allele derived from selfing of the primary transformant. Therefore, this approach requires generation of large numbers of fertile transgenic plants and handling their progenies. As a result, the conventional T-DNA tagging approach is practically infeasible for plant species in which a high efficiency transformation procedure has not been established. In addition, loss-of-function screens are difficult in identifying genes that act redundantly or are essential.

To overcome the limitations of traditional T-DNA insertion mutagenesis, a complementary approach, T-DNA activation tagging, has been developed and used to identify genes involved in a variety of biological pathways. The approach (Fig. 7.1) was initially developed by Walden and colleagues using a tobacco cell culture system, and has recently been modified by Weigel and colleagues to apply it at the whole plant level in *Arabidopsis*.[2-5] Like conventional T-DNA tagging, activation tagging also creates mutations by random insertion of T-DNA into plant genomes. However, the latter uses a T-DNA vector that carries four copies of an enhancer derived from the promoter of the cauliflower mosaic virus 35S gene at its right border. The T-DNA is delivered via *Agrobacterium*-mediated transformation into plant genomes to generate independent transgenic lines. The enhancers (35Se) are expected to cause transcriptional activation of the adjacent plant gene, resulting in a gain-of-function mutation. Since mutations created by activation tagging are usually dominant or semi-dominant, mutant screens can be carried out among the primary transformants. In addition to the 35S enhancers, the 35S promoter and tissue specific promoters have been used to overexpress and/or misexpress genes to create gain-of-function mutations.[5-7] Screening large collections of independent, activation-tagged lines, thus, represents a powerful way of surveying the genome and isolating genes that affect traits of interest. In the last few years, activation tagging has successfully been applied by many research groups in screening various types of mutations and cloning the corresponding genes.[2-6, 8-16] Its ability to activate a biosynthetic pathway is particularly useful for the generation of mutant tissues and plants with enriched accumulation of natural products and for isolating underlying regulatory and biosynthetic genes.

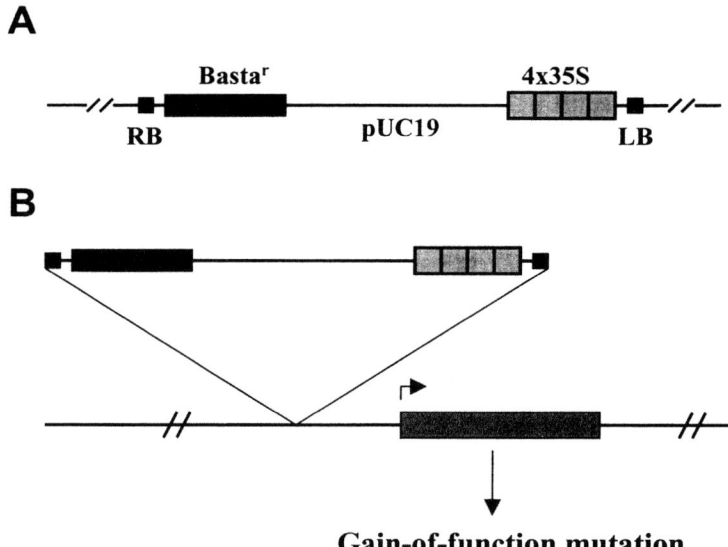

Figure 7.1: Activation tagging.
(A) Structure of an activation tagging vector, pSKI015, that contains four copies of the 35S enhancers next to the left border (LB), the BASTA resistance selectable marker cassette, and the pUC19 plasmid sequence to facilitate cloning T-DNA flanking plant sequences using plasmid rescue. The 35S enhancer sequence can be replaced by other transcriptional enhancers as well as by a constitutive or tissue specific promoter to overexpress or misexpress plant genes. Only the T-DNA region of the vector is shown.
(B) Schematic drawing of generation of gain-of-function mutations resulting from the insertion of 35S enhancers. T-DNA is integrated into the plant genome via *Agrobacterium*-mediated transformation. The 35S enhancers cause transcriptional activation of a nearby gene, resulting in a gain-of-function mutation.

Plants produce an amazing diversity of secondary metabolites that not only play important biological roles in their adaptation to environments but also provide humans with dyes, flavors, drugs, fragrance, and other useful chemicals. However, many of the secondary metabolic pathways are found or are amplified only in limited taxonomic groups. In addition, the compounds are often restricted to a particular tissue and occur at a specific stage of development and in low abundance.[17] Intensive efforts in the last couple of decades to enhance accumulation of some natural products through cell culture for commercial production has essentially failed.[18-20] Transgenic manipulation offers a more promising approach to enhancing natural product biosynthesis. Availability of regulatory genes is paramount for the success of genetic manipulation of biosynthetic pathways. As described here in the case of the identification of PAP1, a MYB transcription factor from *Arabidopsis* that regulates the phenylpropanoid pathway, activation tagging offers a powerful functional genomics approach to override the stringent genetic control over accumulation of specific natural products and to isolate underlying genes.

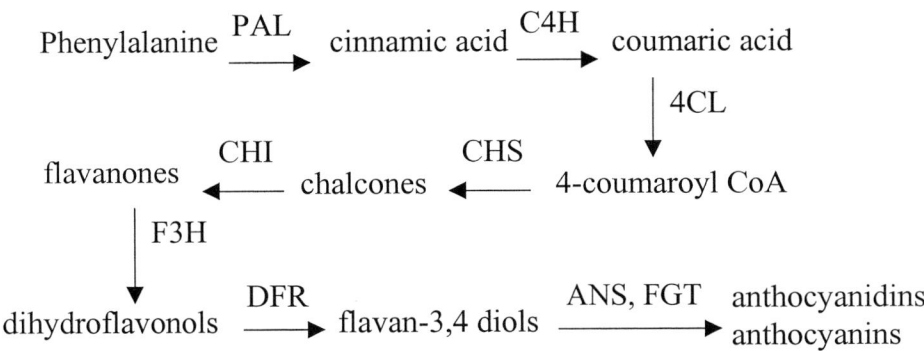

Figure 7.2: Scheme of the anthocyanin pathway. Abbreviations: PAL, phenylalanine ammonia lyase; C4H, cinnamate-4-hydroxylase; 4CL, 4-coumarate:CoA ligase; CHS, chalcone synthase; CHI, chalcone isomerase; F3H, flavanone 3-hydroxylase; DFR, dihydroflavonol 4-reductase; ANS, anthocyanidin synthase; FGT, flavonoid 3-O-glucosyltransferase.

THE *PAP1*-D MUTATION HYPERACTIVATES THE PHENYLPROPANOID PATHWAY

We generated activation tagged *Arabidopsis* lines by transforming ecotype Columbia (Col-0) plants with the activation tagging vector pSKI015 by using the floral dipping procedure.[5,21] Primary transformants were selected for their resistance to the herbicide BASTA. Among a collection of approximately five thousand of primary (T0) transgenic lines, we identified a single line that exhibited a purple phenotype. The mutant was named *pap1-D* (*production of anthocyanin pigment 1-dominant*).[10] The mutant plant was selfed and its T2 progeny segregated 3:1 for the purple phenotype, demonstrating that the mutation is dominant over its wild-type allele. All the purple progeny were found to be resistant to the herbicide BASTA, whereas all wild-type progeny were sensitive to BASTA, indicating that the primary transformant contains a single T-DNA insertion that co-segregates with the mutant allele. The enhanced pigmentation was observed throughout the entire developmental programs of the plant although more purple pigments were accumulated during later stages. The roots and petals, which normally do not accumulate anthocyanins in wild-type plants, also exhibited purple pigmentation. The mutant phenotype was more pronounced under high light intensity and biotic and abiotic stresses, when wild type plants also show a slight pigmentation.

The anthocyanin pathway is a branch from the biosynthesis of flavonoids that are derived from the phenylpropanoid skeleton.[22] Nearly all enzymes involved in the pathway have been characterized, and most of the genes encoding the enzymes have been isolated (Fig. 7.2).[22,23] To examine whether the genes involved in the overall phenylpropanoid pathway are transcriptionally activated in the *pap1*-D mutant, RNA gel blot analysis was conducted to determine the transcript levels of some of the biosynthetic genes, including *PAL1* encoding the first enzyme of the overall phenylpropanoid pathway, *CHS* encoding the entry point enzyme into the flavonoid branch, and *DFR* encoding an enzyme in the anthocyanin branch. The transcript levels of all three genes were at least 50 fold higher in the *pap1*-D plants than those in wildtype plants. The level of *GST*, which is involved in transporting anthocyanins into vacuoles,[22, 24] was about 20 fold higher in the mutant plants.

We then used HPLC to determine effects of the *pap1*-D mutation on accumulation of anthocyanins and other phenylpropanoid-derived compounds. The level of anthocyanins in *pap1*-D was 6.5 fold higher than that in wild-type plants. Two wall-bound phenolic compounds, sinapic acid and coumaric acid, were increased 3 and 20 fold, respectively, in the mutant. The levels of four soluble phenolproponoid compounds, Glc-Rha-quercetin (Glc=glucose, Rha=rhamnose), Glc-Rha-kaempferol, and unidentified conjugates of quercetin and kaempferol, were 9, 8, 10 and 4 fold higher, respectively, in the *pap1*-D mutant. The above findings revealed that the *pap1-D* mutation is able to overcome the stringent genetic control of the overall pathway.

PAP1 ENCODES A MYB-LIKE TRANSCRIPTION FACTOR

Southern blotting analysis was conducted by using the 35S enhancer sequence as a probe against the *Eco*RI and *Kpn*I-digested genomic DNA from the *pap1-D* mutant. Consistent with the genetic data indicating that the mutant contains a single T-DNA insertion, a single 10 kb *Eco*RI and 12 kb *Kpn*I fragment was detected. The 12 kb *Kpn*I fragment that contains the flanking plant sequence and a portion of T-DNA including the entire pUC19 sequence and the 35S enhancers (Fig. 7.3A) was cloned by plasmid rescue.[5] Partial sequencing of the rescued plasmid showed that it contains a 7 kb plant sequence, which was used as a probe to screen a cDNA library, and a 1 kb cDNA clone was isolated. Sequence analysis revealed that the cDNA is derived from a genomic region approximately 3 kb upstream of the integrated 35Se sequences (Fig. 7.3A). This gene encodes a 27 kD protein that shares significant sequence similarity to MYB-like transcription factors. No other candidate genes were found in this 7 kb fragment. In RNA gel blot analysis using the cDNA as a probe, a single 1 kb faint band was detected from wild-type plants; however, its transcript level in the *pap1-D* mutant was massively enhanced as the result of the 35S enhancer insertion (Fig. 7.3B).

To confirm that it is the overexpression of the MYB-like gene that causes the *pap1-D* phenotype, wild-type *Arabidopsis* was transformed with pMN-PAP1. This carries T-DNA containing a 3 kb genomic fragment of the MYB gene fused at its 3' end with two-copies of the 35 enhancers to overexpress the gene (Fig. 7.3A). Multiple transgenic *Arabidopsis* lines containing the construct exhibited a similar phenotype to that in the original *pap1-D* mutant, confirming that the cloned MYB-like gene is indeed the *PAP1* gene.

Database search and sequence alignment analysis revealed that the deduced PAP1 protein is a member of the R2, R3 MYB family, which is estimated to have about 180 members in *Arabidopsis*.[25,26] It is most similar to the other MYB transcriptional activators regulating the anthocyanin pathway, such as maize C1 and petunia AN2 (Fig. 7.3C).[27-30] The MYB domains of those transcription factors have been shown to be sufficient and necessary to bind the *cis* elements of the genes they regulate.[23,31,32] PAP1 is 50% and 40% identical overall to AN2 and C1, respectively. The high level of sequence similarity and the overexpression phenotype suggest that *PAP1* might be the ortholog of those MYB genes. Although some MYB factors such

Figure 7. 3: Molecular characterization of *PAP1*.
(A) Organization of the *pap1-D* allele and the structure of pMN-PAP1. Basta^r, Basta resistance; Kan^r, kanamycin resistance; LB, left border; pUC19, pUC 19 plasmid; RB, right border; 4 x 35S denotes four copies of 35Se.
(B) RNA gel blot hybridization revealed massive enhancement of the *PAP1* expression. Total RNA was isolated from 4-week-old *pap1-D* and wildtype plants.
(C) Partial sequence alignment of the regions containing the R2 and R3 MYB domains from the MYB transcription factors involved in the anthocyanin pathway. The sequences were aligned using the ClustalW program. Black shading denotes matching residues with PAP1. GenBank accession numbers: PAP1 (AF325123), PAP2 (AF325124), AN2 (AAF66727), C1 (AAA33482).

ACTIVATION TAGGING

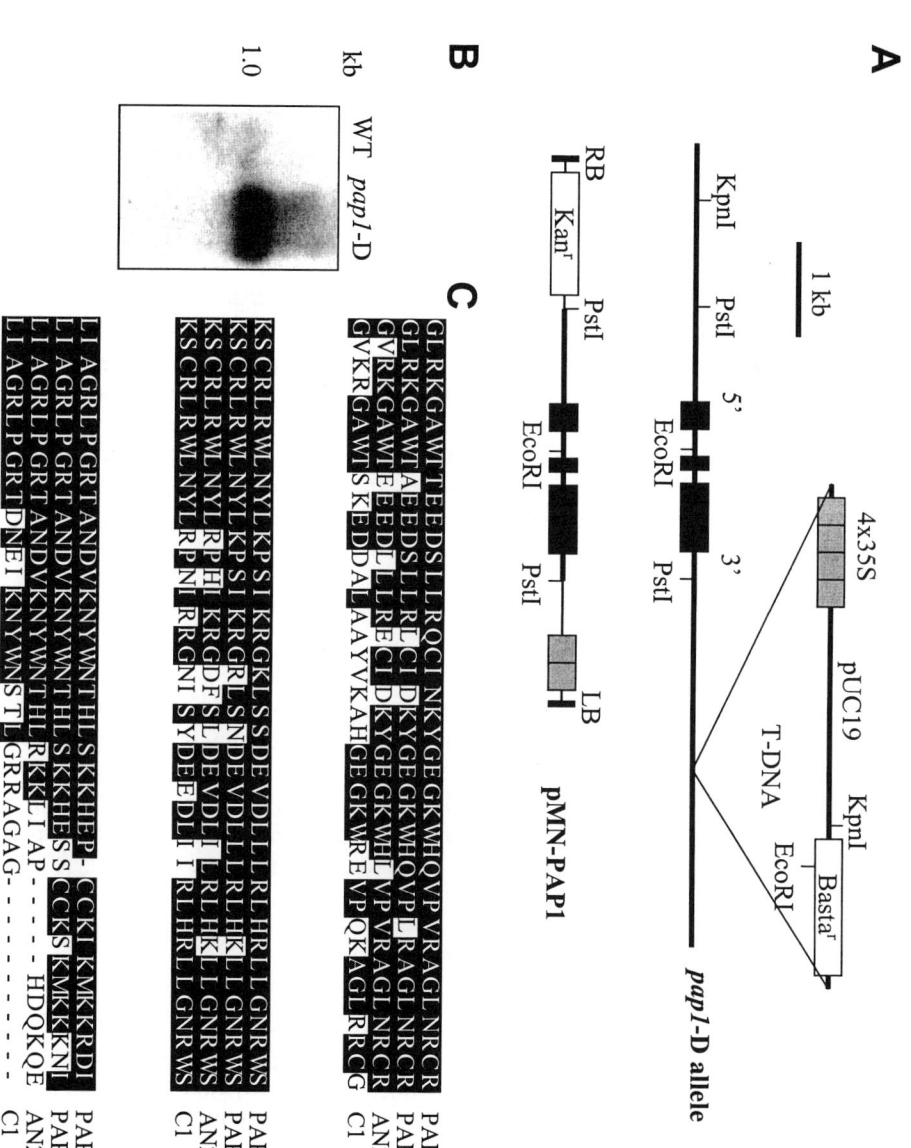

as C1 regulate the expression of phenylpropanoid biosynthetic genes from *CHS* onward, other MYBs such as the snapdragon myb305 and myb340 activate *PAL* and *CHI* but not *CHS* and *DFR*.[23,33-37] The fact that overexpression of *PAP1* is sufficient to activate *PAL* as well as the other downstream biosynthetic genes in *Arabidopsis* suggests that all those *Arabidopsis* biosynthetic genes might contain appropriate PAP1-binding *cis*-elements. However, we cannot rule out that PAP1 might not directly regulate the biosynthetic genes but instead switch on expression of other transcription factor(s) that in turn activate the biosynthetic genes.

OVER-EXPRESSION OF *PAP2*, ANOTHER *MYB* GENE, ENHANCES ANTHOCYANIN ACCUMULATION

Database searching identified another gene in the *Arabidopsis* genome that encodes a MYB protein that shares high sequence identity with PAP1: 93% identity in the MYB domain and 77% overall (Fig. 7.3C). The gene was named *PAP2*. Both *PAP1* and *PAP2* are located on chromosome 1 at a distance of approximately 9cM apart. To test if PAP1 and PAP2 have a similar biological function, we overexpressed *PAP2* by transforming wildtype *Arabidopsis* with a construct in which the expression of the *PAP2* cDNA is under the control of the 35S promoter. Transgenic lines containing the construct also exhibited enhanced pigmentation, a phenotype similar to that of *pap1*-D (data not shown).

OVEREXPRESSION OF *PAP1* AND *PAP2* ENHANCES PIGMENTATION IN TOBACCO

To examine if *PAP1* and *PAP2* can hyperactivate the anthocyanin pathway in another plant species, we generated transgenic tobacco plants that carry 35S promoter-*PAP1* cDNA and 35S promoter-*PAP2* cDNA transcriptional fusion constructs. Overexpression of either gene was found to enhance production of anthocyanin pigments in tobacco. Purple pigmentation was observed in both plant and flower tissues of the transgenic tobacco lines (data not shown).

POTENTIAL OF ACTIVATION TAGGING

Like other secondary metabolic pathways, flavonoid biosynthesis is stringently regulated at the transcriptional level. Expression patterns of the regulatory genes specify the spatial and temporal accumulation patterns of the pathway products.[23] Therefore, genetic manipulation of regulatory genes offers an effective way of enhancing production of natural products. A number of regulatory genes controlling the production of anthocyanin have been isolated from maize, petunia, and snapdragon.[23] They represent two major gene families: one encoding MYB transcription factors such as C1, and the other encoding proteins with a basic helix-loop-helix (bHLH) domain such as the maize *R* gene. *TTG8*, a gene involved in regulation of anthocyanin production in *Arabidopsis*, has recently been cloned and was shown to encode a new member of the bHLH family.[38] However, none of the *Arabidopsis* MYB genes had been found to be involved in regulating the biosynthesis of anthocyanins prior to the cloning of *PAP1* and *PAP2*. Our study has indicated that *PAP1* and *PAP2* could be the orthologs of the MYB genes regulating anthocyanin biosynthesis in other species. Loss-of-function mutant screens have identified 21 loci in *Arabidopsis* that are involved in the anthocyanin pathway, however, none of them maps to either *PAP1* or *PAP2*.[22,39] It is plausible that PAP1 and PAP2 function redundantly and, therefore, have eluded identification through loss-of-function screen efforts.

Our findings have proved that activation tagging is a valuable means to generate gain-of-function mutations that affect complex biosynthetic pathways. Activation tagged mutant plants or tissues could provide enriched sources for isolation of bioactive components. The approach also offers a powerful tool for isolating the regulatory genes involved in biosynthesis of plant natural products, which are not readily identified by biochemical approaches. Like the anthocyanin pathway, genetic activation of secondary metabolism could be scored by visual inspection for other pathways that lead to accumulation of colored products, such as the alkaloid sanguinarine, or isoprenoids such as carotenoids. Another screening strategy is to do activation tagging in a transgenic background that expresses an easily screened marker gene under the control of a promoter from a biosynthetic gene involved in the pathway of interest. High-throughput metabolic profiling also offers a promising method of screening activation tagged lines.[40] (see Summer et al., this volume).

Because activation tagging creates dominant mutations, screens for desired phenotypes can be conducted among the primary transformants. Therefore, it is not a prerequisite to generate a large collection of fertile transgenic plants. Instead, this allows selection directly from among transformed cells or explants for mutations that affect many biochemical and physiological pathways including secondary metabolism. This opens a door to applying activation tagging approaches to those plant species that are recalcitrant to regeneration of transgenic plants. Indeed, the approach was initially developed to generate tobacco cell lines that were able to grow without exogenous auxin or cytokinin supplement.[2] Recently, Memelink and colleagues used the activation tagging approach in *Catharanthus* cell cultures to isolate ORCA3, a transcription factor regulating indole alkaloid production.[41] An alternative approach is to generate and screen activation tagged hairy-roots via *Agrobacterium rhizogenes*-mediated delivery of T-DNA into plant genomes.

SUMMARY

Extensive gene duplication in plants and the potential of functional redundancy among homologous sequences represent a hurdle in understanding physiological and biochemical functions of a vast number of plant genes by using conventional loss-of-function mutant screen approaches. Gain-of-function mutations created by activation tagging might provide important clues about functions of the activated genes. The identification of *PAP1* demonstrates the usefulness of activation tagging for generating massively enriched tissue and plant sources for production of specific natural products and isolating key genes regulating their biosynthetic pathways. Activation tagging has successfully been applied to *Arabidopsis* as well as to other plant species for cloning genes affecting various biological processes and represents a powerful approach for functional genomics.

ACKNOWLEDGMENTS

We thank the following (at the Salk Institute unless otherwise noted): Tsegaye Dabi for help with transgenic tobacco; Mary Anderson at the Nottingham University Stock Centre for help with restriction fragment length polymorphism mapping; Michael Neff, Christian Fankhauser, Kim Hanson, and Joanne Chory for pMN20 and pCHF3 vectors; and Igor Kardailsky, Sioux Christensen, and Detlef Weigel for pSKI015. J.O.B. thanks Joanne Chory for providing laboratory facilities for completion of this work.

REFERENCES

1. KRYSAN, P.J., YOUNG, J.C., SUSSMAN, M.R., T-DNA as an insertional mutagen in *Arabidopsis, Plant Cell*, 1999, **11**, 2283-90.
2. HAYASHI, H. CZAJA, I., LUBENOW, H., SCHELL, J., WALDEN, R., Activation of a plant gene by T-DNA tagging: auxin-independent growth in vitro, *Science,* 1992, **258**, 1350-1353.
3. WALDEN, R., FRITZE, K., HAYASHI., H., MIKLASHEVICHS, E., HARLING, H., SCHELL, J., Activation tagging: a means of isolating genes implicated as playing a role in plant growth and development, *Plant Mol. Biol.*, 1994, **26**, 1521-1528.
4. KARDAILSKY, I., SHUKLA, V., AHN, J.H., DAGENAIS, N., CHRISTENSEN, S.K., NGUYEN, J.T., CHORY, J., HARRISON, M.J., WEIGEL, D., A pair of related genes with antagonistic roles in floral induction, *Science*, 1999, **286**, 1962-1965.
5. WEIGEL, D., AHN, J.H., BLAZQUEZ, M.A., BOREVITZ, J.O., CHRISTENSEN, S.K., FANKHAUSER, C., FERRANDIZ, C., KARDAILSKY, I., MALANCHARUVIL, E.J., NEFF, M.M., NGUYEN, J.T., SATO, S., WANG, Z.Y., XIA, Y., DIXON, R.A., HARRISON, M.J., LAMB, C.J., YANOFSKY, M.F., CHORY, J., Activation tagging in *Arabidopsis, Plant Physiol.*, 2000, **122**, 1003-1014.

6. WILSON, K., LONG, D., SWINBURNE, J., COUPLAND, G., A dissociation insertion causes a semidominant mutation that increases expression of TINY, an *Arabidopsis* gene related to APETALA2, *Plant Cell,* 1996, **8**, 659-671.
7. SUZUKI, Y., UEMURA, S., SAITO, Y., MUROFUSHI, N., SCHMITZ, G., THERES, K., YAMAGUCHI, I., A novel transposon tagging element for obtaining gain-of-function mutants based on a self-stabilizing Ac derivative, *Plant Mol. Biol.,* 2001, **45**, 123-131.
8. KAKIMOTO, T., CKI1, A histidine kinase homolog implicated in cytokinin signal transduction, *Science,* 1996, **274**, 982-985.
9. NEFF, M.M., NGUYEN, S.M., MALANCHARUVIL, E.J., FUJIOKA, S., NOGUCHI, T., SETO, H., TSUBUKI, M., HONDA, T., TAKATSUTO, S., YOSHIDA, S., CHORY, J., *BAS1*: A gene regulating brassinosteroid levels and light responsiveness in *Arabidopsis, Proc. Natl. Acad. Sci.* USA, 1999, **96**, 15316-15323.
10. BOREVITZ, J.O., XIA, Y., BLOUNT, J, DIXON, R.A, LAMB, C., Activation tagging identifies a conserved MYB regulator of phenylpropanoid biosynthesis, *Plant Cell,* 2000, **12**, 2383-2393.
11. ITO, T., MEYEROWITZ, E.M., Overexpression of a gene encoding a cytochrome P450, CYP78A9, induces large and seedless fruit in Arabidopsis, *Plant Cell,* 2000, **12**, 1541-1550.
12. van der GRAAFF, E., DULK-RAS, A.D., HOOYLAAS, P.J.J., KELLER, B., Activation tagging of the LEAFY PETIOLE gene affects leaf petiole development in *Arabidopsis thaliana, Development,* 2000, **127**, 4971-4980.
13. LEE, H., SUH, S.S., PARK, E., CHO, E., AHN, J.H., KIM, S.G., LEE, J.S., KWON, Y.M., LEE, I., The AGAMOUS-like 20 MADS domain protein integrates floral inductive pathways in Arabidopsis, *Genes & Development,* 2000, **14**, 2366-2377.
14. LI, J., LEASE, K.A., TAX, F.E., WALKER, J.C., BRS1, a serine carboxypeptidase, regulates BRI signaling in *Arabidopsis thaliana, Proc Natl Acad Sci* USA, 2001, **98**, 5916-5921.
15. HUANG, S., CERNY, R.E., BHAT, D.S., BROWN, S.M., Cloning of an *Arabidopsis* patatin-like gene *STURDY*, by activation T-DNA tagging, *Plant Physiol.,* 2001, **125**, 573-584.
16. ZHAO, Y., CHRISTERSEN, S.K., FANKHAUSER, C., CASHMAN, J., COHEN, J., WEIGEL, D., CHORY, J., A role for flavin monooxygenase-like enzymes in auxin biosynthesis, *Science,* 2001, **291**, 306-309.
17. PICHERSKY, E., GANG, D.R., Genetics and biochemistry of secondary metabolites in plants: an evolutionary perspective, *Trends Plant Sci.,* 2000, **5**, 439-445.
18. FACCHINI, D., DELUCA, V., Phloem-specific expression of tyrosine dopa decarboxylase genes and the biosynthesis of isoquinoline alkaloids in opium poppy, *Plant Cell,* 1995, **7**, 1811-1821.
19. MCCASKILL, D., CROTEAU, R., Some caveats for bioengineering terpenoid metabolism in plants, *Trends Biotechnol.,* 1998, **16**, 349-355.
20. FOWLER, M.W., Plant-cell cultures: Fact and fantasy, *Biochem. Soc. Trans.,* 1983, **11**, 23-28.
21. CLOUGH, S.J., BENT, A.F., Floral dip: A simplified method fro *Agrobacterium*-mediated transformation of *Arabidopsis thaliana, Plant J.,* 1998, **16**, 735-743.

22. WINKEL-SHIRLEY, B., Flavonoid biosynthesis. A colorful model for genetics, biochemistry, cell biology, and biotechnology, *Plant Physiol.*, 2001, **126**, 485-493.
23. MOL, J., GROTEWOLD, E., KOES, R., How genes paint flowers and seeds, *Trends Plant Sci.,* 1998, **3**, 212-217.
24. ALFENITO, M.R., SOUER, E., GOODMAN, C.D., BUELL, R., MOL, J., KOES, R., WALBOT, V., Functional complementation of anthocyanin sequestration in the vacuole by widely divergent glutathione *S*-transferases, *Plant Cell*, 1998, **10,** 1135-1150.
25. KRANZ, H.D., DENEKAMP, M., GRECO, R., JIN, H., LEYVA, A., MEISSNER, R.C., PETRONI, K., URZAINQUI, A., BEVAN, M., MARTIN, C., SMEEKENS, S., TONELLI, C., PAZ-ARES, J., WEISSHAAR, B., Towards functional characterization of the members of the R2R3-*MYB* gene family from *Arabidopsis thaliana, Plant J.,* 1998, **16**, 263-276.
26. RIECHMANN, J.L., RATCLIFFE, O.J., A genomic perspective on plant transcription factors, *Curr. Opinions Plant Biol.*, 2000, **3**, 423-434.
27. QUATTROCCHIO, F., WING, J.F., VAN DER WOUDE, K., SOUER, E., de Vetten, N., MOL, J.N.M., KOES, R., Molecular analysis of the anthocyanin2 gene of petunia and its role in the evolution of flower color, *Plant Cell,* 1999, **11**, 1433-1444.
28. QUATTROCCHIO, F., WING, J.F., VAN DER WOUDE, K., MOL, J.N.M., KOES, R., Analysis of bHLH and *MYB* domain proteins: Species-specific regulatory differences are caused by divergent evolution of target anthocyanin genes, *Plant J.*, 1998, **13**, 475-488.
29. CONE, K.C., BURR, F.A., BURR, B., Molecular analysis of the maize anthocyanin regulatory locus *C1, Proc Natl Acad Sci* USA, 1986, **83**, 9631-9635.
30. PAZ-ARES, J., GHOSAL, D., WIENAND, U., PETERSON, P.A., SAEDLER, H., The regulatory c1 locus of *Zea mays* encodes a protein with homology to myb proto-oncogene products and with structural similarities to transcriptional activators, *EMBO J.,* 1987, **6**, 3553-3558.
31. SAINZ, M., GROTEWOLD, E., CHANDLER, V., Evidence for direct activation on an anthocyanin promoter by the maize C1 protein and comparison of DNA binding by related Myb-domain proteins, *Plant Cell,* 1997, **9**, 611-625.
32. WILLIAMS, C.E., GROTEWOLD, E., Differences between plant and animal Myb domains are fundamental for DNA binding activity and chimeric Myb domains have novel DNA-binding specificities, *J. Biol. Chem.*, 1997, **272**, 563-571.
33. CONE, K.C., COCCIOLONE, S.M., BURR, F.A., BURR, B., Maize anthocyanin regulatory gene *pl* is a duplicate of *c1* that functions in the plant, *Plant Cell,* 1993, **5**, 1795-1805.
34. CONE, K.C., COCCIOLONE, S.M., MOEHLENKAMP, C.A., WEBER, T., DRUMMOND, B.J., TAGLIANI, L.A., BOWEN, B.A., PERROT, G.H., Role of the regulatory gene *pl* in the photocontrol of maize anthocyanin pigmentation, *Plant Cell,* 1993, **5**, 1807-1816.
35. MOL, J., JENKINS, G., SCHAFER, E., WEISS, D., Signal perception, transduction, and gene expression involved in anthocyanin biosynthesis, *Crit. Rev. Plant Sci.,* 1996, **15**, 525-557.
36. SABLOWSKI, R.W.M., BAULCOME, D.C., BEVAN, M., Expression of a flower-specific Myb protein in leaf cells using a viral vector causes ectopic activation of a target promoter, *Proc Natl Acad Sci* USA, 1996, **92**, 6901-6905.

37. MOYANO, E., MART'NEZ-GARCIA, J.F., MARTIN, C., Apparent redundancy in myb gene function provides gearing for the control of flavonoid biosynthesis in *Antirrhium* flowers, *Plant Cell,* 1996, **8**, 1519-1532.
38. NESI, N., DEBEAUJON, I., JOND, C., PELLETIER, G., CABOCHE, M., LEPINIEC, L., The *TTG8* gene encodes a basic helix-loop-helix domain protein required for expression of DFR and BAN genes in *Arabidopsis* siliques, *Plant Cell*, 2000, **12**, 1863-1878.
39. SHIRLEY, B.W., LUBASEK, W., STORZ, G., BRUGGEMANN, E., KOORNNEEF, M., AUSUBEL, F.M., GOODMAN, H., Analysis of *Arabidopsis* mutants deficient in flavonoid biosynthesis, *Plant J.,* 1995, **8**, 659-671.
40. TRETHEWEY, R.N., Gene discovery via metabolic profiling, *Curr Opinion Biotechnol.,* 2001, **12**, 135-138.
41. VAN DER FITS, L., MEMELINK, J., ORCA3, a jasmonate-responsive transcriptional regulator of plant primary and secondary metabolism, *Science,* 2000, **289**, 295-297.

Chapter Eight

FUNCTIONAL GENOMICS OF CYTOCHROMES P450 IN PLANTS

Kenneth A. Feldmann,[1]* Sunghwa Choe,[2] Hobang Kim,[2] and Joon-Hyun Park[1]

[1]*Ceres, Inc.,*
3007 Malibu Canyon Rd.,
Malibu, CA 90265,

[2]*School of Biological Sciences,*
College of Natural Sciences,
Seoul National University,
Seoul 151-747, Korea

Author for correspondence, e-mail: kfeldmann@ceres-inc.com

Introduction	126
Genomics of P450s	126
Strategies for Identifying Function for P450s	129
Forward Genetics	129
Reverse Genetics	130
PCR-Based Reverse Genetics	130
Ectopic Misexpression	130
Expression Analyses Using Microarrays	131
Hybrid Approach of Forward and Reverse Genetics	131
Examples of Recent Advances in Plants	131
Brassinosteroids and Sterols (CYP72B1, CYP85A1, CYP90A1, and CYP90B1)	131
Glucosinolate and IAA Biosynthesis	134
Phenylpropanoid Pathway	135
Oxylipin Pathway (CYP74A1 and CYP74B2)	137
CYP78A Subfamily	137
Summary	139

INTRODUCTION

Cytochromes P450 (P450s) have many attributes that make them an especially exciting group of proteins for which to elucidate function and utility in plants. First, they play important roles in plant development, protection against biotic and abiotic stresses, and the production of commercially important molecules. In plant development, they are involved in the biosynthesis of phytohormones, *e.g.*, brassinosteroids and gibberellins, the lack of which causes plants to be dwarfed.[1,2] In plant protection, they are critical in the biosynthesis of jasmonates, and other secondary metabolites, *e.g.*, glucosinolates,[3] terpenoids,[4] and anthocyanins.[5] They are also involved in the production of taxol and morphine, important as anti-cancer and anesthetic molecules, respectively.[6,7] Second, P450s are important in the detoxification of xenobiotics such as herbicides into non-toxic substances.[8] Finally, P450s belong to a large superfamily of genes making them relatively easy to identify, compare, and clone.

P450s are heme-thiolate monoxygenases that utilize a flavoprotein system to transfer electrons from NADPH and/or NADH to a substrate.[9] P450s are membrane-anchored in eukaryotes, the majority are likely to be anchored to the endoplasmic reticulum, whereas in prokaryotes they exist as soluble proteins. In *Arabidopsis*, the vast majority of P450s are in the microsomal fraction. Electron transfer to microsomal P450s occurs via a NADPH-dependent reductase. In *Arabidopsis*, there are two P450 reductases, ATR1 and ATR2.[10]

Plants contain hundreds of distinct P450 genes, while only three are found in *Saccharomyces cerevisiae*, 80 in *Drosophila melanogaster*, and 85 in *Caenorhabditis elegans*, three eukaryotes whose genomes are sequenced. With the exception of *Saccharomyces*, the function for the vast majority of the P450s in any species is unknown. However, incredibly rapid progress is being made in plants and in particular in *Arabidopsis*. It is likely that in the next few years, most of the P450s will be assigned to a biochemical pathway, and for many we will have elucidated the exact function and ascertained how the gene is regulated in the plant. In this review, we focus on the various approaches that are being used to understand *Arabidopsis* P450s but include research from other species as an aid to identifying putative function for P450s in this model plant system.

GENOMICS OF P450s

P450s can be recognized by several conserved motifs (Fig. 8.1). The heme-binding domain near the C-terminus of the protein contains a conserved Cys residue that serves as the axial ligand to the heme. The PERF and K-helix domains are located 5' of the Cys residue. The Arg in the PERF domain and the Glu and Arg in the K-helix domain are hypothesized to form a salt bridge that locks the Cys pocket into position to provide heme association with the protein.[11] The I helix contributes to proton transfer and oxygen activation by the binding and activation of dioxygen, which is essential for oxygen incorporation into substrates. Most P450s contain a hydrophobic region at the amino terminal end, which anchors the protein to the membrane. In between the hydrophobic domain and the catalytic domain is a proline-rich region that acts as a type of hinge region. While the primary amino acid sequence can be variable, the secondary and tertiary structures of P450s are similar because of these conserved domains.[11] Another distinctive feature is their length; P450s contain approximately 500 amino acids.

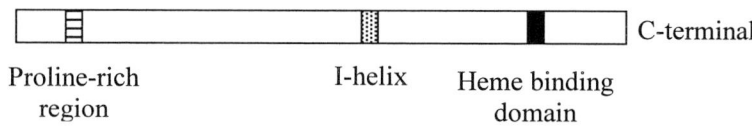

Proline-rich region   I-helix   Heme binding domain

Figure 8.1: Schematic of a typical cytochrome P450 protein showing the approximate locations for the heme binding domain, the proline-rich region and the I helix.

Recently the genome of *Arabidopsis* has been sequenced, and homology searches revealed 273 P450 genes (see http://www.biobase.dk/P450/p450.shtml http://www.*Arabidopsis*.org/, and http://drnelson.utmem.edu/CytochromeP450.html). The P450 genes are more or less randomly distributed on the chromosomes except for deficiencies on the lower arm of chromosome 3 and the upper arm of chromosomes 2 and 4 and the clustering of related P450s (Fig.8.2). Phylogenetic clustering of the functional P450s (see below) showed that 155 of them belong to a single clade, the A-type or plant-specific, while 91 belong to the non-A type, those found across kingdoms. Among these P450s, 21 A-type and 6 non-A-type genes are hypothesized to be pseudogenes (map positions for 25 of these are shown on Fig. 8.2). These putative pseudogenes are promoterless or contain premature stop codons. The A-type P450s are important in the biosynthesis of many plant-specific secondary metabolites, while the non-A-type carry out conserved reactions such as sterol biosynthesis in many life forms. A-type P450s are likely derived from a common ancestor,[12,13] while the non-A-type P450s are diverse, sharing more homology to non-plant P450s than to other plant P450s. The A-type P450s have a simple genomic organization generally containing only a single conserved intron; a small number (n = 17) are intronless. Not surprisingly, closely related A-type P450s are often clustered in the genome, *e.g.*, 19 members of the CYP71B family are clustered together on chromosome 3. Most of the non-A-type P450s contains multiple introns; one exception is the clade that contains the two CYP51A genes that contain only a

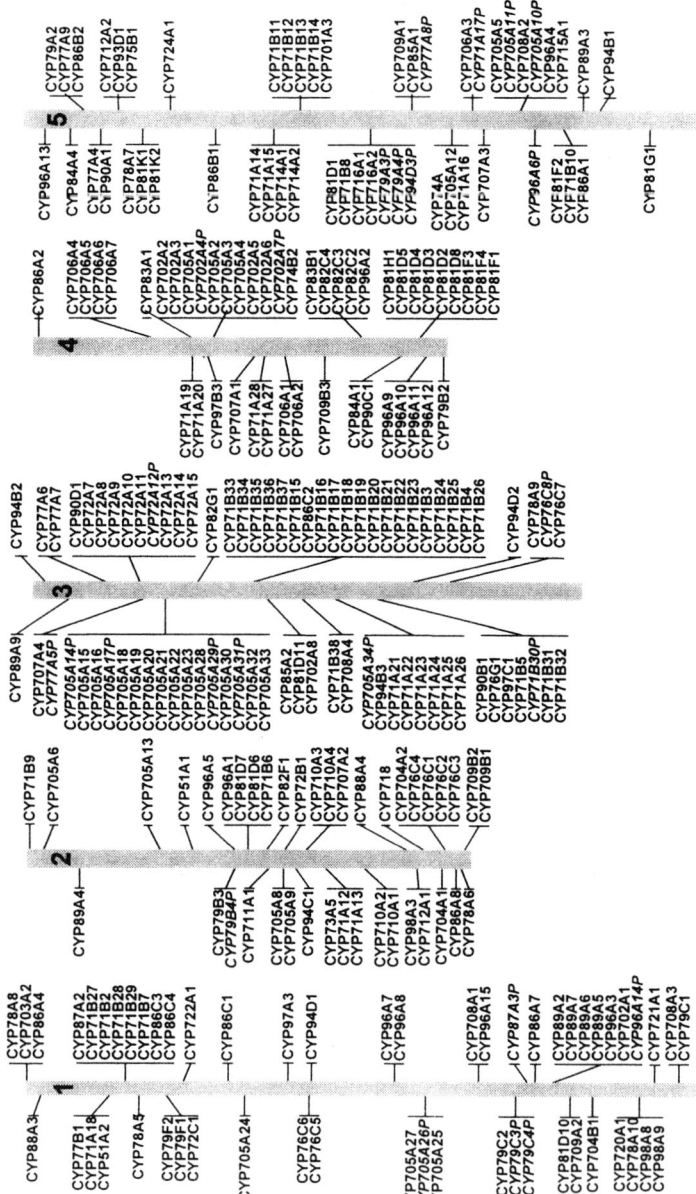

Figure 8.2: The location of the cytochrome P450s on the *Arabidopsis* chromosomes. Putative pseudogenes are italicized and followed by a P (see http://www.biobase.dk/P450/p450.shtml).

single intron. Two other clades containing the CYP96A (ten members) and CYP710A (four members) subfamilies are intronless. Many members of these two subfamilies are clustered on chromosomes 4 and 2, respectively.

P450s are named according to the degree of amino acid sequence identity that they share with other P450s. P450s that have greater than 40% amino acid identity are placed in the same family, and those that share more than 55% identity are placed in the same subfamily; if they possess >97% identity, they are referred to as allelic variants. *Arabidopsis* has 45 P450 families and 72 subfamilies, with 13 families containing a single member, while one family, CYP71, contains 54 members in just two subfamilies. Eight families contain just two members each, both belonging to the same subfamily, with the exception of CYP74 (see below). Amino acid identities from different organisms can be as low as 20%.[14] There remain unsequenced regions of the *Arabidopsis* genome, mostly at centromeric regions, so additional P450s may be added to this list in the coming months.

STRATEGIES FOR IDENTIFYING FUNCTION FOR P450s

Until recently, the characterization of plant P450s has lagged behind animal systems because 1) P450s are generally difficult to purify, due to their low abundance, 2) there are a large number of homologs, and 3) the proteins are unstable. Recently, plant molecular biologists, biochemists, and geneticists have joined forces to begin to make rapid progress toward identifying functions for a growing number of P450s. While a mutation resulting in a mutant phenotype in one of the P450s does not in itself yield the function of a P450, the mutants can be used in a comparative analysis to the wildtype to give valuable clues about function. As such, both forward and reverse genetics approaches in plants have proven useful.[15]

Forward Genetics

In forward genetics, a P450 mutant is identified, either from phenotype prediction or by luck, and the mutation is used to clone the gene. There are generally two types of mutant populations used in *Arabidopsis*, chemical (EMS) and insertion (T-DNA predominantly, but also transposons). Chemically-mutagenized populations offer the advantage of having more mutations per genome than insertion populations.[16] However, it has generally been much faster to clone a gene from an insertion mutant than to clone a gene positionally from a point mutation. Mutants for a number of P450s have been isolated from T-DNA insertion populations in *Arabidopsis*, including CYP90A1,[17] CYP90B2,[18] and CYP72B1.[19] With the complete sequencing of the genome and thousands of molecular markers available (http://www.*Arabidopsis*.org/cgi-bin/maps/Schrom, http://www.*Arabidopsis*.org/servlets/mapper, and http://www.*Arabidopsis*.org/), it is becoming routine to clone a gene based on its map position.[20]

Reverse Genetics

Reverse genetic approaches start with a sequence and attempt to identify a function for the gene. This can be done by using PCR-based screens on DNA from insertionally mutagenized populations, ectopic misexpression of the gene, and expression studies, *e.g.*, with RT-PCR or microarray analyses.

PCR-Based Reverse Genetics

In the past few years, the predominant reverse genetics approach in *Arabidopsis* has been to use the gene sequence in a PCR-based approach.[21-23] In this approach, primers are made for the gene of interest and used in combination with a T-DNA specific primer on DNA from as many as 5,000 transformed lines at a time. The pooling strategy for the population is done in a manner that allows the fewest PCRs to be performed in order to identify the disrupted line. An efficient PCR-based system requires large populations of DNA from insertion mutants and sequence information for all genes of interest; *Arabidopsis* meets both of these needs. In fact, reverse genetics methods on T-DNA populations have lead to many mutants for specific P450s, *e.g.*, CYP83B1,[3] CYP74A1, CYP74B2 (J.-H. Park, unpublished data), CYP51A1, CYP51A2 (H. Kim, unpublished data), and CYP78A5-CYP78A9 (H. Kim, unpublished data) to name a few.

Ectopic Misexpression

The misexpression of endogenous or heterologous genes can be a valuable tool toward identifying a putative function for the gene. However, ectopic misexpression has not been utilized to its full advantage to date, probably because of the lack of full-length cDNAs. With the complete genome sequence, it is relatively easy to make primers for the various P450s and to PCR-clone full-length cDNAs from appropriate libraries. Alternatively, full-length cDNA sequences and clones are being made available in the public sector through the ABRC, TIGR, and TAIR (http://www.*Arabidopsis*.org/servlets/Order?state=checkLogin, ftp://ftp.tigr.org/pub/data/a_thaliana/ceres/, and http://www.*Arabidopsis*.org/, respectively). Still, even with misexpression plants for every P450, it takes considerable biochemistry to ascertain function. For example, Zondlo and Irish[24] and Ito and Meyerowitz[25] overexpressed CYP78A5 and CYP78A9 and found floral phenotypes (see below). While leading to phenotypes that hint at the function of the CYP78A family, the phenotypes have thus far not yielded the biochemical reaction that is taking place.

GENOMICS: CYTOCHROMES P450

Expression Analyses Using Microarrays

Considerable insight can be gained by examining the expression patterns of the P450s in various tissues, mutants, and environments. Xu et al.[26] made a P450 microarray chip representing 142 different P450s. They used different tissues as targets against these probes and ascertained expression levels as well as ratios among tissue sources. These data confirmed what we already knew about some P450s and as such offered valuable clues about the function of others. For example, Xu et al.[26] showed that CYP74A is expressed much more highly in the leaves than in roots, siliques, and flowers, which is expected as CYP74A (allene oxide synthase) is involved in jasmonate biosynthesis, and is localized to the chloroplast.

Hybrid Approach of Forward and Reverse Genetics

A hybrid of forward and reverse genetics is now being utilized to identify gene knock-outs in a high-throughput manner. Large populations of insertion lines are being generated and the plant DNA flanking the insertion is being cloned from each line. Again, T-DNA tagged populations have been the most widely employed. Generally, single transformed lines are used in this approach. TAIL-PCR is conducted on DNA from tissues (leaves or seeds) of each transformant. When the TAIL product matches a P450 sequence, the line can be screened for a phenotype, morphological and/or biochemical. Two such populations are generally available to researchers (http://www.*Arabidopsis*.org/abrc/tdna_ecker.html and http://www.nadii.com/pages/collaborations/garlic_files/GarlicDescription.html). J. Ecker has already generated and distributed to the ABRC >20,000 lines, 60% of these with associated plant flanking sequences. He is going to generate 140,000 independent T-DNA insertion lines by 2003. The TMRI site contains ~70,000 plant flanking sequences from attempting to TAIL 100,000 transformed lines. The sequences can be screened and the seeds ordered for a specific P450 insertion mutant. An alternative to this approach is to phenotype the T-DNA population simultaneous with TAIL-PCR; this is the approach Ceres, Inc. (unpublished) has taken. Integrating the two data sets quickly reveals the consequence of the knock-out. The beauty of this approach is that as transformed populations become larger, second and third knock-out alleles will be found for each gene allowing co-segregation of the mutant phenotype with the disruption *in silico*.

EXAMPLES OF RECENT ADVANCES IN PLANTS

Brassinosteroids and Sterols (CYP72B1, CYP85A1, CYP90A1, and CYP90B1)

Brassinosteroids (BR) are a group of plant steroid hormones that help to regulate many aspects of plant growth and development. Several P450s have been identified that are important in the brassinosteroid biosynthetic pathway. Szekeres et al.[17] identified the first P450, CYP90A1. A T-DNA knock-out allele of this gene, constitutive photomorphogenesis and dwarfism (cpd), causes severe dwarfism. In fact, the CYP90A1 loss-of-function alleles are smaller than loss-of-function mutants at any other step in the brassinosteroid biosynthetic pathway. This gene catalyzes the C-23α hydroxylation of 6-deoxocathasterone and cathasterone to 6-deoxoteasterone and teasterone, respectively (Fig. 8.3). This is the most downstream step in the pathway for which a mutant has been identified in *Arabidopsis*. Mathur et al.[27] revealed that the CPD promoter is feed-back downregulated specifically by the end product brassinolide (BL).

Choe et al.[18] identified a T-DNA knock-out allele of CYP90B1 that hydroxylates campestanol and 6-oxocampestanol at C-22α to form 6-deoxocathasterone and cathasterone, respectively. CYP90B1 is the rate-limiting step in BL biosynthesis and as such represents one of the most important enzymes in brassinosteroid biosynthesis. The intermediates upstream of CYP90B1 have little biological activity and they accumulate, while intermediates downstream start to have dramatic bioactivity, probably as a result of being quickly metabolized to BL. They are scarce in the plant. Not surprisingly, overexpression of CYP90B1 leads to greater biomass and seed set in several dicotolydenous species.[28] Another member of the CYP90 family, CYP90C1 (ROT3),[29] when disrupted, results in a rounder leaf phenotype that is also observed in other BR dwarfs. There is no biochemical evidence that CYP90C1 is involved in BR biosynthesis, but given that the leaves have a round phenotype, suggesting a perturbation in cell elongation, it seems likely that it is somehow involved in BR biosynthesis.

In tomato, Bishop et al.[30,31] identified a transposon tagged dwarf. Cloning of the disrupted gene, Dwarf, resulted in the first member of a new family of P450s, CYP85. Bishop et al.[30] showed that CYP85 is responsible for the C-6 oxidation of 6-deoxocasterasterone to castasterone. Shimada et al.[32] cloned an *Arabidopsis* ortholog of the tomato CYP85 gene, AtBR6ox. By expressing these genes, tomato CYP85 and AtBR6ox, in yeast they were able to show that both enzymes catalyzed the C-6 oxidation of 6-deoxytyphasterol and 6-deoxocastasterone, and probably 6-deoxoteasterone and 3-deydro-6-deoxoteasterone, to their respective oxidized forms (Fig. 8.3). These data show conclusively that both genes encode steroid 6-oxidases with broad substrate specificity. Sequence analysis of AtBR6ox showed that it is the CYP85A1 gene according to the nomenclature of Nelson et al.[14] Dwarf and CYP85A1 share 68% amino acid identity. CYP85A1 clusters with the CYP90 family, P450s that are involved in brassinosteroid biosynthesis.

Using a yeast two hybrid screen with Pra2, a small G protein that plays an essential role in the phytochrome-mediated light signal transduction pathway, Kang et al.[33] found a cytochrome P450, DDWF1, that is involved in BL biosynthesis in pea. This P450 shares high sequence identity to P450s involved in brassinosteroid

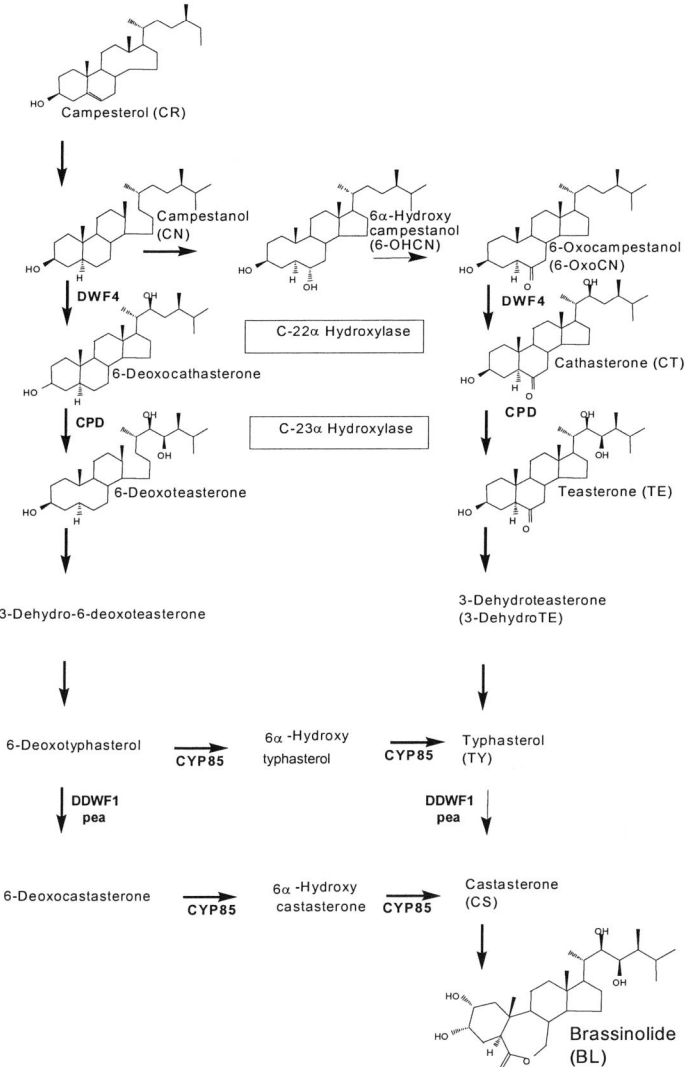

Figure 8.3: Brassinosteroid biosynthetic pathway. DWF4 and CPD catalyze the 22α- and 23α-hydroxylation reaction, respectively, in both the early and late C-6 oxidation pathways. CYP85 catalyzes the C-6 oxidation of 6-deoxytyphasterol and 6-deoxocastasterone and tentatively the C-6 oxidation of 6-deoxocathasterone and 6-deoxoteasterone (arrows not shown). *DDWF1*: dark-induced *DWF*-like protein 1; DWF4: DWARF4; CPD: CONSTITUTIVE PHOTOMORPHOGENESIS AND DWARF.

biosynthesis. They were able to show that DDWF1 catalyzes the C-2 hydroxylation of 6-deoxytyphasterol and typhasterol to 6-deoxocastasterone and castasterone, respectively (Fig. 8.3). The *Arabidopsis* ortholog has not yet been identified.

Finally, Neff et al.[19] identified CYP72B1 in *Arabidopsis* as being important in the degradation of BL. Using T-DNA activation tagging of a phytochrome B mutant (phyB-4), an activation line was observed that had a short hypocotyl (dwarf) phenotype (bas1-D phyB-4). The activation tag was just upstream of the CYP72B1 gene. The mutant has reduced levels of BL and accumulates 26-hydroxyBL.

Glucosinolate and IAA Biosynthesis

The cloning and characterization of several P450 genes in the pathway that leads from tryptophan to auxin and glucosinolates has been accomplished recently (Fig. 8.4; reviewed[34]). Glucosinolates are amino acid-derived natural products containing a sulfate and a thioglucose moiety that are stored in the vacuoles of cruciferous plants. The amino acids are converted to unstable aldoximes by P450s of the CYP79 family. The aldoximes are hydroxylated by a different P450 and eventually metabolized to a glucosinolate (Fig. 8.4). Auxins are growth-promoting substances that are important in maintaining apical dominance, initiating roots, vascular differentiation, and certain tropic behaviors. Though the structure of the primary plant auxin, indole-3-acetic acid (IAA), has been known for decades, the pathway has been difficult to elucidate.

Hull et al.[35] used a different approach to identify CYP79B2 and CYP79B3 from *Arabidopsis* as enzymes for the conversion of tryptophan to indole-3-acetaldoxime (IAOx) (Fig. 8.4). They took advantage of the fact that an analog of indole, 5-fluoroindole, is toxic to yeast, and they screened for yeast colonies containing *Arabidopsis* cDNAs that were resistant to the analog.

Using a PCR-based reverse genetics screen, Bak et al.[3] identified a T-DNA knock-out allele, *runt1* (*rnt1*), of CYP83B1 that has an auxin accumulating phenotype (strong apical dominance, short roots, and proliferation of callus at root tips). Seedlings of *rnt1* contained reduced levels of indole glucosinolates, while overexpression lines contained elevated levels of indole glucosinolates and showed reduced apical dominance. The knock-out of CYP83B1 shunts IAOx to IAA, thus producing an auxin accumulation phenotype (Fig. 8.4).

Glucosinolates can also be derived from phenylalanine. Wittstock and Halkier[36] have identified CYP79A2 from *Arabidopsis* and shown that it catalyzes the conversion of phenylalanine to phenylacteladoxime with the subsequent accumulation of benzylglucosinolate.

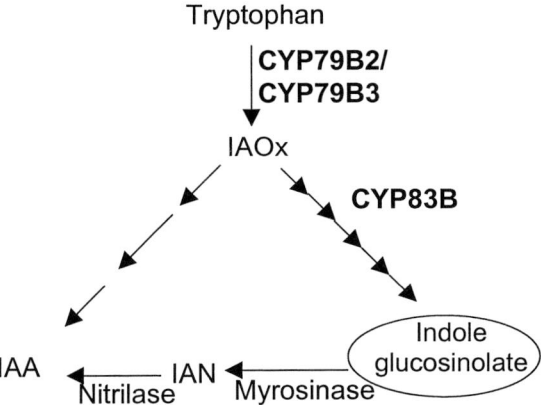

Figure 8.4: Glucosinolate and IAA pathways. IAA: indole-3-acetic acid; IAN: indole-3-acetonitrile; IAOx: indole-3-acetaldoxime. A loss of function of CYP83B drives the IAOx to IAA and an auxin-overproducing phenotype is the result.

Phenylpropanoid Pathway

Another pathway where significant progress has been made is the latter part of the phenylpropanoid pathway leading to isoflavonoids (Fig. 8.5). Isoflavonoids are found almost exclusively in leguminous species where they play a role in the response of microorganisms to the plant, *e.g.*, defense against pathogen attack and establishment of the symbiotic relationship between roots and rhizobial bacteria. In addition, the isoflavonoids genistein and daidzein are being studied for their beneficial effects on human health, including reducing the risk of cancer and coronary heart disease, and improvement of cholesterol levels.[37] Liquiritigenin and naringenin are first converted to 2-hydroxy isoflavanones by 2-hydroxyisoflavanone synthase (IFS; CYP93C2) via a 1,2-aryl migration. Flavanone 2-hydroxylase (F2H; CYP93B1) competes for the same substrate as IFS to generate a 2-hydroxy-flavanone as a precursor to retrochalcone[38] (Fig. 8.5). IFS was recently identified by several sets of researchers.[37,39,40] Steele et al.[39] first identified a cDNA encoding IFS by showing that liquiritigenin could be converted to daidzein by isolated microsomes from insect cells expressing a soybean cDNA. These cells also converted naringenin to genistein. Akashi et al.[40] isolated a full-length cDNA for IFS (CYP93C2) from cultured cell lines of licorice (*Glycyrrhiza echinata*) that were producing an isoflavonoid-derived phytoalexin, medicarpin, and showed that it could convert liquiritigenin and naringenin to daidzein and genistein, respectively (Fig. 8.5).

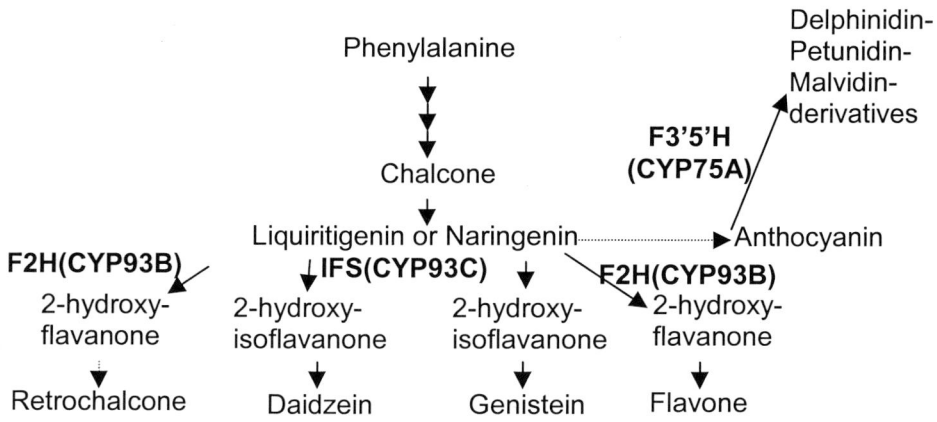

Figure 8.5: Phenylpropanoid biosynthetic pathway. IFS: 2-hydroxyisoflavanone synthase; F2H: flavanone-2-hydroxylase; F3'5'H: flavnonoid-3',5'-hydroxylase

As the P450s in this pathway all accumulate transiently upon elicitation of the cells, Jung et al.[37] were able to isolate IFS from a soybean cDNA library made from leaves infected with the fungus *Sclerotinia sclerotiorum* by functional expression in yeast. They transformed IFS into *Arabidopsis*, where the substrate naringenin is present as an intermediate in the anthocyanin pathway, and were able to get transgenic plants containing genistein. The authors isolated IFS from a number of leguminous species as well as sugarbeet and showed that, remarkably, all of the enzymes shared >95% amino acid similarity.[37]

In another branch of the phenylpropanoid pathway, Shimada et al.[41] isolated the flavonoid-3'5'-hydroxylase gene (F3'5'H) from prairie gentian (*Enstoma russelianum*) using heterologous hybridization to the *Petunia* F3'5'H cDNA (Fig. 8.5). This protein is responsible for the blue and purple pigments in flowers. The *Petunia* F3'5'H gene was transferred to *Petunia* and tobacco varieties, which do not normally contain the gene. While the flowers in transgenic *Petunia* went from pink to magenta, tobacco flowers showed much less change in color. However, both accumulated 3',5'-hydroxylated anthocyanidins, products of the F3'5'H gene.[41]

Oxylipin Pathway (CYP74A1 and CYP74B2)

The CYP74 genes are non-classical P450s in that they do not use NADPH or the NADPH-dependent P450 reductase but instead use already activated substrates, fatty acid hydroperoxides.[9] This phenomenon occurs only in this P450 family.[13,42] The CYP74 family members are also unusual among P450s in that they appear to contain chloroplast-targeting sequences.[13] The fact that the CYP74 genes do not require oxygen and NADPH, along with their very reactive peroxide substrates, indicates that these enzymes have very high activity with over 1000 turnovers per second for CYP74A.[43]

Fatty acid hydroperoxides (lipoxygenase products) are metabolized to allene oxide by CYP74A (allene oxide synthase) or to C-6 volatiles and 12-oxo-dodecenoic acid (ODA) by CYP74B (hydroperoxy lyase), two P450s that have been detected in many plant species[44-46] (Fig. 8.6). The C-6 volatiles are important as flavorings in fruits and vegetables, while C-12 ODA is important in plant-insect interactions.[47] Because the two genes have been cloned, the biochemical nature and molecular structure of their proteins have been elucidated.[48-50]

Utilizing PCR-based reverse genetics we have identified knock-out mutants in both of the CYP74 genes in *Arabidopsis*. From analysis of a CYP74A1 knock-out mutant, which shows an impaired wound signal and severe male sterility, it is clear that CYP74A1 is essential, and a major modulating factor, in the production of jasmonates in *Arabidopsis*. (J.-H. Park, unpublished data).

In contrast to the sterility phenotype in the CYP74A mutant, the knock-out allele of CYP74B had no detectable visible alteration in phenotype. Matsui et al.[50] showed the expression level of CYP74B is relatively low in inflorescences and flowers of *Arabidopsis* compared to other organs. Forty-eight hours after wounding, the accumulation of CYP74B mRNAs was increased.[50] Chemical analyses of the endogenous oxylipins should give us some clues as to the importance of this pathway in the plant, and generate plants that can be tested in bioassays with various insect pests. Double mutant analyses with the knock-out alleles of CYP74A and CYP74B will be important tools for understanding the regulation and interactions between these two pathways.

CYP78A Subfamily

The approaches taken to understand the CYP78A subfamily typify the approach that will need to be taken with other P450 families. Early data on the CYP78A family suggested that these genes were important in flower development. For example, CYP78A1 from maize and CYP78A2 from *Phalaenopsis* are expressed specifically in the tassel primordia and pollen tube, respectively.[51,52]

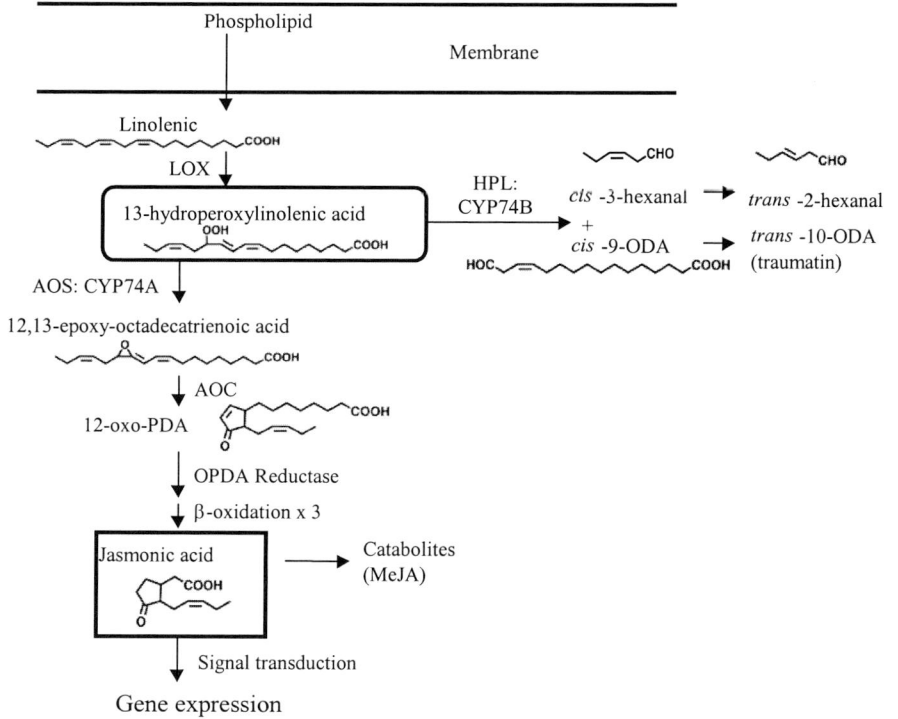

Figure 8.6: A diagram of the oxylipin pathway. LOX: lipoxygenase; AOS: allene oxide synthase; HPL: hydroperoxide lyase; AOC: allene oxide cyclase; OPDA Reductase: 12-oxo-phytodienoic acid-10,11-reductase.

In *Arabidopsis*, the CYP78A subfamily contains six genes, CYP78A5-CYP78A10, located around the genome. The first lead in this family came from Zondlo and Irish.[24] They were studying genes that were important in the flowering process by subtracting cDNAs from cauliflower heads that were induced to flower with cDNAs from leaves. They identified a CYP78A clone that was most similar to CYP78A5 in *Arabidopsis*. CYP78A5 was expressed in the peripheral regions of the vegetative and reproductive shoot meristems.[24] Most interesting in this study, they were not able to recover any antisense expression plants, suggesting that the loss of function of CYP78A5, and related genes, causes embryo lethality. The loss of expression of CYP78A5 by itself does not explain this phenotype, as we have isolated knock-out alleles of this gene and do not observe lethality in the homozygous condition (H. Kim, unpublished data). Overexpression of CYP78A5 caused stems to be twisted and kinked.[24] In the flowers, the buds were slow to open, and the petals and stamens were stunted. The anthers were also disorganized with undeveloped locules. The overexpression lines produced very few seeds, and the reduced fecundity could not be reversed with wild-type pollen, suggesting that the ovules were also abnormal. One interesting note is that the more acropetal flowers of the OE lines were more wildtype.[24]

Ito and Meyerowitz[25] used an activation tag screen in *Arabidopsis* to identify a plant that made large seedless fruits. The activation tagged gene was CYP78A9. The OE plants shared many characteristics of the CYP78A5 OE plants.[24] In the CYP78A9 OE plants, pollen production was reduced and dehiscence delayed, stamens were short, few seeds were produced, and siliques were 10-20% longer and considerably wider than wildtype. Also, the OE plants displayed fruit growth independent of fertilization.

Thus, it appears that 4 different members of the CYP78A family in three different species are expressed in the flower. We are taking a more thorough approach to understanding this gene family in *Arabidopsis* by identifying a knock-out for each member, conducting northern analyses on all family members on a variety of tissues, ectopic over- and under-expression of all members, and by making warranted double mutants. Initial results show that a knock-out in CYP78A5 results in a subtle flower phenotype, but a double mutant between CYP78A5 and CYP8A7, a gene that shares the closest expression pattern to CYP78A5, results in a double mutant that has a severe dwarf phenotype and abnormal flowers (H. Kim, unpublished data). Knock-out alleles in the other CYP78A genes yield no clues about their possible function. Analyses of the overexpression lines and additional double mutants will shed some light on the possible function of this gene family.

SUMMARY

We have illustrated several examples of methods for identifying the function for plant P450s. Most of these P450s are the sole member of a subfamily such that gene redundancy is not an *a priori* factor. We also presented some work on the CYP78A family where there is significant gene redundancy and discussed how we used expression analysis coupled with double mutant analysis to deduce gene function. Double mutant analysis is possible in the CYP78A family because the family members are not linked. However, most of the P450 genes in *Arabidopsis* are redundant and clustered. The function encoded by these P450 genes will be more difficult to ascertain. Still, the biological resources, knock-outs and resources for ectopic expression in *Arabidopsis*, should prove useful for making progress on these genes.

To generate a saturated knock-out population in *Arabidopsis*, we estimate that 92,000 transformants (138,000 inserts) are required to have a 95% probability of an insert in any specific P450 gene. This is based on the hypotheses that P450 genes are average size genes (2.8 kb including 5' and 3' UTRs with a minimal promoter), the *Arabidopsis* genome is ~130,000 kbp in length (containing 27,000 genes), and T-DNA insertion appears to be random at the gene level. Larger populations exist in both public and private holdings. The advantage of large populations of T-DNA lines is that multiple alleles will be isolated. Knock-outs for the remaining lines are isolated via directed PCR-based screening on DNA from the same population as described in the section above. Our experiences tell us that it will be more difficult to obtain full-length cDNAs for all P450 genes. Our large-scale sequencing program suggests that many P450 transcripts are under-represented even in diverse tissues.

REFERENCES

1. KOORNNEEF, M., VAN DER VEEN, J.H., Induction and analysis of gibberellin sensitive mutants in *Arabidopsis thaliana* (L.) Heynh, *Theor. Appl. Genet.*, 1980, **58**, 257-263.
2. CLOUSE, S.D., FELDMANN, K.A., Molecular genetics of brassinosteroid action. In: *Brassinosteroids: Steroidal Plant Hormones,* (A. Sakurai, T. Yokota, and S.D. Clouse, eds,), Springer Verlag, Tokyo, 1999, pp 163-190.
3. BAK, S., TAX, F.E., FELDMANN, K.A., GALBRAITH, D.W., FEYEREISEN, R., CYP83B1, A cytochrome P450 at the metabolic branch point in auxin and indole glucosinolate biosynthesis in *Arabidopsis, Plant Cell,* 2001, **13**, 101-111.
4. DOONER H.K., ROBBINS, T.P., JORGENSEN, R.A., Genetic and developmental control of anthocyanin biosynthesis, *Annu. Rev. Genet.*, 1991, **125,** 173-99.
5. TOGURI, T., UMEMOTO, N., KOBAYASHI, O., OHTANI, T., Activation of anthocyanin synthesis genes by white light in eggplant hypocotyl tissues, and identification of an inducible P-450 cDNA, *Plant. Mol. Biol.*, 1993, **23,** 933-46.
6. JENNEWEIN, S., RITHNER, C.D., WILLIAMS, R.M., CROTEAU, R.B., Taxol biosynthesis: Taxane 13alpha-hydroxylase is a cytochrome P450-dependent monooxygenase, *Proc. Natl. Acad. Sci. USA,* 2001, **98**, 13595-600.
7. MUKHERJEE, A.K, BASU, S., SARKAR, N., GHOSH, A.C., Advances in cancer therapy with plant based natural products, *Curr. Med. Chem.,* 2001, **8**, 1467-86.
8. WERCK-REICHHART, D., HEHN, A., DIDIERJEAN, L., Cytochromes P450 for engineering herbicide tolerance, *Trends Plant. Sci.,* 2000, **3**, 116-123.
9. SCHULER, M.A., Plant cytochrome P450 monooxygenases, *Crit. Rev. in Plant Sci.*, 1996, **15**, 235-284.
10. URBAN, P., MIGNOTTE, C., KAZMAIER, M., DELORME, F., POMPON, D., Cloning, yeast expression, and characterization of the coupling of two distantly related *Arabidopsis thaliana* NADPH-cytochrome P450 reductases with P450 CYP73A5, *Bio. Chem.,* 1997, **272,** 19176-86.
11. HASEMANN, C.A., KURUMBAIL R.G., BODDUPALLI, S.S., PETERSON, J.A., DEISENHOFER, J., Structure and function of cytochromes P450: A comparative analysis of three crystal structures, *Structure,* 1995, **3**, 41-62.

12. DURST, F., NELSON, D.R., Diversity and evolution of plant P450 and P450-reductases, *Drug Metabol. Drug Interact.,* 1995, **12**, 189-206.
13. PAQUETTE, S.M., BAK, S., FERYEREISEN, R., Intron-exon organization and phylogeny in a large superfamily, the paralogous cytochrome P450 genes of *Arabidopsis thaliana, DNA Plant Biol.,* 2000, **5**, 307-317.
14. NELSON, D.R., KOYMANS, L., KAMATAKI, T., STEGEMAN, J.J., FEYEREISEN, R., WAXMAN, D.J., WATERMAN, M.R., GOTOH, O., COON, M.J., ESTABROOK, R.W., GUNSALUS, I.C., NEBERT, D.W., P450 superfamily: Update on new sequences, gene mapping, accession numbers and nomenclature, *Pharmacogenetics,* 1996, **6**, 1-42.
15. AZPIROZ, R., FELDMANN, K.A., T-DNA insertion mutagenesis in *Arabidopsis*: Going back and forth, 1997, *Trends Genet.,* **13**:152-156.
16. FELDMANN, K.A., MALMBERG, R., DEAN, C., Mutagenesis in *Arabidopsis*. In: *Arabidopsis,* E. Meyerowitz and C. Somerville (eds.) Cold Spring Harbor, NY, 1994, pp. 137-172.
17. SZEKERES, M., NEMETH, K., KONCZ-KALMAN, Z., MATHUR, J., KAUSCHMANN, A., ALTMANN, T., REDEI, G.P., NAGY, F., SCHELL, J., KONCZ, C., Brassinosteroids rescue the deficiency of CYP90, a cytochrome P450, controlling cell elongation and de-etiolation in *Arabidopsis, Cell,* 1996, **85**, 171-182.
18. CHOE, S., DILKES, B.P., FUJIOKA, S., TAKATSUTO, S., SAKURAI, A., FELDMANN, K.A., The DWF4 gene of *Arabidopsis* encodes a cytochrome P450 that mediates multiple 22-hydroxylation steps in brassinosteroid biosynthesis, *Plant Cell,* 1998, **10**, 231-243.
19. NEFF, M. M., NGUYEN, M.S., BAS1: A gene regulating brassinosteroid levels and light responsiveness in *Arabidopsis, Proc. Natl. Acad. Sci.* USA, 1999, **96**, 15316-15323.
20. SHIRLEY, A.M., McMICHAEL, C.M., CHAPPLE, C., The sng2 mutant of *Arabidopsis* is defective in the gene encoding the serine carboxypeptidase-like protein sinapoylglucose: Choline sinapoyltransferase, *Plant J.,* 2001, **28**, 83-94.
21. McKINNEY, E.C., ALI, N., TRAUT, A., FELDMANN, K.A., BELOSTOTSKY, D.A., McDOWELL, J.A., MEAGHER, R.B., Sequence based identification of T-DNA insertion mutations in *Arabidopsis*: Actin mutants *act2-1* and *act4-1, Plant J.,* 1995, **8**, 613-622.
22. WINKLER, R., FRANK, M., GALBRAITH, D.W., FEYEREISEN, R., FELDMANN, K.A., Systematic reverse genetics of the P450 superfamily in *Arabidopsis, Plant Physiol.,* 1998, **118**, 743-749.
23. KRYSAN P.J., YOUNG, J.C., TAX, F., SUSSMAN, M.R., Identification of transferred DNA insertions within *Arabidopsis* genes involved in signal transduction and ion transport, *Proc. Natl. Acad. Sci. USA,* 1996, **93**, 8145-50.
24. ZONDLO, S.C., IRISH, V.F., CYP78A5 encodes a cytochrome P450 that marks the shoot apical meristem boundary in *Arabidopsis, Plant J.,* 1999, **19**, 259-268.
25. ITO, T., MEYEROWITZ, E.M., Overexpression of a gene encoding a cytochrome P450, CYP78A9, induces large and seedless fruit in *Arabidopsis, Plant Cell,* 2000, **12**, 1541-1550.

26. XU, W., BAK, S., DECKER, A., PAQUETTE, S.M., FEYEREISEN, R., GALBRAITH, D.W., Microarray-based analysis of gene expression in very large gene families: The cytochrome P450 gene superfamily of *Arabidopsis thaliana*, *Gene*, 2001, **272**, 61-74.
27. MATHUR, J., MOLNAR, G., FUJIOKA, S., TAKATSUTO, S., SAKURAI, A., YOKOTA, T., ADAM, G., VOIGT, B., NAGY, F., MAAS, C., SCHELL, J., KONCZ, C., SZEKERES, M., Transcription of the *Arabidopsis* CPD gene, encoding a steoidogenic cytochrome P450, is negatively controlled by brassinosteroids, *Plant J.*, 1998, **14**, 593-602.
28. CHOE, S., FUJIOKA, S., NOGUCHI, T., TAKATSUTO, S. YOSHIDA, S., FELDMANN, K.A., Overexpression of DWARF4 in the brassinosteroid biosynthetic pathway results in increased vegetative growth and seed yield in *Arabidopsis*, *Plant J.*, 2001, **26**, 1-11.
29. KIM, G.T., TSUKAYA, H., UCHIMIYA, H., The ROTUNDIFOLIA3 gene of *Arabidopsis thaliana* encodes a new member of the cytochrome P-450 family that is required for the regulated polar elongation of leaf cells, *Genes Dev.*, 1998, **12**, 2381-91.
30. BISHOP, G.J., HARRISON, K., JONES, J.D., The tomato Dwarf gene isolated by heterologous transposon tagging encodes the first member of a new cytochrome P450 family, *Plant Cell*, 1996, **8**, 959-969.
31. BISHOP, G.J., NOMURA, T., YOKOTA, T., HARRISON, K., NOGUCHI, T., FUJIOKA, S., TAKATSUTO, S., JONES, J.D., KAMIYA, Y., The tomato DWARF enzyme catalyses C-6 oxidation in brassinosteroid biosynthesis, *Proc. Natl. Acad. Sci. USA*, 1999, **96**, 1761-6.
32. SHIMADA, Y, FUJIOKA, S., MIYAUCHI, N., KUSHIRO, M., TAKATSUTO, S., MONURA, T., YOKOTA, T., KAMIYA, Y., BISHOP, G.J., YOSHIDA, S., Brassinosteroid-6-Oxidases from *Arabidopsis* and tomato catalyze multiple C-6 oxidations in brassinosteroid biosynthesis, *Plant Physiol.*, 2001, **126**, 770-779.
33. KANG, J.G., YUN, J., KIM, D.H., CHUNG, K.S., FUJIOKA, S., KIM, J.I., DAE, W.H., YOSHIDA, S., TAKATSUTO, S., SONG, P.S., PARK, C.M., Light and brassinosteroid signals are integrated via a dark-induced small G protein in etiolated seedling growth, *Cell*, 2001, **105**, 625-636.
34. FELDMANN, K.A., Cytochrome P450s as genes for crop improvement, *Curr. Opin. Plant Biol.*, 2001, **4**, 165-167.
35. HULL, A.K., VIJ, R., CELENZA, J.L., *Arabidopsis* cytochrome P450s that catalyze the first step of tryptophan-dependent indole-3-acetic acid biosynthesis, *Proc. Natl. Acad. Sci. USA*, 2000, **97**, 2379-2384.
36. WITTSTOCK, U., HALKIER, B.A., Cytochrome P450 CYP79A2 from *Arabidopsis thaliana* L. catalyzes the conversion of L-phenylalanine to phenylacetaldoxime in the biosynthesis of benzylglucosinolate, *J. Biol. Chem.*, 2000, **275**, 14659-14666.
37. JUNG, W., YU, O., LAU, S.M., O'KEEFE, D.P., ODELL, J., FADER, G., McGONIGLE, B., Identification and expression of isoflavone synthase, the key enzyme for biosynthesis of isoflavones in legumes, *Nature Biotech.*, 2000, **18**, 208-212.
38. OTANI, K., TAKAHASHI, T., FURUYA, T., AYABE, S., Licondione synthase, a cytochrome P450 monooxygenase catalyzing 2-hydroxylation of 5-deoxyflavanone, in cultured *Glycyrrhiza enchinata* L. cells, *Plant Physiol.*, 1994, **105**, 1427-1432.

39. STEELE, C.L., GIJZEN, M., QUTOB, D., DIXON, R.A., Molecular characterization of the enzyme catalyzing the aryl migration reaction of isoflavonoid biosynthesis in soybean, *Arch. of Biochem. and Biophys.*, 1999, **367**, 146-150.
40. AKASHI, T., AOKI, T., AYABE, S., Cloning and functional expression of a cytochrome P450 cDNA encoding 2-hydroxyisoflavanone synthase involved in biosynthesis of the isoflavonoid skeleton in licorice, *Plant Physiol.*, 1999, **121**, 821-828.
41. SHIMADA, Y., NAKANO-SHIMADA, R., OHBAYASHI, M., OKINAKA, Y., KIYOKAWA, S., KIKUCHI, Y., Expression of chimeric P450 genes encoding flavonoid-3',5'-hydroxylase in transgenic tobacco and petunia plants, *FEBS Lett.*, 1999, **461**, 241-245.
42. WERCK-REICHHART, D., FEYEREISEN, R., Cytochromes P450: A success story, *Genome Biol.* 2000, **1**, 3003.
43. CHAPPLE, C., Molecular-genetic analysis of plant cytochrome P450-dependent monooxygenases, *Annu. Rev. Plant Physiol., Plant Mol. Bio.*, 1989, **49**, 311-343.
44. MATSUI, K., SHIBUTANI, M., HASE, T., KAJIWARA, T., Bell pepper fruit fatty acid hydroperoxide lyase is a cytochrome P450 (CYP75B) in many plant tissues, *FEBS Lett.*, 1996, **394**, 21-24.
45. LAU, S.M., HARDER, P.A., O'KEEFE, D.P., Low carbon monoxide affinity allene oxide synthase is the predominant cytochrome P450 in many plant tissues, *Biochemistry*, 1993, **32**, 1945-1950.
46. PAN, Z., DURST, F., WERCK-REICHHART, D., GARDNER, H.W., CAMARA, B., CORNISH, K., BACKHAUS, R. A., The major protein of guayule rubber particles is a cytochrome P450. Characterization based on cDNA cloning and spectroscopic analysis of the solubilized enzyme and its reaction products, *J. Bio. Chem.*, 1995, **270**, 8487-8494.
47. PARE, P.W., TUMLINSON, J.H., Plant volatiles as a defense against insect herbivores, *Plant Physiol.*, 1999, **121**, 325-332.
48. SONG, W.C., FUNK, C.D., BRASH, A.R., Molecular cloning of an allene oxide synthase: A cytochrome P450 specialized for the metabolism of fatty acid hydroperoxides, *Proc. Natl. Acad. Sci. USA*, 1993, **90**, 8519-8523.
49. LAUDERT, D., PFANNSCHMIDT, U., LOTTSPEICH, F., HOLLANDER-CZYTKO, H., WEILER, E.W., Cloning, molecular and functional characterization of *Arabidopsis thaliana* allene oxide synthase (CYP74) the first enzyme for the octadecanoid pathway to jasmonates, *Plant Mol. Biol.*, 1996, **31**, 323-335.
50. MATSUI, K., WILKINSON, J., HIATT, B., KNAUF, V., KAJIWARA, T., Molecular cloning and expression of *Arabidopsis* fatty acid hydroperoxide lyase, *Plant Cell Physiol.*, 1999, **40**, 477-481.
51. NADEAU, J.A., ZHANG, X.S., LI, J., O'NEILL, S.D., Ovule development: Identification of stage-specific and tissue-specific cDNAs, *Plant Cell*, 1996, **8**, 213-239.
52. LARKIN, J., Isolation of a cytochrome P450 homologue preferentially expressed in the developing inflorescences of Zea mays, *Plant Mol. Biol.*, 1994, **25**, 343-353.

Chapter Nine

FUNCTIONAL GENOMICS APPROACHES TO UNRAVEL ESSENTIAL OIL BIOSYNTHESIS

Bernd Markus Lange,[1,2]* Raymond E.B. Ketchum[1]

[1] *Washington State University, Pullman, WA 99164-6340*

[2] *present address: Torrey Mesa Research Institute,*
Syngenta Genomics Research and Technology,
3115 Merryfiled Row,
San Diego, CA 92121-1125

*Author for correspondence, e-mail: mark.lange@syngenta.com

Introduction... 146
Sequencing of ESTs from a Peppermint Secretory Cell cDNA Library............ 147
Characterization of Genes Involved in Essential Oil Biosynthesis.................. 149
Metabolite Profiling of Oil gland Secretory Cells....................................... 154
Metabolic Engineering of Essential Oil Yield... 158
Summary... 158

INTRODUCTION

Essential oils have been valued for thousands of years for their culinary, fragrant and medicinal properties. Some of the most prominent essential oil plants include members of the mint family, such as basil, lavender, peppermint, rosemary, sage, and thyme. A common feature of essential oil biosynthesis and storage is the localization to specialized anatomical structures, which, in the case of peppermint, are composed of modified epidermal hairs termed peltate glandular trichomes (Fig. 9.1A). These oil glands, which consist of a single basal cell, a stalk cell, and a cluster of secretory cells (Fig. 9.1B), accumulate a complex mixture of predominantly monoterpenes, with sesquiterpenes and methylated flavonoids being minor constituents.[1-3]

Owing to the technical difficulties in purifying enzymes involved in essential oil biosynthesis, traditional biochemical approaches to gene discovery have presented a considerable challenge.[4] Peppermint oil gland secretory cells express the complete enzymatic machinery necessary to produce monoterpenes from imported sucrose and can be isolated in high yield and purity from leaves.[5,6] This encouraged us to use an alternative approach to gene cloning: the secretory cell purification protocol was modified for the isolation of intact mRNA and the subsequent generation of a cDNA library, thus providing a highly enriched source of transcripts involved in the accumulation of monoterpenoid essential oils in these specialized structures.[7] The partial sequencing of random clones (expressed sequence tags, ESTs) from this oil gland secretory cell cDNA library was followed by a bioinformatic analysis to predict their enzymatic functions. Full-length cDNA clones putatively involved in essential oil biosynthesis and storage, if necessary generated by RACE (rapid amplification of cDNA ends)-PCR or obtained by screening the peppermint secretory cell cDNA library with gene-specific probes, were further evaluated by functional expression in a heterologous host and characterization of the recombinant enzymatic activities.[8-13]

The data obtained as part of this effort to establish the identity and functional characteristics of genes and derived enzymes involved in monoterpenoid essential oil biosynthesis were complemented by analyses of the early intermediates of this pathway by using a combination of high performance liquid chromatography (HPLC) and mass spectrometry (MS),[14] which indicated that 1-deoxy-D-xylulose 5-phosphate reductoisomerase (DXR) may catalyze a slow step of this route. The dose-dependent decrease in the ability of isolated oil gland secretory cells to convert biosynthetic precursors into monoterpenes in the presence of fosmidomycin, a specific inhibitor of DXR, provided additional evidence that this enzyme activity may be a rate-limiting step in monoterpene biosynthesis. Transgenic plants overexpressing DXR under control of a strong constitutive promoter were shown to accumulate increased levels of monoterpenes, thus providing further evidence for the regulatory role of this enzyme in the biosynthesis of monoterpenoid essential oils in peppermint.[15]

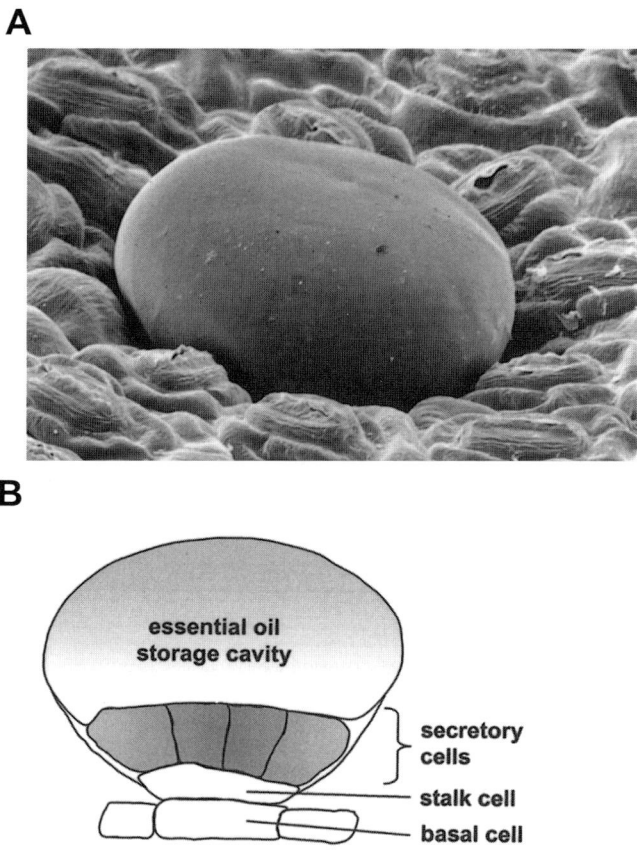

Figure 9.1: Peppermint peltate glandular trichomes; (**A**) scanning electron micrograph (courtesy of Dr. Glenn Turner) and (**B**) anatomical scheme of a cross-section.

SEQUENCING OF ESTs FROM A PEPPERMINT SECRETORY CELL cDNA LIBRARY

The general approach for the functional characterization of ESTs from an oil gland cDNA library is outlined in Fig. 9.2. The isolation of secretory cells was achieved by using a procedure modified after Gershenzon et al.,[6] which was optimized to afford intact mRNA from a tissue rich in monoterpenes and phenolic compounds. Roughly 1,300 anonymous clones from this cDNA library were partially sequenced, and these sequences were - after a BLAST search and additional comparisons that used algorithms to predict functional motifs (Prosite [http://ca.expasy.org/prosite]) and posttranslational modifications (ChloroP [http://www.cbs.dtu.dk/services/ChloroP], MitoProt [http://www.mips.biochem.mpg.de/cgi-bin/proj/med-gen/mitofilter], PSORT [http://psort.nibb.ac.jp], SignalP [http://www.cbs.dtu.dk/services/SignalP]) - subdivided into functional classes (Fig. 9.3). The majority of ESTs were predicted to code for enzymes involved in different metabolic processes (48 % of total clones), with isoprenoid biosynthetic genes constituting the most abundant class of transcripts (roughly 30 %). A second notable class of genes was predicted to code for enzymes involved in the transport of proteins and metabolites (7 %).

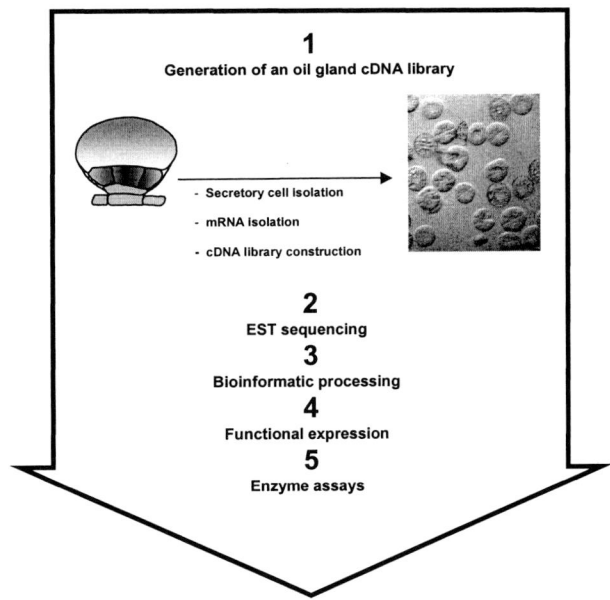

Figure 9.2: Outline of a functional genomics approach to evaluate ESTs from a peppermint oil gland secretory cell cDNA library.

GENOMICS -- ESSENTIAL OIL BIOSYNTHESIS

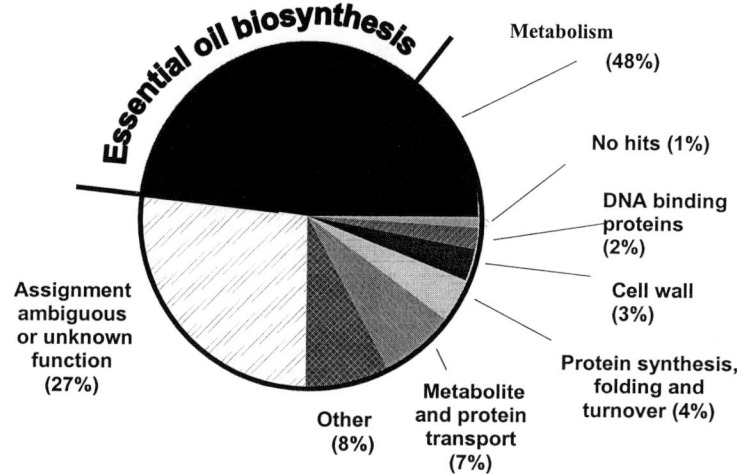

Figure 9.3: Functional classification of ESTs from a peppermint oil gland secretory cell cDNA library.

CHARACERIZATION OF GENES INVOLVED IN ESSENTIAL OIL BIOSYNTHESIS

The biosynthesis of monoterpenes, the major components of peppermint essential oils, can be divided into four stages (Fig. 9.4). Stage 1 includes the formation of isopentenyl diphosphate (IPP) and dimethylallyl alcohol (DMAPP). In plants, two separate pathways are utilized for the synthesis of these universal C5 intermediates, with the cytosolic mevalonate pathway being responsible for the formation of sterols and certain sesquiterpenes, and the plastidial mevalonate-independent pathway being involved in the biosynthesis of isoprene, monoterpenes, certain sesquiterpenes, diterpenes, tetraterpenes, as well as the side chains of chlorophyll and plastoquinone.[16] In peppermint oil gland secretory cells, however, the mevalonate pathway is blocked and the biosynthesis of monoterpenoid essential oils proceeds exclusively via the mevalonate-independent pathway.[17,18] Stage 2 comprises the isomerization of IPP to DMAPP, which then serves as the reactive starter molecule for a subsequent condensation reaction with IPP to form geranyl diphosphate (GPP, C_{10}).[19] It is still a matter of debate whether IPP is the end-product of the mevalonate-independent pathway, as suggested by McCaskill and Croteau[20] and Arigoni et al.,[21] or if this pathway branches and yields IPP and DMAPP independently as shown for *E. coli*.[22] In stage 3, GPP undergoes cyclization reactions, catalyzed by terpene synthases, to form primarily 1,8-cineole and limonene. Stage 4 involves a series of substitution reactions of the limonene precursor to yield the numerous *p*-menthane end products.[7]

Figure 9.4: Monoterpene biosynthesis in peppermint oil gland secretory cells. The enzymes involved in this pathway are (1) 1-deoxy-D-xylulose 5-phosphate synthase, (2) 2-C-methyl-D-erythritol 4-phosphate reductoisomerase, (3) 2-C-methyl-D-erythritol 4-phosphate cytidyltransferase, (4) 4-(cytidine 5'-diphospho)-2-C-methyl-D-erythritol kinase, (5) 2-C-methyl-D-erythritol 2,4-cyclodiphosphate synthase, (6) isopentenyl diphosphate isomerase, (7) geranyl diphosphate synthase, (8) 1,8-cineole synthase, (9) (-) limonene synthase, (10) (-) limonene 3-hydroxylase, (11) (-)-*trans*-isopiperitenol dehydrogenase, (12) (-)- isopiperitenone-$\Delta^{1,2}$-reductase, (13) (+)-*cis*-isopulegone isomerase, (14) (+)-pulegone-$\Delta^{4,8}$-reductase, (15) (-)-menthone reductase, (16) (-)-menthofuran synthase. In peppermint secretory cells, the mevalonate pathway (boxed) is blocked as indicated by a gray shading.

Based on sequence homology to characterized gene products related to monoterpenoid essential oil biosynthesis from peppermint or other organisms, functional classifications were made for a certain groups of ESTs, including genes putatively coding for IPP isomerase, prenyl transferases, terpene synthases, and monoterpene-specific cytochrome P450 monooxygenases. In most cases, however, assignments to specific enzymatic activities were not possible when the sequencing results of the EST project became available. Functional expression of candidate genes derived from the EST effort in *E. coli* established the identity of genes encoding the two subunits of geranyl diphosphate synthase,[10] a (-)-limonene synthase (A. Crowell and R. Croteau, unpublished results), a (-)-limonene 3-hydroxylase,[9] a (+)-methofuran synthase,[13] and several oxidoreductases involved in the synthesis of (-)-menthol and (-)-menthone from the (-)-*trans*-isopiperitenol intermediate (M. McConkey, E. Davis, T. Ringer, R. Croteau, unpublished results).

In addition to providing candidate genes for the steps specific to peppermint essential oils, the EST project also proved to be a valuable resource to characterize genes involved in the plastidial mevalonate-independent pathway of general isoprenoid biosynthesis, which, when the peppermint EST project was initiated, had been shown to occur in some eubacteria, a green alga, and plant plastids.[23-27] The individual steps of this pathway to IPP and DMAPP, however, were unknown. A hypothetical mechanism had been suggested in which the condensation of D-glyceraldehyde 3-phosphate and "activated acetaldehyde" derived from pyruvate would yield 1-deoxy-D-xylulose 5-phosphate (DXP) as the first intermediate.[24,25,28] Following the hypothesis that the formation of DXP was catalyzed by a transketolase-like enzyme (Fig. 9.5A), we used the putative thiamin diphosphate binding motif of the transketolase involved in the pentose phosphate pathway (E.C.

2.2.1.1) to query peppermint EST sequences. Interestingly, the transketolase-like EST sequences fell into two groups, one of which shared high sequence identity with the pentose phosphate pathway transketolase, whereas a second group exhibited high homology to a gene family of unknown function in eubacteria and *Arabidopsis thaliana*. A probe derived from the latter group of transketolase-like peppermint sequences was used to obtain a full-length clone (Fig. 9.5B) from the peppermint secretory cell cDNA library, which was shown, by heterologous expression in *E. coli*, to encode the predicted 1-deoxy-D-xylulose 5-phosphate synthase (DXPS) activity.[8]

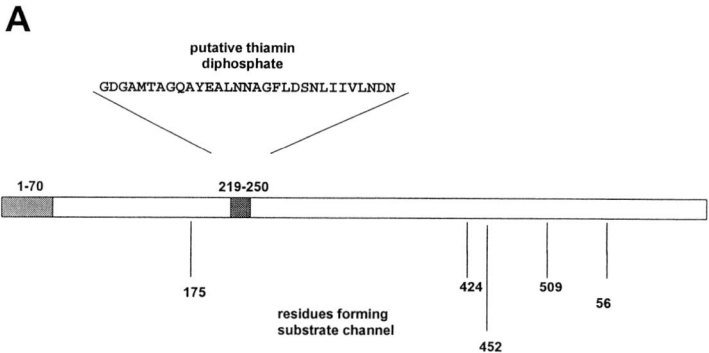

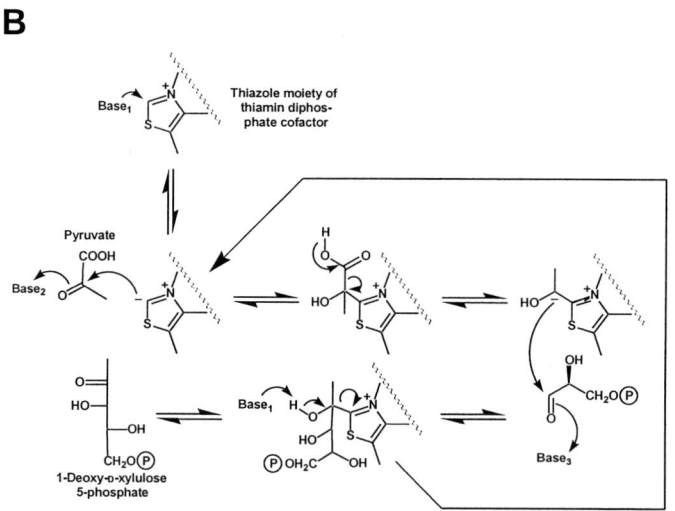

Figure 9.5: 1-Deoxy-D-xylulose 5-phosphate synthase; (**A**) schematic representation of peptide sequence features and (**B**) hypothetical enzymatic mechanism of this transketolase.

It had been suggested that the second step of the mevalonate-independent pathway involved an intramolecular rearrangement and subsequent reduction of DXP to yield 2-*C*-methyl-D-erythritol 4-phosphate (Fig. 9.6A).[29-31] Seto and co-workers reported the isolation, by using a mutant complementation approach, and characterization of such a reductoisomerase gene from *E. coli*.[32] Based on likely

Figure 9.6: 1-Deoxy-D-xylulose 5-phosphate reductoisomerase; (**A**) schematic representation of peptide sequence features and (**B**) hypothetical enzymatic mechanism.

conserved regions of this reductoisomerase gene, degenerate PCR primers were designed and employed with the peppermint oil gland secretory cell cDNA library to amplify a 223 bp fragment with significant homology to the *E. coli* reductoisomerase. By screening the peppermint secretory cell cDNA library with a probe derived from this PCR product, a full-length clone was obtained the gene product of which was evaluated by expression in *E. coli* and was shown to catalyze the rearrangement and pyridine nucleotide-dependent reduction of DXP to 2-*C*-methyl-D-erythritol 4-phosphate (Fig. 9.6B).[11]

The end products of the mevalonate-independent pathway are the prenyl diphosphates IPP and DMAPP, and because the two known intermediates both represent carbohydrate monophosphates, a phosphorylation step had to occur at some point during the reaction sequence. A database search for ATP-binding domain motifs revealed several peppermint ESTs, two of which showed significant homology to a chromoplast-directed protein of unknown function from ripening tomato fruits and also to several hypothetical proteins from eubacteria, all of which shared the signature motif of the GHMP family of kinases.[33] The heterologous expression of a full-length version of these clones in *E. coli* resulted in a gene product capable of catalyzing the phosphorylation of isopentenyl monophosphate to IPP.[12] However, it was subsequently shown that the orthologous gene products from *E. coli* and tomato (abbreviated with CMK) phosphorylated 4-(cytidine 5'-diphospho)-2-*C*-methyl-D-erythritol at much higher rates than isopentenyl monophosphate,[34-36] indicating that the phosphorylation of isopentenyl monophosphate is most likely metabolically irrelevant. To date, two additional gene products involved in the mevalonate-independent pathway have been characterized, and for both 2-*C*-methyl-D-erythritol 4-phosphate cytidyltransferase (MCT)[36-38] and 2-*C*-methyl-D-erythritol 2,4-cyclodiphosphate synthase (MECPS),[40,41] putative peppermint orthologues were detected in our EST database, an additional indication of the enrichment of the oil gland cDNA library in genes related to isoprenoid biosynthesis.

METABOLITE PROFILING OF OIL GLAND SECRETORY CELLS

Although the nature of the individual biosynthetic steps of the mevalonate-independent pathway were unfolding and recombinant enzymes had been characterized in some detail, no information was available on putatively rate-limiting steps in plant plastids. Again, the isolated oil gland secretory cells appeared to be a suitable experimental model system due to the fact that these cells are dedicated to monoterpene biosynthesis via the mevalonate-independent pathway and amenable to feeding with exogenous precursors for labeling experiments.[5,14,17]

As a first step, a method for the rapid and reproducible analysis of intermediates of the mevalonate-independent pathway was established. For all of our experiments, the HPLC-MS instrumentation consisted of an Agilent Series 1100 HPLC with diode array and model 1946 A mass detector, with the mass detector being operated in electrospray (ES) negative ion mode. An extensive array of HPLC chromatographic columns, buffers, ion-pairing reagents, and MS conditions were tested to develop a routine method for the separation and detection of intermediates of the mevalonate-independent pathway. Mixtures of compounds were used as standards to test chromatographic separation and mass detector response (Table 9.1). Columns tested for separation of these compounds were Phenomenex Prodigy ODS3 (5 μm, 250 x 4.6 mm), Phenomenex Luna C18(2) (5 μm, 150 x 4.6 mm), Amersham Source 5RPC ST (5 μm, 150 x 4.6 mm), Metachem Taxsil (3 μm, 250 x 4.6 mm, and Dionex OmniPac PAX-500 (5 μm, 150 x 4.0 mm). To avoid plugging the ionization and source components, and thus causing a loss of analyte signal, a separation method based on a volatile mobile phase was sought. Ammonium formate and ammonium acetate were tested either alone or in combination with volatile ion-pairing reagents. Tested concentrations of the buffers ranged from 0.1 mM to 90 mM and pH 4 to pH 8 adjusted with either glacial acetic acid or formic acid. Ion-pairing reagents were tested at concentrations of 0.01 mM to 10 mM and included triethylamine, trihexylamine, tributlyamine, heptafluorobutyrate, and pentadecafluorooctanoic acid. For the development of methods for the separation of polar metabolites to be analyzed in the metabolic profiling experiments, a delicate balance between compound resolution and detector sensitivity had to be established. However, there was no combination of buffers and ion pair reagents that gave both satisfactory separation on conventional reversed phase chromatography columns as well as a good mass detector response. For example, with an isocratic elution on a Phenomenex Prodigy ODS3 column using a buffer of 90 mM ammonium formate with 10 mM tetrabutylamine (pH 6.4) in combination with acetonitrile (80 : 20, buffer : acetonitrile) a baseline resolution of AMP, ADP, and ATP was achieved. However, after only a few runs, the mass detector electronics shorted out due to incomplete buffer volatilization. If the flow rate was lowered to avoid deposition of the residue, the analyte signal became unacceptably low. A similar problem occurred with regard to signal strength in the mass detector: excellent intensity of molecular ions could be obtained for any of the tested metabolites without the use of ion pairing reagents, but these compounds all eluted within a time range of two minutes with little separation. A breakthrough in analytical methods was achieved when we began to test a Macherey Nagel Nucleodex ß-OH cyclodextrin column (5 μm, 200 x 4 mm), which had been sucessfully used to analyze phosphorylated carbohydrates.[42] Using 10 mM ammonium acetate (pH 6.5) and acetonitrile as mobile phases, a gradient elution was developed that resulted in an excellent separation of the carbohydrate standard mixture (Table 9.1) and of the prenyl diphosphates IPP, DMAPP, and GPP.[14]

Table 9.1: Standard mixtures for LC-MS methods development.

Carbohydrate phosphates	Adenosine phosphates	Intermediates of the mevalonate-independent pathway
D-Erythrose 4-phosphate	AMP	1-Deoxy-D-xylulose
D-Fructose 1,6-diphosphate	ADP	1-Deoxy-D-xylulose 5-phosphate
D-Galactose 1-phosphate	ATP	2-C-Methyl-D-erythritol
D-Glucose 6-phosphate		2-C-Methyl-D-erythritol 4-phosphate
D,L-Glyceraldehyde 3-phosphate		
D-Ribose 5-phosphate		
Pyruvate		

Intermediates of the mevalonate-independent pathway to be used as standards for the metabolite profiling analysis by LC-MS were generated biosynthetically. Thus, the genes for DXS, DXR, MCT, and CMK were PCR-amplified from *E. coli* genomic DNA, and amplicons were inserted into the pBAD TOPO TA vector (Invitrogen, Carlsbad, CA) for expression as a 6 x His fusion protein. Recombinant proteins were expressed in *E. coli*, partially purified by using Ni^{2+} affinity columns, and were utilized in consecutive enzyme assays to produce DXP from pyruvate and D-glyceraldehyde 3-phosphate (catalyzed by recombinant DXPS), MEP from DXP (catalyzed by recombinant DXR, 4-(cytidine 5'-diphospho)-2-C-methyl-D-erythritol from MEP (catalyzed by recombinant MCT), and 2-phospho-4-(cytidine 5'-diphospho)-2-C-methyl-D-erythritol from 4-(cytidine 5'-diphospho)-2-C-methyl-D-erythritol (catalyzed by recombinant CMK) (B.M. Lange, R.E.B. Ketchum, R. Croteau, unpublished data). Using the newly developed LC-MS method these compounds were readily separated (Fig. 9.7). Subsequently, isolated oil gland secretory cells were incubated under various conditions with [2,3-$^{13}C_2$] pyruvate as a precursor. Metabolites were isolated and analyzed by LC-MS. IPP, DMAPP, and geranyl diphosphate were found to accumulate in several assays, whereas DXP was the only intermediate of the mevalonate-independent pathway to accumulate in detectable amounts (data not shown), indicating that the reaction catalyzed by DXR may be a slow step of this route. To test the hypothesis that the DXR reaction indeed constitutes a rate-limiting step in this pathway, the effect of fosmidomycin, a specific inhibitor of DXR,[43,44] on monoterpene accumulation in isolated peppermint oil gland secretory cells was studied. Preliminary co-incubation experiments of these cells with [2-^{14}C] pyruvate and fosmidomycin resulted in a dose-dependent reduction of monoterpene accumulation (Table 9.2), thus providing additional evidence for a regulatory role of DXR in peppermint monoterpene biosynthesis.

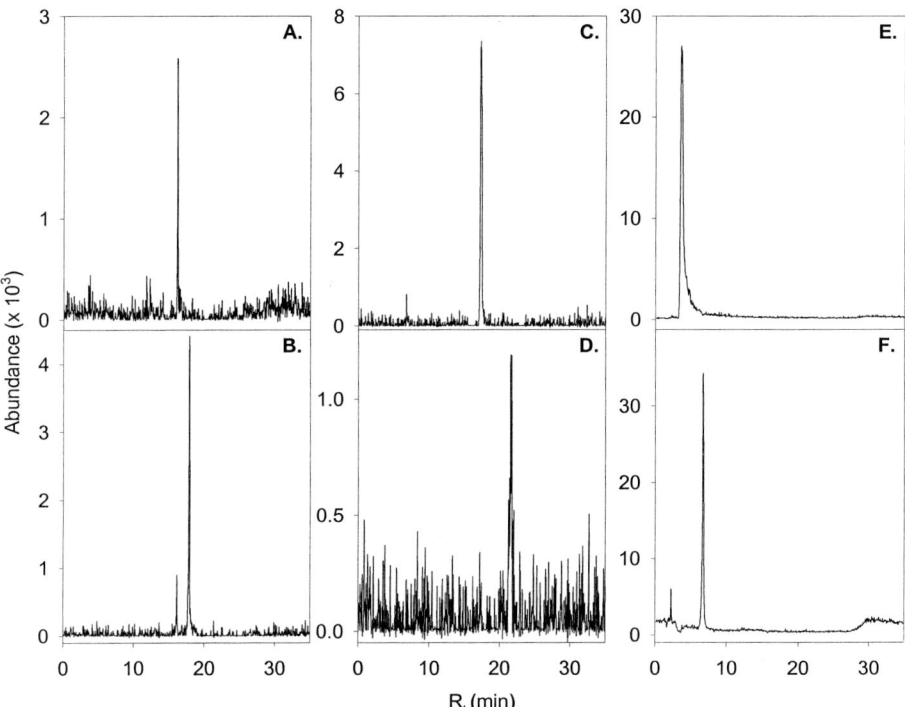

Figure 9.7: Separation and detection of intermediates of the mevalonate-independent pathway of isoprenoid biosynthesis by LC-MS; extracted ion chromatograms at (**A**) *m/z* 213 (1-deoxy-D-xylulose 5-phosphate), (**B**) *m/z* 215 (2-*C*-methyl-D-erythritol 4-phosphate), (**C**) *m/z* 520 (4-(cytidine 5'-diphospho)-2-*C*-methyl-D-erythritol), (**D**) *m/z* 600 (2-phospho-4-(cytidine 5'-diphospho)-2-*C*-methyl-D-erythritol), (**E**) *m/z* 133 (1-deoxy-D-xylulose), (**F**) *m/z* 135 (2-*C*-methyl-D-erythritol).

Table 9.2: Incorporation rate of [2-^{14}C]-pyruvate into monoterpenes of isolated peppermint oil gland secretory cells in the presence of fosmidomycin, a specific inhibitor of 1-deoxy-D-xylulose 5-phosphate reductoisomerase, an enzyme of the mevalonate-independent pathway of isoprenoid biosynthesis.

Inhibitor concentration	Incorporation rate [% of control]
1 mM	81.7 ± 9.2
2 mM	51.0 ± 11.4
5 mM	33.3 ± 4.5

METABOLIC ENGINEERING OF ESSENTIAL OIL YIELD

To determine whether flux through the mevalonate-independent pathway is regulated at the transcriptional level by limiting precursor supply, a transgenic approach was taken. Based on the results described above, DXR appeared to be a good target for metabolic engineering. Thus, peppermint was transformed with a homologous DXR cDNA under the control of a strong constitutive promoter using an established protocol,[45] and the resulting transformants were analyzed for DXR transcript abundance, DXR enzyme activity, and essential oil production and composition.[15] In summary, co-suppressed plants accumulated less essential oil, whereas the transgenic up-regulation of DXR resulted in an increase in essential oil formation without affecting oil composition, indicating that DXR catalyzes a slow step of the mevalonate-independent pathway.

SUMMARY

The peppermint oil gland secretory cell cDNA library has proven to provide a highly enriched source of candidate genes involved in essential oil biosynthesis. A functional genomics approach has successfully been employed to clone genes involved in the mevalonate-independent pathway of isoprenoid biosynthesis and in the peppermint-specific steps producing (-)-menthol and (-)-menthone. The optimization of LC-MS technology to profile phosphorylated carbohydrates and prenyl diphosphates has helped to establish that the reaction catalyzed by DXR is a slow step of the mevalonate-independent pathway in peppermint. Transgenic plants with either increased or decreased steady-state DXR transcript accumulation and enzyme activity have confirmed the correlation of DXR expression rate and essential oil yield, thus providing evidence for the utility of profiling biosynthetic intermediates to predict flux control through metabolic pathways.

ACKNOWLEDGEMENTS

The work reviewed in this article was performed in the laboratory of Dr. Rodney Croteau (Institute of Biological Chemistry, Washington State University) and we thank our mentor for helpful discussions and continuous support.

REFERENCES

1. AMELUNXEN, F., WAHLIG, T., ARBEITER, H., Über den Nachweis des ätherischen Öls in isolierten Drüsenhaaren und Drüsenschuppen von *Mentha* x *piperita* L., *Z. Pflanzenphysiol.*, 1969, **61**, 68-72.
2. GASIC, O., MIMICA-DUDIC, N., ADAMOVIC, D., BOROJEVIC, K., Variability of content and composition of essential oil in different genotypes of peppermint, *Biochem. System. Ecol.*, 1987, **15**, 335-340.
3. VOIRIN, B., BAYET, C., Developmental variations in leaf flavonoid aglycones of *Mentha* x *piperita*, *Phytochemistry*, 1992, **31**, 2299-2304.
4. COLBY, S.M., ALONSO, W.R., KATAHIRA, E.J., McGARVEY, D.J., CROTEAU, R., 4S-Limonene synthase from the oil glands of spearmint (*Mentha spicata*), *J. Biol. Chem.*, 1993, **268**, 23016-23024.
5. McCASKILL, D., GERSHENZON, J., CROTEAU, R., Morphology and monoterpene biosynthetic capabilities of secretory cell clusters isolated from glandular trichomes of peppermint (*Mentha x piperita* L.), *Planta*, 1992, **187**, 445-454.
6. GERSHENZON, J., McCASKILL, D., RAJAONARIVONY, J.I.M., MIHALIAK, C., KARP, F., CROTEAU, R., Isolation of secretory cells from plant glandular trichomes and their use in biosynthetic studies of monoterpenes and other gland products, *Anal. Biochem.*, 1992, **200**, 130-138.
7. LANGE, B.M., WILDUNG, M.R., STAUBER, E.J., SANCHEZ, C., POUCHNIK, D., CROTEAU, R., Probing essential oil biosynthesis by functional evaluation of expressed sequence tags from mint glandular trichomes, *Proc. Natl. Acad. Sci. USA*, 2000, **97**, 2934-2939.
8. LANGE, B.M., WILDUNG, M.R., McCASKILL, D., CROTEAU, R., A family of transketolases that directs isoprenoid biosynthesis via a mevalonate-independent pathway, *Proc. Natl. Acad. Sci. USA*, 1998, **95**, 2100-2104.
9. LUPIEN, S., KARP, F., WILDUNG, M., CROTEAU, R., Regiospecific cytochrome P450 limonene hydroxylases from mint (*Mentha*) species: cDNA isolation, characterization, and functional expression of (-)-4S-limonene 3-hydroxylase and (-)-4S-limonene 6-hydroxylase, *Arch. Biochem. Biophys.*, 1999, **368**, 181-192.

10 BURKE C.C., WILDUNG, M.R., CROTEAU, R., Geranyl diphosphate synthase: cloning, expression, and characterizatin of this prenyltransferase as a heterodimer, *Proc. Natl. Acad. Sci. USA*, 1999, **96**, 13062-13067.
11 LANGE, B.M., CROTEAU, R., Isoprenoid biosynthesis via a mevalonate-independent pathway in plants: cloning and heterologous expression of 1-deoxy-D-xylulose 5-phosphate reductoisomerase from peppermint, *Arch. Biochem. Biophys.*, 1999, **365**, 170-174.
12 LANGE, B.M., CROTEAU, R., Isopentenyl diphosphate biosynthesis via a mevalonate-independent pathway: isopentenyl monophosphate kinase catalyzes the terminal enzymatic step, *Proc. Natl. Acad. Sci. USA*, 1999, **96**, 13714-13719.
13 BERTEA, C.M., SCHALK, M., KARP, F., MAFFEI, M., CROTEAU, R., Demonstration that menthofuran synthase of mint (*Mentha*) is a cytochrome P450 monooxygenase: cloning, functional expression, and characterization of the responsible gene, *Arch. Biochem. Biophys.*, 2001, **390**, 279-286.
14 LANGE, B.M., KETCHUM, R.E.B., CROTEAU, R., Isoprenoid biosynthesis: metabolite profiling of peppermint oil gland secretory cells and application to herbicide target analysis, *Plant Physiol.*, 2001, **127**, 305-314.
15 MAHMOUD, S.S., CROTEAU, R., Metabolic engineering of essential oil yield and composition in mint by altering expression of deoxyxylulose phosphate reductoisomerase and menthofuran synthase, *Proc. Natl. Acad. Sci. USA*, 2001, **98**, 8915-8920.
16 EISENREICH, W., ROHDICH, F., BACHER, A., Deoxyxylulose phosphate pathway to terpenoids, *Trends Plant Sci.*, 2001, **6**, 78-84.
17 McCASKILL, D., CROTEAU, R., Monoterpene and sesquiterpene biosynthesis in glandular trichomes of peppermint (*Mentha* x *piperita*) rely exclusively on plastid-derived isopentenyl diphosphate, *Planta*, 1995, **197**, 49-56.
18 EISENREICH, W., SAGNER, S., ZENK, M.H., BACHER, A., Monoterpenoid essential oils are not of mevalonoid origin, *Tetrahedron Lett.*, 1997, **38**, 3889-3892.
19 OGURA, K., Isomerase and prenyl transferases. In: Comprehensive Natural Products Chemistry, Vol 2: Isoprenoids Including Carotenoids and Steroids (D.E. Cane, ed.) Pergamon Press, Oxford, 1999, pp. 69-96.
20 McCASKILL, D., CROTEAU, R., Isopentenyl diphosphate is the terminal product of the deoxyxylulose 5-phosphate pathway for terpenoid biosynthesis in plants, *Tetrahedron Lett.*, 1999, **40**, 653-656.
21 ARIGONI, D., EISENREICH, W., LATZEL, C., SAGNER, S., RADYKEWICZ, T., ZENK, M.H., BACHER, A., Dimethylallyl pyrophosphate is not the committed precursor of isopentenyl pyrophosphate during terpenoid biosynthesis from 1-deoxyxylulose in higher plants, *Proc. Natl. Acad. Sci. USA*, 1999, **96**, 1309-1314.
22 RODRIGUEZ-CONCEPTION, M., CAMPOS, N., MARIA LOIS, L., MALDONADO, C., HOEFFLER, J.F., GROSDEMANGE-BILLIARD, C., ROHMER, M., BORONAT, A., Genetic evidence of branching in the isoprenoid pathway for the production of isopentenyl diphosphate and dimethylallyl diphosphate in *Escherichia coli, FEBS Lett.*, 2000, **473**, 328-332.

23 ROHMER, M., KNANI, M., SIMONIN, P., SUTTER, B., SAHM, H., Isoprenoid biosynthesis in bacteria: a novel pathway for the early steps leading to isopentenyl diphosphate, *Biochem. J.*, 1993, **295**, 517-524.
24 BROERS, S.T.J., Ph.D. thesis, 1994, Eidgenössische Technische Hochschule, Zürich, Switzerland.
25 SCHWARZ, M.C., Ph.D. thesis, 1994, Eidgenössische Technische Hochschule, Zürich, Switzerland.
26 SCHWENDER, J., SEEMANN, M., LICHTENTHALER, H.K., ROHMER, M., Biosynthesis of isoprenoids (carotenoids, sterols, prenyl side-chains of chlorophylls and plastoquinone) via a novel pyruvate/glyceraldehyde 3-phosphate non-mevalonate pathway in the green alga *Scenedesmus obliquus*, *Biochem. J.*, 1996, **316**, 73-80.
27 LICHTENTHALER, H.K., SCHWENDER, J., DISCH, A., ROHMER, M., Biosynthesis of isoprenoids in higher plant chloroplasts proceeds via a mevalonate-independent pathway, *FEBS Lett.*, 1997, **400**, 271-274
28 ROHMER, M., SEEMANN, M., HORBACH, S., BRINGER-MEYER, S., SAHM, H., Glyceraldehyde 3-phosphate and pyruvate as precursors of isoprenic units in an alternative non-mevalonate pathway for terpenoid biosynthesis, *J. Am. Chem. Soc.*, 1996, **118**, 2564-2566.
29 DUVOLD, T., BRAVO, J.-M., PALE-GROSDEMANGE, C., ROHMER, M., Biosynthesis of 2-*C*-methyl-D-erythritol, a putative C_5 intermediate in the mevalonate independent pathway for isoprenoid biosynthesis, *Tetrahedron Lett.*, 1997, **38**, 4769-4772.
30 DUVOLD, T., CALÍ, P., BRAVO, J.-M., ROHMER, M., Incorporation of 2-*C*-methyl-D-erythritol, a putative isoprenoid precursor in the mevalonate-independent pathway, into ubiquinone and menaquinone of *Escherichia coli*, *Tetrahedron Lett.*, 1997, **38**, 6181-6184.
31 SAGNER, S., EISENREICH, W., FELLERMEIER, M., LATZEL, C., BACHER, A., ZENK, M.H., Biosynthesis of 2-*C*-methyl-D-erythritol in plants by rearrangement of the terpenoid precursor, 1-deoxy-D-xylulose 5-phosphate, *Tetrahedron Lett.*, 1998, **39**, 2091-2094.
32 TAKAHASHI, S., KUZUYAMA, T., WATANABE H., SETO, H., A 1-deoxy-D-xylulose 5-phosphate reductoisomerase catalyzing the formation of 2-*C*-methyl-D-erythritol 4-phosphate in an alternative nonmevalonate pathway, *Proc. Natl. Acad. Sci. USA*, 1998, **95**, 9879-9884.
33 TSAY, Y.H., ROBINSON, G.W., Cloning and characterization of ERG8, an essential gene of *Saccharomyces cerevisiae* that encodes phosphomevalonate kinase, *Mol. Cell. Biol.*, 1991, **11**, 620-631.
34 LÜTTGEN, H., ROHDICH, F., HERZ, S., WUNGSINTAWEEKUL, J., HECHT, S., SCHUHR, C.A., FELLERMEIER, M., SAGNER, S., ZENK, M.H., BACHER, A., EISENREICH, W., Biosynthesis of terpenoids: ychB protein of *Escherichia coli* phosphorylates the 2-hydroxy group of 4-diphosphocytidyl-2-*C*-methyl-D-erythritol, *Proc. Natl. Acad. Sci. USA*, 2000, **97**, 1062-1067.

35 KUZUYAMA, T., TAKAGI, M., KANEDA, K., WATANABE, H., DAIRI, T., SETO, H., Studies on the nonmevalonate pathway: conversion of 4-(cytidine 5'-diphospho)-2-C-methyl-D-erythritol to its 2-phospho derivative by 4-(cytidine 5'-diphospho)-2-C-methyl-D-erythritol kinase, *Tetrahedron Lett.*, 2000, **41**, 2925-2928.

36 ROHDICH, F., WUNGSINTAWEEKUL, J., LÜTTGEN, H., FISCHER, M., EISENREICH, W., SCHUHR, C.A., FELLERMEIER, M., SCHRAMEK, N., ZENK, M.H., BACHER, A., Biosynthesis of terpenoids: 4-diphosphocytidyl-2-C-methyl-D-erythritol kinase from tomato, *Proc. Natl. Acad. Sci. USA*, 2000, **97**, 8251-8256.

37 ROHDICH, F., WUNGSINTAWEEKUL, J., FELLERMEIER, M., SAGNER, S., HERZ, S., KIS, K., EISENREICH, W., BACHER, A., ZENK, M.H., Cytidine 5'-triphosphate-dependent biosynthesis of isoprenoids: ygbP protein of *Escherichia coli* catalyzes the formation of 4-diphosphocytidyl-2-C-methylerythritol, *Proc. Natl. Acad. Sci. USA*, 1999, **96**, 11758-11763.

38 KUZUYAMA, T., TAKAGI, M., KANEDA, K., WATANABE, H., DAIRI, T., SETO, H., Formation of 4-(cytidine 5'-diphospho)-2-C-methyl-D-erythritol from 2-C-methyl-D-erythritol 4-phosphate by 2-C-methyl-D-erythritol 4-phosphate cytidyltransferase, a new enzyme in the nonmevalonate pathway, *Tetrahedron Lett.*, 2000, **41**, 703-706.

39 ROHDICH, F., WUNGSINTAWEEKUL, J., EISENREICH, W., RICHTER, G., SCHUHR, C.A., ZENK, M.H., BACHER, A., Biosynthesis of terpenoids: 4-diphosphocytidyl-2-C-methyl-D-erythritol synthase of *Arabidopsis thaliana*, *Proc. Natl. Acad. Sci. USA*, 2000, **97**, 6451-6456.

40 HERZ, S., WUNGSINTAWEEKUL, J., SCHUHR, C.A., HECHT, S, LÜTTGEN, H., SAGNER, S., FELLERMEIER, M., EISENREICH, W., ZENK, M.H., BACHER, A., ROHDICH, F., Biosynthesis of terpenoids: ygbB protein converts 2-phospho-4-(cytidine 5'-diphospho)-2-C-methyl-D-erythritol to 2-C-methyl-D-erythritol 2,4-cyclodiphosphate, *Proc. Natl. Acad. Sci. USA*, 2000, **97**, 2486-2490.

41 TAKAGI, M., KUZUYAMA, T., KANEDA, K., WATANABE, H., DAIRI, T., SETO, H., Studies on the nonmevalonate pathway: formation of 2-C-methyl-D-erythritol 2,4-cyclodiphosphate from 2-phospho-4-(cytidine 5'-diphospho)-2-C-methyl-D-erythritol, *Tetrahedron Lett.*, 2000, **41**, 3395-3398.

42 FEURLE, J., JOMAA, H., WILHELM, M., GUTSCHE, W.B., HERDRICH, M., Analysis of phosphorylated carbohydrates by high-performance liquid chromatography-electrospray ionization tandem mass spectrometry utilising β-cyclodextrin bonded stationary phase, *J. Chromatogr.*, 1998, **803**, 111-119.

43 KUZUYAMA, T., SHIMIZU, T., TAKAHASHI, S., SETO, H., Fosmidomycin, a specific inhibitor of 1-deoxy-D-xylulose 5-phosphate reductoisomerase in the nonmevalonate pathway for terpenoid biosynthesis, *Tetrahedron Lett.*, 1998, **39**, 7913-7916.

44 ZEIDLER, J., SCHWENDER, J., MÜLLER, C., WIESNER, J., WEIDEMEYER, C., BECK, E., JOMAA, H., LICHTENTHALER, H.K., Inhibition of the non-mevalonate 1-deoxy-D-xylulose 5-phosphate pathway of plant isoprenoid biosynthesis by fosmidomycin, *Z. Naturforsch.*, 1998, **53c**, 980-986.

45 NIU, X., LIN, K., HASEGAWA, P.M., BRESSAN, R.A., WELLER, S.C., Transgenic peppermint (*Mentha* x *piperita*) plants obtained by cocultivation with *Agrobacterium tumefaciens*, *Plant Cell Rep.*, 1998, **17**, 165-171.

Chapter Ten

SEQUENCE-BASED APPROACHES TO ALKALOID BIOSYNTHESIS GENE IDENTIFICATION

Toni M. Kutchan

Leibniz-Institut für Pflanzenbiochemie,
Weinberg 3,
06120 Halle (Saale),
Germany

e-mail: kutch@ipb-halle.de

Introduction...	164
Strictosidine Synthase...	164
Berberine Bridge Enzyme...	165
Berbamunine Synthase..	167
(S)-N-Methylcoclaurine 3'-Hydroxylase...	169
Coclaurine 6-O-Methyltransferase..	170
Codeinone Reductase..	172
Salutaridinol 7-O-Acetyltransferase..	173
Summary..	176

INTRODUCTION

Alkaloids are pharmacologically active, nitrogen-containing, basic compounds originally believed to be only of plant origin. Since the isolation of the first alkaloid morphine, more than 12,000 alkaloids have been defined. Approximately 20% of flowering plants produce alkaloids. Each species accumulates alkaloids in a unique and defined patterned. The role of alkaloids in plants has been a longstanding question, but a picture emerges that supports an ecochemical function for these compounds. Alkaloid-containing plants were also mankind's original *materia medica*. Many of these plants are still used today as sources of prescription drugs, for example the analgesics morphine and codeine are isolated from the opium poppy *Papaver somniferum*. Alkaloid biosynthetic pathways are attractive targets for molecular biology because of their role in plant chemical ecology and the biotechnological potential for the production of commercially important compounds. Good progress has been made since the late 1980's towards isolating genes involved in plant alkaloid biosynthesis.

Since the isolation of the first cDNA of alkaloid biosynthesis, that encoding strictosidine synthase from *Rauwolfia serpentina* cell suspension cultures in 1988,[1] the methods that we use to clone have changed, reflecting improvements in the field of molecular genetics in general. Efforts in early years concentrated on protein purification followed by amino acid sequence determination or antibody production.[2] The amount of purified protein that barely sufficed for an N-terminus sequence determination by Edman degradation in the late 1980s is today more than enough to generate multiple internal peptide sequences. These cDNAs isolated early on showed very little, if any, homology to those few sequences present in the data bases at the time. The advent of the polymerase chain reaction and automated sequencing dramatically changed the way that we work in that speed and sensitivity were both greatly improved. EST sequencing programs, in particular in plant science, and genome sequencing efforts in general have generated a wealth of information publicly available to scientists. This has lead us to think differently about the magnitude of projects that we can undertake and about how we go about isolating cDNAs that encode biosynthetic enzymes of interest.

This chapter is an historical perspective with selected examples from our own laboratory, first at the University of Munich, and now at the Leibniz Institute of Plant Biochemistry in Halle, that reflect the changes in how we define and exploit "sequence-based approaches to alkaloid biosynthesis gene identification."

STRICTOSIDINE SYNTHASE

The alkaloidal glucoside strictosidine was recognized in 1968 as the biosynthetic precursor of monoterpenoid indole alkaloids.[3] The enzyme that catalyzes the formation of strictosidine from tryptamine and the iridoid secologanin is called strictosidine synthase (Fig. 10.1). In immobilized form, this enzyme is stable and can be used to produce gram quantities of strictosidine.[4] The only limiting factor at the time in the application of this process was the availability of strictosidine synthase. In order to overcome this limitation, and in an attempt to apply the techniques of molecular biology to the study of alkaloid biosynthesis, strictosidine synthase was purified to apparent homogeneity from cell suspension cultures of *R. serpentina*.[5] From 2.5 kg fresh weight of cells, 1.9 mg pure protein were obtained. This was a sufficient quantity to obtain the amino acid sequences of eight internal peptides. Oligodeoxynucleotides were synthesized, based upon sequences of two of these peptides, but the high degree of degeneracy in the primers precluded a successful screening of an *R. serpentina* cDNA library.

A new technique was developed for the indirect sequencing of RNA using degenerate oligodeoxynucleotides.[6] This resulted in a non-degenerate sequence of a portion of the strictosidine synthase gene transcript. A new, non-degenerate 52 nucleotide-long oligodeoxynucleotide was synthesized and used to screen the cDNA library. This resulted in quick success, and the cDNA was sequenced and functionally expressed in *Escherichia coli*.[7] Strictosidine synthase-like sequences have now been reportedly expressed in a number of organisms ranging from *Arabidopsis* to *Caenorhabditis elegans* and in human brain.[8,9]

BERBERINE BRIDGE ENZYME

The berberine bridge enzyme [(*S*)-reticuline:oxygen oxidoreductase (methylene bridge forming)] is a vesicular plant enzyme that catalyzes the oxidative cyclization of the *N*-methyl group of (*S*)-reticuline into the berberine bridge carbon of (*S*)-scoulerine (Fig. 10.2).[10] (*S*)-Scoulerine then serves as biosynthetic precursor to a multitude of protoberberine, berberine, and benzo[*c*]phenanthridine alkaloids. Benzo[*c*]phenanthridine alkaloids function in selected species of the Papaveraceae and Fumariaceae in the adaptive response to fungal infection. The elicitation of alkaloid accumulation is a phenomenon that was recognized in plant cell cultures,[11] and has been mainly used to improve the biosynthetic capacity of cell culture systems for secondary products. Elicited plant cell suspension cultures can provide a rich source of the enzymes that catalyze the formation of secondary metabolites, and this was the case for the enzymes of benzo[*c*]phenanthridine alkaloid biosynthesis in *Eschscholzia californica*.

The berberine bridge enzyme was purified to apparent electrophoretic homogeneity from elicited cell suspension cultures of *E. californica*.[12] Fifty micrograms of pure protein sufficed in 1988 for amino acid sequence determination of tryptic peptides. From the five peptide sequences obtained, the indirect

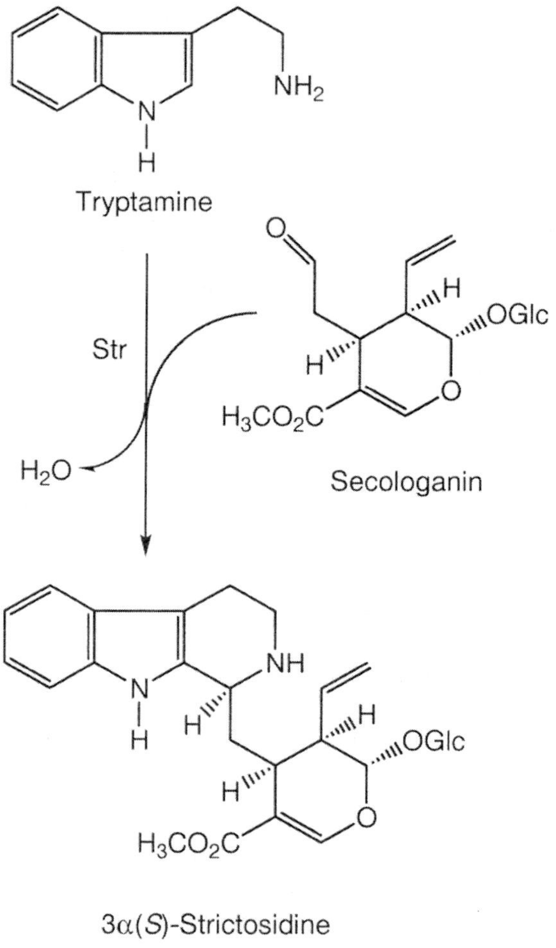

Fig. 10.1: Reaction catalyzed by strictosidine synthase (Str) in monoterpenoid indole alkaloid formation.

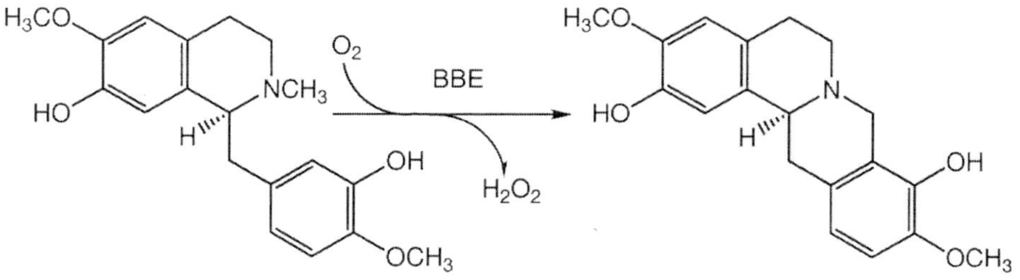

(S)-Reticuline → BBE (O₂, H₂O₂) → (S)-Scoulerine

Fig. 10.2: Reaction catalyzed by the berberine bridge enzyme (BBE) along the biosynthetic pathway leading to protoberberine, berberine and benzo[c]phenanthridine alkaloids.

sequencing of RNA by extension of degenerate oligodeoxynucleotides was used, as described above. A twelve-fold degenerate 40mer was then used to screen a cDNA library to isolate the cDNA *bbe1* that encodes the berberine bridge enzyme. Heterologously expressed *bbe1* was not active in *E.coli*, but was active in *Saccharomyces cerevisiae* protein extracts after partial purification. Due to the difficulties encountered with low levels of expression in yeast, a successful baculovirus expression system was adapted for *bbe1* functional expression.[13,14] This system has proven extremely useful for most of the alkaloid genes that were subsequently cloned and expressed.

BERBAMUNINE SYNTHASE

Bisbenzylisoquinoline alkaloids are dimeric benzyltetrahydroisoquinoline alkaloids that are known for their pharmacological activities. A well-described example is the muscle relaxant (+)-tubocurarine, which in crude form serves as an arrow poison for South American Indian tribes. In the biosynthesis of this broad class of dimeric alkaloids, it has been postulated that the mechanism of phenol coupling proceeds by generation of phenolate radicals followed by radical pairing to form either an inter- or intramolecular C – O or C – C bond. Enzyme studies on the formation of bisbenzylisoquinoline alkaloids indicated that a cytochrome P-450-dependent oxidase catalyzes C – O bound formation in the biosynthesis of berbamunine in *Berberis* cell suspension culture.[15] This enzyme, berbamunine synthase (CYP80A1), is one of the few cytochromes P-450 that can be purified to apparent electrophoretic homogeneity from plants (Fig. 10.3). Other examples include allene oxide synthase from flax,[16] cinnamate hydroxylase from Jerusalem artichoke,[17] and P450tyr from sorghum.[18]

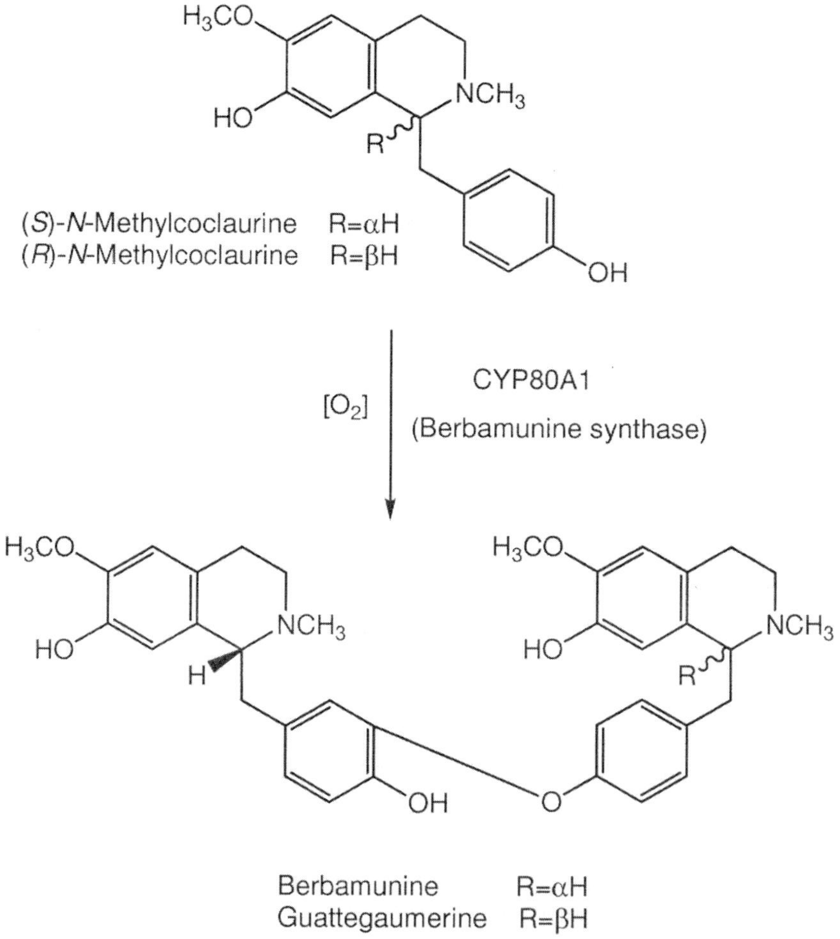
H3CO
HO
NCH3
R
OH
(S)-N-Methylcoclaurine R=αH
(R)-N-Methylcoclaurine R=βH
[O2]
CYP80A1
(Berbamunine synthase)
H3CO
HO
NCH3
H
OH
O
H3CO
HO
NCH3
R
Berbamunine R=αH
Guattegaumerine R=βH

Fig. 10.3: Reaction catalyzed by the cytochrome P-450-dependent oxidase berbamunine synthase (CYP80A1). This enzyme creates a branchpoint in the (*S*)-reticuline biosynthetic pathway to form the bisbenzylisoquinoline alkaloids.

Berbamunine synthase was solubilized from *Berberis stolonifera* microsomes and was purified to homogeneity.[19] From 1.6 kg fresh weight *B. stolonifera* suspension cells, 450 mg microsomal protein were obtained after solubilization. After five column chromatography steps, 4 µg of pure berbamunine synthase were achieved, representing a yield of 0.2 %. This amount of protein was sufficient to determine the sequence of the first 24 amino acids from the amino terminus. Expression levels of *cyp80a1* were low enough that the indirect sequencing of RNA by extension of degenerate oligodeoxynucleotides did not yield a contiguous, clear sequence. The relatively new method of RT-PCR was then used to amplify a DNA fragment that encoded the berbamunine synthase amino terminus.[19] This cDNA fragment was used to screen a cDNA library. *Cyp80a1* was then functionally expressed in insect cell culture using a baculovirus expression vector.[19]

(*S*)-*N*-METHYLCOCLAURINE 3'-HYDROXYLASE

Alkaloids derived from the tetrahydrobenzylisoquinoline alkaloid (*S*)-*N*-methylcoclaurine represent a vast and varied structural array of physiologically active molecules. These compounds range from the dimeric bisbenzylisoquinoline (+)-tubocurarine, mentioned above, to the powerful anesthetic opiate morphine. The 3'-hydroxylation of (*S*)-*N*-methylcoclaurine is a branch point that is the penultimate biosynthetic step in the formation of (*S*)-reticuline. The hydroxylation was long thought to be catalyzed by a phenolase. Molecular genetic studies on benzo[*c*]phenanthridine formation in *E. californica* cell suspension cultures identified the catalysts as a cytochrome P-450-dependent monooxygenase.[20] The first cDNAs encoding plant cytochromes P-450 of known function were isolated using either polyclonal antibodies or internal amino acid sequences obtained from purified protein. As primary protein sequences for plant cytochromes P-450 accumulated, conserved domains could be exploited for direct cloning of cDNAs, which eliminated the necessity to first purify the proteins.[21]

Two alleles encoding a new enzyme (*S*)-*N*-methylcoclaurine 3'-hydroxylase (CYP80B1) (Fig. 10.4) were isolated from a cDNA library prepared from mRNA isolated from methyl jasmonate-induced cell suspension cultures of *E. californica*. Partial clones that were generated by RT-PCR with cytochrome P-450-specific primers were used as hybridization probes. Partial clones that hybridized to blotted transcripts that accumulated in response to addition of methyl jasmonate to the culture medium were obtained as full-length clones by screening a cDNA library. We knew from our earlier work on the berberine bridge enzyme from *E. californica* that the enzyme activity of the berberine bridge enzyme and the cytochromes P-450 of benzo[*c*]phenanthridine alkaloid formation were inducible by methyl jasmonate.[12] At least for the berberine bridge enzyme, this appeared to be a result of transcriptional activation. By analogy, the cytochrome P-450 genes of benzo[*c*]phenanthridine alkaloid formation should also be transcriptionally activated by methyl jasmonate treatment. Three full-length cDNAs encoding cytochromes P-450 were functionally expressed in *Spodoptera frugiperda* Sf9 cells. Two of these were found to be alleles that encode a highly substrate specific (S)-N-methylcoclaurine 3'-hydroxylase. This was the first time that we discovered a new enzyme of alkaloid biosynthesis by employing molecular genetic methods. The cDNA *cyp80b1* was subsequently isolated from the opium poppy *Papaver somniferum*.[22]

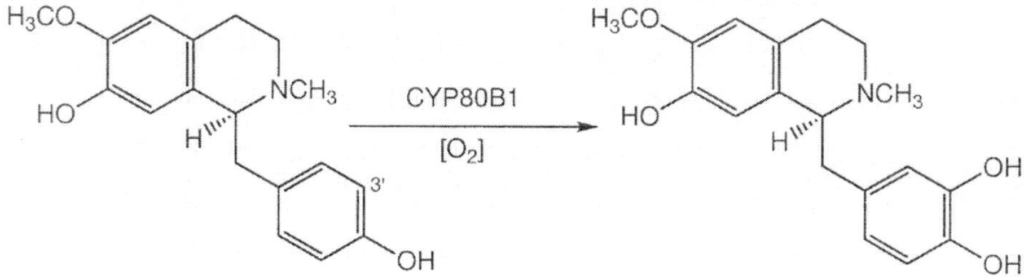

(S)-N-Methylcoclaurine → CYP80B1 [O_2] → (S)-3'-Hydroxy-N-methylcoclaurine

Fig. 10.4: Reaction catalyzed by the cytochrome P-450-dependent monooxygenase (S)-N-methylcoclaurine 3'-hydroxylase (CYP80B1) along the (S)-reticuline biosynthetic pathway. (S)-N-Methylcoclaurine 3'-hydroxylase acts upon the same substrate as berbamunine synthase (CYP80A1).

COCLAURINE 6-*O*-METHYLTRANSFERASE

In analogy to the isolation of the cytochrome P-450-dependent monooxygenase CYP80B1 of (S)-reticuline biosynthesis by RT-PCR with primers based upon highly conserved sequences in the coding region of cytochromes P-450, we attempted to isolate cDNAs encoding methyltransferases of (S)-reticuline biosynthesis by RT-PCR. In cell suspension cultures of the meadow rue *Thalictrum tuberosum*, biosynthesis of the anti-microbial alkaloid berberine was induced by addition of methyl jasmonate to the culture medium. The activities of the four methyltransferases involved in the formation of berberine are increased in response to elicitor addition. Partial clones were generated by RT-PCR with methyltransferase specific primers. These clones were used as hybridization probes to isolate four cDNAs encoding apparent methyltransferases from a cDNA prepared from methyl jasmonate-treated suspension cells.[23] As for the cytochrome P-450 encoding cDNAs from *E. californica*, only those partial clones that hybridized to inducible transcripts were used to isolate full-length cDNAs.

The biosynthesis of the central isoquinoline alkaloid intermediate (S)-reticuline requires the action of two O-methyltransferases, norcoclaurine 6-O-methyltransferase and 3'-hydroxy-N-methylcoclaurine 4'-O-methyltransferase, and one N-methyltransferase. The four cDNAs that were functionally expressed in *S. frugiperda* Sf9 cells were tested for activity in an assay for each of these three enzymes. Only two of the four nearly identical (93.2 – 99.7 % identity) recombinant enzymes were found to O-methylate norcoclaurine at the 6-position (Fig. 10.5). This surprising result initiated a more thorough investigation of the substrate specificity of the recombinant enzymes. As for alkaloid biosynthesis, O-methyltransferases are central also to flavonoid biosynthesis. A common structural feature, a catechol moiety, is found between the substrate of early isoquinoline alkaloid biosynthesis, norcoclaurine, and the substrate of early phenylpropanoid biosynthesis, caffeic acid.[23] A series of thirty-eight potential substrates were tested with the four recombinant methyltransferases of *T. tuberosum*. All four enzymes were demonstrated to be functional O-methyltransferases with overlapping, but not identical, substrate specificity. The O-methyltransferases were surprisingly permissive with respect to the broad range of substrates that could be methylated. The structures tested varied from simple catechols, such as catechol and guaiacol, to complex tetrahydrobenzylisoquinoline alkaloids. The most unexpected result for us was that all four recombinant methyltransferases could O-methylate caffeic acid. This was the first time that we observed a functional overlap between an enzyme of alkaloid and of phenylpropanoid biosynthesis.[23]

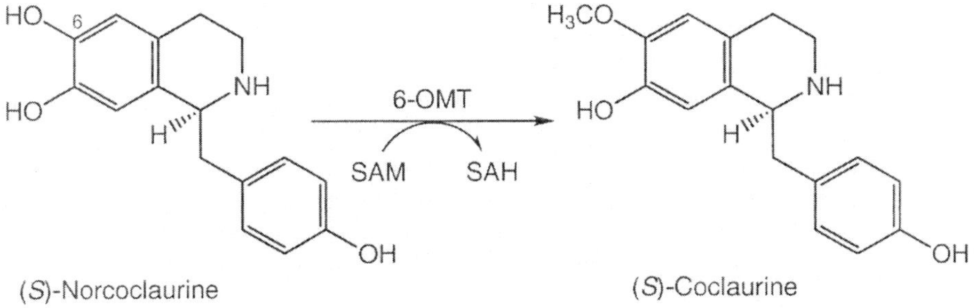

Fig. 10.5: Reaction catalyzed by norcoclaurine 6-O-methyltransferase (6-OMT) in the (S)-reticuline biosynthetic pathway. SAM, S-adenosyl-L-methionine; SAH, S-adenosylhomocysteine.

CODEINONE REDUCTASE

The narcotic analgesic morphine is the major alkaloid of opium poppy *P. somniferum*. Its biosynthetic precursor codeine is the most widely used and effective antitussive agent. Along the biosynthetic pathway that leads to morphine in *P. somniferum*, codeinone reductase catalyzes the NADPH-dependent reduction of codeinone to codeine, the penultimate precursor to morphine (Fig. 10.6). At the time that this cloning project was initiated, not enough information was known about the nature of NADPH-dependent oxidoreductases in alkaloid biosynthesis to facilitate a direct molecular genetic approach to a cDNA encoding codeinone reductase. Native codeinone reductase was, therefore, purified to electrophoretic homogeneity from suspension cells of *P. somniferum*.[24] This time, less than five micrograms of pure protein were sufficient to determine the amino acid sequences of seven internal peptides.[25] In effect, this was ten times less protein than was necessary to determine peptide sequences for the berberine bridge enzyme in 1988.[12]

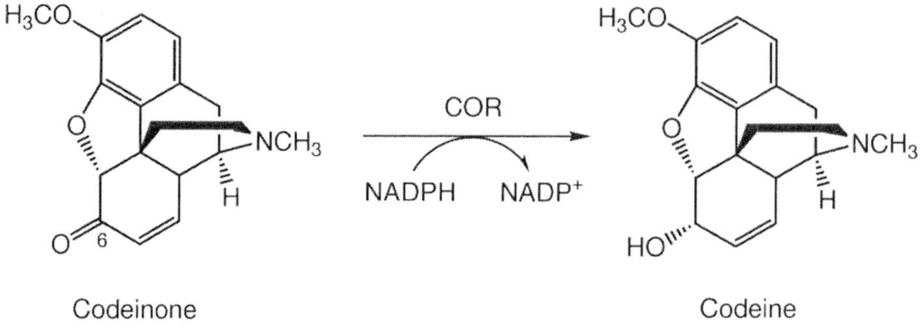

Fig. 10.6: Reaction catalyzed by codeinone reductase (COR), the penultimate step in morphine biosynthesis.

The peptide sequences obtained for codeinone reductase aligned well with the amino acid sequences for 6'-deoxychalcone synthase (chalcone reductase) from alfalfa, *Glycerrhiza,* and soybean. Knowledge of the relative positions of the peptides allowed for a quick RT-PCR based isolation of cDNAs encoding codeinone reductase from *P. somniferum*. The codeinone reductase isoforms are 53 % identical to chalcone reductase from soybean.[25] By sequence comparison, both codeinone reductase and chalcone reductase belong to the aldo/keto reductase family, a group of structurally and functionally related NADPH-dependent oxidoreductases, and thereby possibly arise from primary metabolism. Six alleles encoding codeinone

ALKALOID BIOSYNTHESIS -- GENE IDENTIFICATION

reductase were found to be expressed in *P. somniferum*, although no biochemical differences or differences in the expression patterns could be detected.[25]

With a morphine biosynthetic gene in hand, we believed we could begin to address the question why only *P. somniferum* produces morphine, while other *Papaver* species such as *P. rhoeas*, *P. orientale,* and *P. bracteatum* do not. Unexpectedly, we found that the codeinone reductase transcript was present to some degree in all four species investigated. A review of the literature revealed no alkaloids reported in *P. rhoeas* for which codeinone reductase should participate in the synthesis. Similarly, *P. orientale* accumulates the alternate morphine biosynthetic precursor oripavine, but codeinone reductase is not involved in the biosynthesis of oripavine, acting instead after this alkaloid along the biosynthetic pathway to morphine.[22] *P. bracteatum* produces the morphine precursor thebaine as a major alkaloid. As for oripavine in *P. orientale*, codeinone reductase would act in *P. bracteatum* after thebaine formation on the pathway to morphine. It appears, therefore, that the reason that *P. rhoeas*, *P. orientale,* and *P. bracteatum* do not produce morphine is not related to the absence of the transcript of the morphine biosynthesis-specific gene codeinone reductase. The expression of codeinone reductase may simply be an evolutionary remnant in these species.

SALUTARIDINOL 7-*O*-ACETYLTRANSFERASE

Salutaridinol 7-*O*-acetyltransferase catalyzes the conversion of the phenanthrene alkaloid salutaridinol to salutaridinol-7-*O*-acetate, the immediate precursor of thebaine along the morphine biosynthetic pathway in *P. somniferum* (Fig. 10.7).[26] Acetyl CoA-dependent acetyltransferases have an important role in plant alkaloid metabolism. They are involved in the synthesis of monoterpenoid indole alkaloids in medicinal plant species such as *Rauwolfia serpentina*. In this plant, the enzyme vinorine synthase transfers an acetyl group from acetyl CoA to 16-epi-vellosimine to form vinorine. This acetyl transfer is accompanied by a concomitant skeletal rearrangement from the sarpagan- to the ajmalan-type (reviewed in[2]). An acetyl CoA-dependent acetyltransferase also participates in vindoline biosynthesis in *Catharanthus roseus*, the source of the chemotherapeutic dimeric indole alkaloid vinblastine (reviewed in[2]). Acetyl CoA:deacetylvindoline 4-*O*-acetyltransferase catalyzes the last step in vindoline biosynthesis. A cDNA encoding acetyl CoA:deacetylvindoline 4-*O*-acetyltransferase was recently successfully isolated.[27]

Salutaridinol 7-*O*-acetyltransferase was purified to apparent electrophoretic homogeneity from *P. somniferum* cell suspension cultures and the amino acid sequence of ten endoproteinase Lys-C-generated peptides was determined.[28] A comparison of these amino acid sequences with those available in the GenBank/EMBL sequence databases indicated no relevant similarity to known proteins. The first attempt to isolate a cDNA encoding salutaridinol 7-*O*-

acetyltransferase from *P. somniferum* was based upon RT-PCR. A series of PCR primer pairs based on salutaridinol 7-*O*-acetyltransferase peptide combinations yielded only DNA fragments of irrelevant sequence. During the course of the initial RT-PCR experiments, sequence comparison information appeared in the literature for acetyl CoA:deacetylvindoline 4-*O*-acetyltransferase.[27]

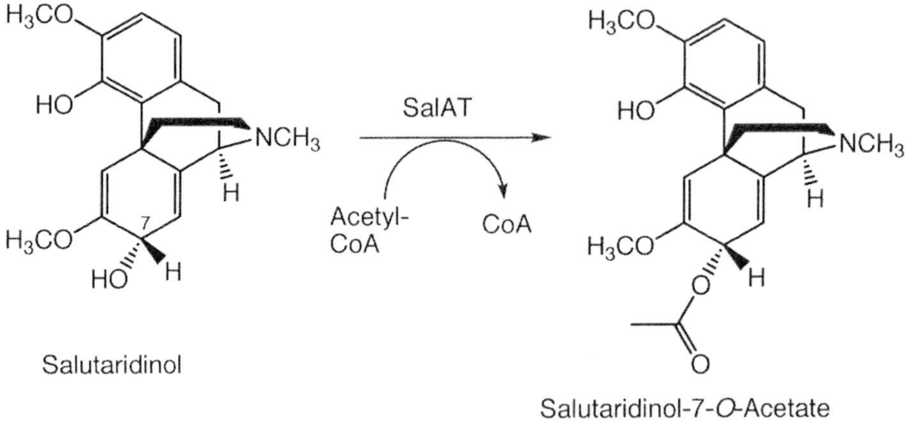

Fig. 10.7: Reaction catalyzed by salutaridinol 7-*O*-acetyltransferase (SalAT) along the morphine biosynthetic pathway.

The translation of the sequence of the cDNA encoding deacetylvindoline 4-*O*-acetyltransferase compared to other putative plant acetyltransferases revealed a conserved region near the carboxy terminus of the proteins. This sequence was used to design a degenerate antisense oligodeoxynucleotide primer for PCR. The sense primer was based upon an internal peptide sequence of salutaridinol 7-*O*-acetyltransferase. This approach finally yielded a partial cDNA that encoded salutaridinol 7-*O*-acetyltransferase. The full-length clone was obtained by RACE-PCR and was functionally expressed in *S. frugiperda* Sf9 cells.[28] The amino acid sequence of salutaridinol 7-*O*-acetyltransferase is most similar (37% identity) to that of deacetylvindoline acetyltransferase of *C. roseus*.[27]

RNA gel blot analysis of several members of the genus *Papaver* demonstrated that salutaridinol 7-*O*-acetyltransferase transcript accumulated in three-week-old seedlings of *P. orientale* and *P. bracteatum*, though not in *P. atlanticum* or *P. nudicaule*.[28] *P. orientale* accumulates the alternate biosynthetic precursor oripavine, and *P. bracteatum* accumulates the morphine biosynthetic precursor thebaine, both structures contain the oxide bridge formed by action of salutaridinol 7-O-acetyltransferase. It was, therefore, expected that these two species should contain the hybridizing salutaridinol 7-O-acetyltransferase transcript. Neither *P. atlanticum* nor *P. nudicaule* contain an alkaloid with the morphinan skeleton, consistent with the absence of transcript in these two species.

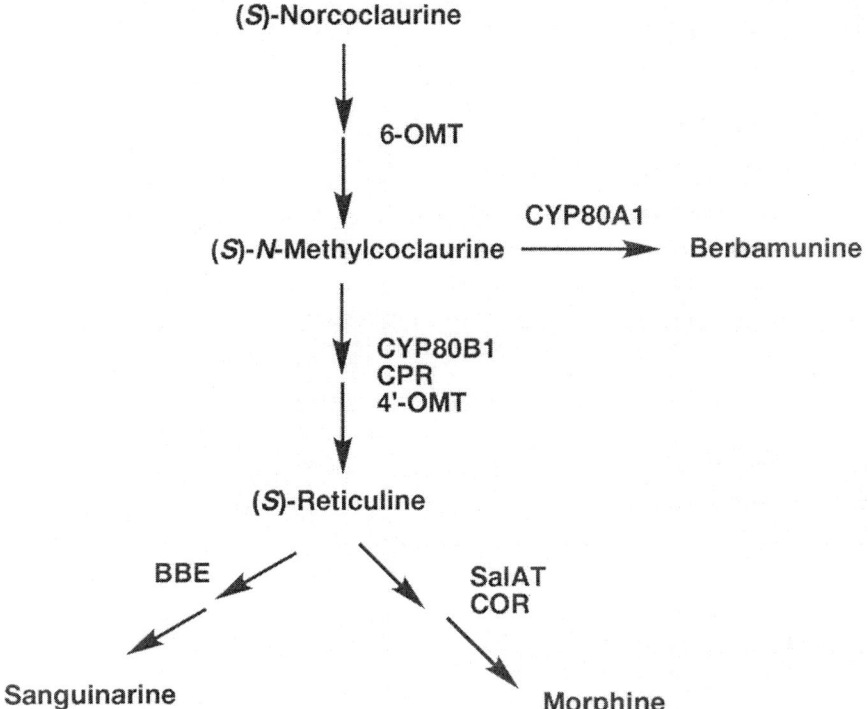

Fig. 10.8: Selected cDNAs isolated in recent years that encode enzymes involved in the biosynthesis of various classes of isoquinoline alkaloids. 6-OMT, norcoclaurine 6-O-methyltransferase;[23] CYP80A1, berbamunine synthase;[19] CYP80B1, (S)-N-methylcoclaurine 3'-hydroxylase;[20] CPR, cytochrome P-450 reductase;[29] 4'-OMT, (S)-3'-hydroxy-N-methylcoclaurine 4'-O-methyltransferase;[30] BBE, berberine bridge enzyme;[12] SalAT, salutaridinol 7-O-acetyltransferase;[28] COR, codeinone reductase.[25]

SUMMARY

Since 1988, the methods that we use to isolate cDNAs of alkaloid biosynthesis have become ever more facile and sensitive, allowing for more efficient cDNA identification. We do not, however, yet understand enough about the cellular localization of alkaloid formation or about the nature of the catalysts to move completely away from enzymology and biochemistry and to use only molecular genetic techniques to dissect these biosynthetic pathways. Even our most recently successful cDNA isolations and identifications involved classical protein purification. We are beginning now to use proteomics and EST sequencing to identify natural product biosynthetic cDNAs, but these approaches are more feasible when a specialized cell/tissue type in which secondary metabolite biosynthetic pathways are active, can be isolated and used as a protein or RNA source.

In summary, the *de novo* isolation of the cDNAs encoding enzymes of alkaloid biosynthesis is still achieved by using a variety of "classical" techniques, such as protein purification followed by partial amino acid sequence determination, and by newer techniques such as proteomics coupled to functional heterologous expression. The current status of cloned cDNAs specifically related to isoquinoline alkaloid biosynthesis is schematically presented in Figure 10.8. New additions to this list will certainly be made in the future as a result of a combination of approaches both new and old.

REFERENCES

1. KUTCHAN, T.M., HAMPP, N., LOTTSPEICH, F., BEYREUTHER, K., ZENK, M.H., The cDNA clone for strictosidine synthase from *Rauvolfia serpentina* – DNA sequence determination and expression in *Escherichia col.*, *FEBS Lett.*, 1988, **237**, 40-44.
2. KUTCHAN, T.M., Molecular genetics of plant alkaloid biosynthesis. In: The Alkaloids Vol. 50, (G. Cordell, ed.), Academic Press, San Diego, 1998, pp. 257-316.
3. SMITH, G.N., Strictosidine: A key intermediate in the biogenesis of indole alkaloids, *Chem. Commun.* 1968, 912-914.
4. PFITZNER, U., ZENK, M.H., Immobilization of strictosidine synthase from *Catharanthus* cell cultures and preparative synthesis of strictosidine, *Planta Med.*, 1982, **46**, 10-14.
5. HAMPP, N., ZENK, M.H., Homogeneous strictosidine synthase from cell suspension cultures of *Rauvolfia serpentina*, *Phytochemistry*, 1988, **27**, 3811-3815.
6. KÖHRER, K., KUTCHAN, T.M., DOMDEY, H., Specific oligodeoxynucleotide probes obtained through RNA sequencing, *DNA*, 1989, **8**, 143-147.
7. KUTCHAN, T.M., Expression of enzymatically active cloned strictosidine synthase from the higher plant *Rauvolfia serpentina* in *Escherichia coli*, *FEBS Lett.*, 1989, **257**, 127-130.

8. FABBRI, M., DELP, G., SCHMIDT, O., THEOPOLD, U., Animal and plant members of a gene family with similarity to alkaloid-synthesizing enzymes, *Biochem. Biophys. Res. Commun.*, 2000, **271**, 191-196.
9. MORITA, M., HARA, Y., TAMAI, Y., ARAKAWA, H., NISHIMURA, S., Genomic construct and mapping of the gene for CMAP (leukocystatin/cystatin F, CST7) and identification of a proximal novel gene, BSCv (C20orf3), *Genomics*, 2000, **67**, 87-91.
10. STEFFENS, P., NAGAKURA, N., ZENK, M.H., Purification and characterization of the berberine bridge enzyme from *Berberis beaniana* cell cultures, *Phytochemistry*, 1985, **24**, 2577-2583.
11. WOLTERS, B., EILERT, U., Accumulation of acridone-epoxides in callus cultures of *Ruta graveolens* increased by coculture with non host-specific fungi, *Z. Naturforsch.*, 1982, **37c**, 575-583.
12. DITTRICH, H., KUTCHAN, T.M., Molecular cloning, expression, and induction of berberine bridge enzyme, an enzyme essential to the formation of benzophenanthridine alkaloids in the response of plants to pathogenic attack, *Proc. Natl. Acad. Sci. USA*, 1991, **88**, 9969-9973.
13. KUTCHAN, T.M., BOCK, A., DITTRICH, H., Heterologous expression of strictosidine synthase and berberine bridge enzyme in insect cell culture, *Phytochemistry*, 1994, **35**, 353-360.
14. KUTCHAN, T.M., DITTRICH, H., Characterization and mechanism of the berberine bridge enzyme, a covalently flavinylated oxidase of benzophenanthridine alkaloid biosynthesis in plants, *J. Biol. Chem.*, 1995, **270**, 24475-24481.
15. STADLER, R., ZENK, M.H., The purification and characterization of a unique cytochrome P-450 enzyme from *Berberis stolonifera* plant cell cultures, *J. Biol. Chem.*, 1993, **268**, 823-831.
16. SONG, W.C., FUNK, C.D., BRASH, A.R., Molecular cloning of an allene oxide synthase: A cytochrome P450 specialized for the metabolism of fatty acid hydroperoxides, *Proc. Natl. Acad. Sci. USA*, 1993, **90**, 8519-8523.
17. TEUTSCH, H.G., HASENFRATZ, M.P., LESOT, A., STOLTZ, C., GARNIER, J.M., JELTSCH, J.M., DURST, F., WERCK-REICHHART, D., Isolation and sequence of a cDNA encoding the Jerusalem artichoke cinnamate 4-hydroxylase, a major plant cytochrome P450 involved in the general phenylpropanoid pathway, *Proc. Natl. Acad. Sci. USA*, 1993, **90**, 4102-4106.
18. KOCH, B.M., SIBBESEN, O., HALKIER, B.A., SVENDSEN, I., MØLLER, B.L., The primary sequence of cytochrome P450tyr, the multifunctional N-hydroxylase catalyzing the conversion of L-tyrosine to *p*-hydroxyphenylacetaldehyde oxime in the biosynthesis of the cyanogenic glucoside dhurrin in *Sorghum bicolor* (L.) Moench, *Arch. Biochem. Biophys.*, 1995, **323**, 177-186.
19. KRAUS, P.F.X., KUTCHAN, T.M., Molecular cloning and heterologous expression of a cDNA encoding berbamunine synthase, a C – O phenol coupling cytochrome P450 from the higher plant *Berberis stolonifera*, *Proc. Natl. Acad. Sci. USA*, 1995, **92**, 2071-2075.
20. PAULI, H.H., KUTCHAN, T.M., Molecular cloning and functional heterologous expression of two alleles encoding (*S*)-*N*-methylcoclaurine 3'-hydroxylase (CYP80B1), a new methyl jasmonate-inducible cytochrome P-450-dependent monooxygenase of benzylisoquinoline alkaloid biosynthesis, *Plant J.*, 1998, **13**, 793-801.

21. UDVARDI, M.K., METZGER, J.D., KRISHNAPILLAI, V., PEACOCK, W.J., DENNIS, E.S., Cloning and sequencing of a full-length cDNA from *Thlaspi arvense* L. that encodes a cytochrome P-450, *Plant Physiol.*, 1994, **105**, 755-756.
22. HUANG, F.-C., KUTCHAN, T.M., Distribution of morphinan and benzo[c]phenanthridine alkaloid gene transcript accumulation in *Papaver somniferum*, *Phytochemistry*, 2000, **53**, 555-564.
23. FRICK, S., KUTCHAN, T.M., Molecular cloning and functional expression of *O*-methyltransferases common to isoquinoline alkaloid and phenylpropanoid biosynthesis, *Plant J.*, 1999, **17**, 329-339.
24. LENZ, R., ZENK, M.H., Purification and properties of codeinone reductase (NADPH) from *Papaver somniferum* cell cultures and differentiated plants, *Eur. J. Biochem.*, 1995, **233**, 132-139.
25. UNTERLINNER, B., LENZ, R., KUTCHAN, T.M., Molecular cloning and functional heterologous expression of codeinone reductase: The penultimate enzyme in morphine biosynthesis in the opium poppy *Papaver somniferum*, *Plant J.*, 1999, **18**, 465-475.
26. LENZ, R., ZENK, M.H., Acetyl coenzyme A:salutaridinol 7-*O*-acetyltransferase from *Papaver somniferum* plant cell cultures, *J. Biol. Chem.*, 1995, **270**, 31091-31096.
27. ST-PIERRE, B., LAFLAMME, P., ALARCO, A.-M., DE LUCA, V., The terminal *O*-acetyltransferase involved in vindoline biosynthesis defines a new class of proteins responsible for coenzyme A-dependent acyl transfer, *Plant J.*, 1998, **14**, 703-713.
28. GROTHE, T., LENZ, R., KUTCHAN, T.M., Molecular characterization of the salutaridinol 7-*O*-acetyltransferase involved in morphine biosynthesis in opium poppy *Papaver somniferum*, *J. Biol. Chem.* 2001, **276**, 30717-30723.
29. ROSCO, A., PAULI, H.H., PRIESNER, W., KUTCHAN, T.M., Cloning and heterologous expression of cytochrome P450 reductases from the Papaveraceae, *Arch. Biochem. Biophys.*, 1997, **348**, 369-377.
30. MORISHIGE, T., TSUJITA, T., YAMADA, Y., SATO, F., Molecular characterization of the S-adenosyl-L-methionine:3'-hydroxy-*N*-methylcoclaurine 4'-*O*-methyltransferase involved in isoquinoline alkaloid biosynthesis in *Coptis japonica*, *J. Biol. Chem.*, 2000, **275**, 23398-23405.

Chapter Eleven

AN INTEGRATED APPROACH TO *MEDICAGO* FUNCTIONAL GENOMICS

Gregory D. May

Plant Biology Division,
The Samuel Roberts Noble Foundation,
Ardmore, OK 73402, USA

e-mail: gdmay@noble.org

Introduction..	180
Sequence Analysis...	181
Expression Analysis..	185
Proteomics..	187
Metabolic Profiling..	188
Medicago truncatula Biological Materials...............................	189
Insertional Mutagenesis...	189
Virus Induced Gene Silencing...	190
Summary..	192

INTRODUCTION

With more than 18,000 types of legumes belonging to the pea family (*Leguminosae*), these plants are second only to grasses in economic importance worldwide. Forage and pasture legumes are an important source of nutrition for animal and dairy production. Seeds of legumes such as peanut, soybeans, chickpeas, and lentils contain from 20 to 50 percent protein – two to three times that of cereal grains and meat. Legumes, therefore, serve as an excellent source of protein and dietary fiber that is often deficient in the diets of individuals in developing nations. Moreover, in comparison with other crops, the production of legumes reduces economic and environmental costs given their ability to fix nitrogen.

Medicago truncatula (commonly known as "barrel medic" because of the shape of its seed pods) is a forage legume commonly grown in Australia. It is an omni-Mediterranean species and closely related to the world's major forage legume, alfalfa. Unlike alfalfa, which is a tetraploid, obligate outcrossing species, *M. truncatula* has a simple diploid genome (two sets of eight chromosomes) and can be self-pollinated.

M. truncatula has been chosen as a model species for genomic studies in view of its small genome, fast generation time (from seed-to-seed), and high transformation efficiency.[1-2] Genes from *M. truncatula* share high sequence identity to their counterparts from alfalfa (*e.g.*, 98.7 and 99.1% at the amino acid levels for isoflavone reductase and vestitone reductase, respectively), so it serves as an excellent genetically tractable model for alfalfa. Studies on syntenic relationships (comparisons of genome content and organization between organisms) are establishing links between *M. truncatula*, alfalfa, and pea, as well as *Arabidopsis*.

As a legume, and unlike the most studied genetic model plant, *Arabidopsis*, *M. truncatula* establishes symbiotic relationships with nitrogen fixing *Rhizobia*. Roots of *M. truncatula* are also colonized by beneficial arbuscular mycorrhizal fungi.[3] The complex interactions of legumes with microorganisms have resulted in the evolution of a rich variety of natural product biosynthetic pathways impacting both mutualistic and disease/defense interactions. Of these, the isoflavonoid pathway, which is not present in *Arabidopsis*, leads to nodulation gene inducers and repressors, pterocarpan phytoalexins involved in host disease resistance, and isoflavones with anticancer and other health promoting effects for humans. This pathway has been well characterized in alfalfa, and in other legumes such a soybean and chickpea, at the metabolic, enzymatic, and genetic levels.[4-6] *Medicago* species are also a rich source of triterpene saponins with a wide range of biological activities (see Osbourn and Haralampidis, this volume). The phenylpropanoid polymer lignin is ubiquitous in monocots and dicots, but is of particular importance in forage legumes such as *Medicago* because of its impact on forage digestibility. Many genes are involved in lignin biosynthesis and deposition. Exploitation of the diverse and complex chemistry of legumes for the benefit of humankind requires in-depth knowledge of the legume genome.

A Center for *Medicago* Genomics Research was established at The Samuel Roberts Noble Foundation in the fall of 1999. We have taken a global approach in studying the genetic and biochemical events associated with the growth, development, and environmental interactions of the model legume *M. truncatula*. Our approach includes: large-scale EST and genome sequencing, gene expression profiling, the generation of *M. truncatula* activation-tagged, promoter trap insertion and gene knockout systems, and high-throughput metabolite and protein profiling. An overview is illustrated in Figure 11.1. The resulting multidisciplinary databases developed in our program will be interfaced to provide scientists with an integrated set of tools to address fundamental questions pertaining to legume biology. The program's bioinformatics needs are currently being met through partnerships with the National Center for Genome Resources (NCGR) and the Virginia Bioinformatics Institute (VBI). Our aim is to develop a program that will make a significant contribution to the areas of legume molecular biology, biochemistry, and genetics research.

SEQUENCE ANALYSIS

The *Medicago* Genome Initiative (MGI) is an EST sequence database of the model legume *M. truncatula*. The database is available to the public and results from a collaborative research effort between the Noble Foundation and the NCGR to investigate the genome of *M. truncatula*. MGI was first reported in the *Nucleic Acids Research* 2001 Database Issue and featured a prototype database, interface, and analysis pipeline.[7] We have since developed an entirely new system that retains the advantages of the prototype, with improvements that make it more portable, modular, flexible, interactive, and reusable.[8] The data model is designed around the concept of an analysis operation (which may run a third-party sequence analysis tool) whose input and output consists of sets of sequences (zero, one, or many). This permits analysis methods that use individual (*e.g.*, similarity search) or multiple (*e.g.*, EST clustering) sequences to interact with the same generalized relational database structure. It also allows for the flexible addition of sequence analysis methods, and the storage and analysis of genomic DNA sequences in the same schema. The analysis pipeline is run automatically upon receipt of new sequences and can be configured to perform any series of available operations. The current suite of operations includes:

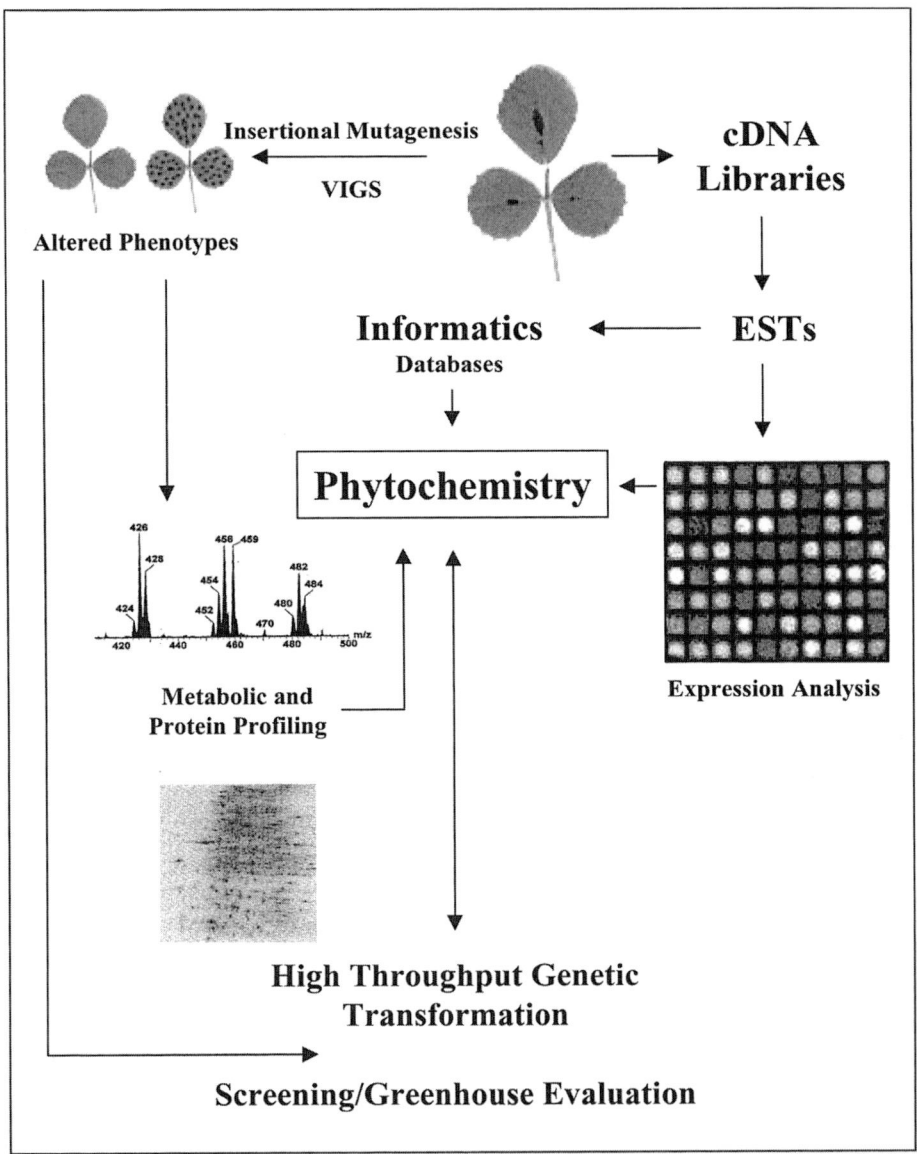

Figure 11.1: A schematic overview of the Noble Foundation *Medicago truncatula* functional genomics program.

Import; Vector Screen; Quality Control; BLASTN search to identify non-mRNA contamination; clustering, multiple sequence alignment, and extraction of a consensus; BLASTX versus a protein database; and Blocks+ (protein motif) search. Annotation is automated by linking high-scoring BLAST and Blocks+ hits to their cognate entries in the Gene Ontology database (http://geneontology.org). Users view, query, and manipulate their data via a WWW browser through an interface running on a secure server. All analysis operations are performed on consensus sequences (gene sequences) resulting from the clustering and assembly operation, rather than on individual ESTs. MGI now incorporates all publicly available *M. truncatula* data available from Genbank combined with public Noble data in clustering and analysis runs. Typically the data are refreshed, including a complete reanalysis with all available new data, four times per year. As of September 2001, MGI contained over 95,000 sequences of which the 65,000 GenBank ESTs grouped into 8,843 clusters and 11,279 singletons resulting in 20,122 total analyzed consensus sequences. Clusters ranged in membership from two ESTs (3585) to 256 ESTs (one). A publicly viewable version of MGI has been deployed (http://xgi.ncgr.org/mgi), which can be accessed by following the login instructions on the main page.

In addition to data from the Noble Foundation and the inclusion of all publicly available *M. truncatula* data from GenBank, the database and analysis system has been designed to present a gene-centric view of ESTs. The new interface improvements include keyword searches, query restriction by library and sequence type, a multiple sequence alignment viewer, and a features and annotation viewer. These additions, coupled with automated assignment of Gene Ontology annotations, have resulted in a vastly improved information resource for model legume research.

As of late 2001, the *M. truncatula* research community has generated approximately 141,000 ESTs from 24 different cDNA libraries. These ESTs separate into more than 126,000 tentative consensus sequences and more than 14,000 singletons. Greater than 29,000 unique sequences have been identified. A breakdown of the types of genes identified, based on Gene Ontology assignments, can be found in Figure 11.2. In addition to MGI, the TIGR *Medicago truncatula* Gene Index (http://www.tigr.org/tdb/mtgi/) and the NSF-sponsored *Medicago truncatula* Consortium (http://www.medicago.org/) are two additional databases of particular interest to the legume research community.

A *M. truncatula* whole genome sequence program has begun at the University of Oklahoma (http://www.genome.ou.edu/medicago.html). The initial goal of the project was to generate an approximately one-fold whole genome shotgun sequence data of the 500 megabase genome from a plasmid-based genomic library and obtain target shotgun clones for additional primer walking-based sequencing. However, preliminary results from the shotgun approach suggest that the *M. trucatula* genome is highly repetitive. As previously predicted, estimates are that approximately 80% of the genome is highly repetitive and that approximately 80% of the gene-rich regions represent only 20% of the total genome. To reduce the amount of redundant sequence, the strategy has been modified now to sequence bacterial artificial chromosome (BAC) clones from a *M. truncatula* BAC library. More than 1,000 BACs will be identified based on DNA markers or gene content and will be sequenced to working draft coverage (four- to five-fold) in the first year by utilizing a BAC-based shotgun sequencing approach.

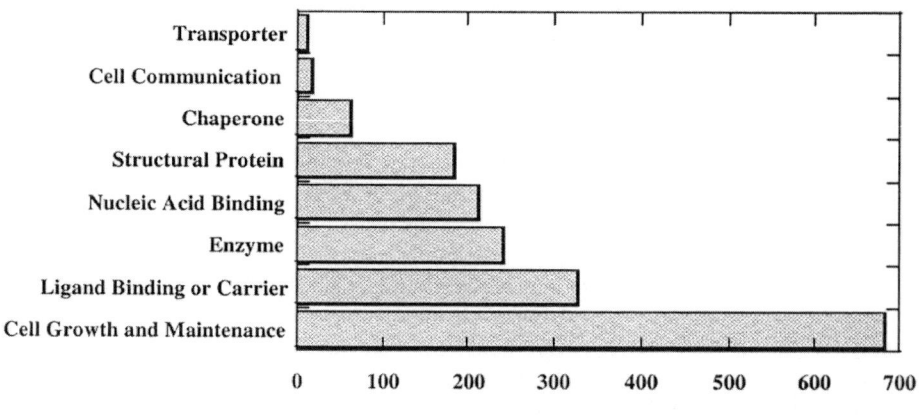

Figure 11.2: Gene Ontology assignments of the *M. truncatula* expressed sequence tags.

The whole genome shotgun approach has already resulted in the sequencing of the *M. truncatula* chloroplast genome, since the total genomic DNA preparation not only contains the nuclear genome, but also a significant level of the chloroplast DNA. The DNA sequence of the *M. truncatula* chloroplast genome has been completed and consists of one contiguous 124,039 base pair circle. Artificially linearizing the sequence at the histidine tRNA prior to the *psbA* gene allows the *Medicago* chloroplast genomic sequence to be co-linear with the *Arabidopsis*, tobacco, and most other chloroplast genomes. The semi-automated annotation of the *M. truncatula* chloroplast genome using Web-

Artemis has been completed, and can be viewed at: http://www.genome.ou.edu/medicago_chloroplast/med_chloro_art.html.

EXPRESSION ANALYSIS

Changes in gene expression underlie many biological phenomena. The use of DNA microarrays provides insight into tissue- and developmental-specific expression of genes and the response of gene expression to environmental stimuli. Oligonucleotide- and cDNA-based microarrays are being generated on the *BioRobotics* Ltd, *Micro*Grid system by using unique cDNA isolates identified in the *M. truncatula* EST programs. These glass slide arrays are hybridized with fluorescently labeled cDNA probes and analyzed with the GSI Lumonics ScanArray 4000 two-color microarray analysis system.[9-11] Arrays will be used to profile transcript levels in 1) elicited cell cultures, 2) developmental comparisons within a particular tissue type (*e.g.*, stems for the study of lignin biosynthesis), and 3) host-symbiont/host-pathogen interactions. Many of the gene features (or probes) on the arrays represent unknown, previously undescribed genes. The correlation of transcript profiles, probe clustering, and expression patterns in a particular tissue or developmental stage may help to assign biological relevance to many of the unknown genes.

Discussions are currently underway within the *M. truncatula* research community to establish a standardized set of controls to be included on arrays generated from all research programs. These experimental controls will increase the likelihood that data sets obtained from each program will be interchangeable. Standardized controls used in the *Arabidopsis* community are also being used.

To supplement the expression analyses data generated by microarrays, we have added an "open system" serial analysis of gene expression (SAGE) to our set of transcript profiling tools. Such open system approaches allow for the identification and analysis of genes not previously characterized. With "closed systems" such as microarrays, analysis is limited to only those species previously identified and assigned to an array. SAGE analyses have already been used to study gene expression in plant systems.[12-13] Among the high-throughput, comprehensive technological methods used to analyze transcript expression levels, array-based hybridization and SAGE are currently the most common approaches. In a recent comparison of SAGE and array-based technologies, the two methods correlated quite well in both absolute expression analyses and comparative analyses during differentiation.[14] The correlation was better for genes with higher expression levels and greater changes in expression.

Plants are continuously exposed to biotic and abiotic (notably harmful UV radiation) stress elements and have evolved mechanisms to reduce their deleterious effects. The UV-inducibility of a number of DNA repair[15] and metabolic pathways has previously been determined.[16-20] To date, no global assessment of the effects of UV irradiation on transcript abundance in plants has been performed. As an example of the application of SAGE to monitor changes in gene expression, we have profiled modulation in transcript accumulation through analysis of libraries generated from mRNAs isolated from UV-, γ-treated, and non-treated *Arabidopsis* plantlets. Ten days post germination, whole plantlets were exposed to either UV-B light at a dosage rate of 20 J $M^{-2}s^{-1}$ and harvested after a 8000 J M^{-2} exposure, or were irradiated with γ-rays at a dose rate of 181cGy min-1 and harvested after 217Gy of γ-rays irradiation. More than 73,000 tags were identified between the control and treatment libraries and were analyzed using SAGE 2000 software (http://www.sagenet.org/sage_protocol.htm).

More than 300 transcripts, identified in treatment comparisons, had a five-fold or greater difference in abundance. The ten most abundant SAGE tags identified in the UV-treated SAGE library are listed in Table 11.1. Greater than 36,000 of the more than 73,000 SAGE tags were unique. The largest differences in tag abundance were between a 16-fold increase and a 13-fold decrease for transcripts in a comparison between untreated and UV-treatment SAGE tag libraries. A comparison of the γ-ray and untreated SAGE tag libraries reveals that the greatest fluctuation in transcript levels are an approximately ten-fold increase or decrease in relative tag abundance. A number of the SAGE tags had identity and mapped to "putative proteins" corresponding to predicted open reading frames (ORFs) in the *Arabidopsis* genome. The identification of SAGE tags that correspond to putative proteins confirms that the predicted ORFs are indeed expressed.

Table 11.1: The ten most abundant SAGE tags identified in the UV-treated SAGE library.

Tag Count	Percent	Tag Sequence	Gene Description	BAC Locus
109	0.6568	GTGCGTTTGT	putative glycine-rich protein	T20G20.13
100	0.6026	GGCCTTCGCC	chlorophyll a/b binding protein	F1N18_23
85	0.5122	TTTCCTTCCT	ash1 protein	F6I18_230
82	0.4941	AGTGTACGAT	unknown protein	F16J10.7
81	0.4881	GCCATTGGAA	glutathione S transferase	T10P11_18
78	0.47	TGATGAGTTT	putative ACC oxidase	F19P19_18
77	0.464	GGCATCGACA	X-Pro dipeptidase-like protein	T16L4_10
74	0.4459	AAGGTGTGGC	RuBisCO small chain precursor	MXI10_15
74	0.4459	TGTTTGTCAT	ESTs	-
60	0.3615	GTGAGTTTGT	delta-cadinene synthase like pro.	Dl3975c

The sequence of the yeast genome has been completed for some time, and analyses of yeast SAGE databases have led to the discovery of new, previously un-annotated ORFs.[21] In these experiments, SAGE tags were identified that did not correspond to known or predicted genes in the yeast genome. Likewise, our comparison of the UV, γ-ray and untreated SAGE tag libraries has led to the discovery of new ORFs in the *Arabidopsis* genome. SAGE tag counts in the non-treated control library reflected transcript quantitation results similar to those in previous studies reported in the literature that utilized other profiling technologies. SAGE analysis of the UV- and gamma-treated library demonstrated an increase in stress-related transcripts. Genes for many of these transcripts are in the phytochrome signaling and flavonoid biosynthetic pathways, both of which were previously described as UV-inducible.[22] These studies illustrate the power of SAGE technology as a tool for both transcript profiling and gene discovery and its use in examining global changes in plant gene expression patterns. As additional plant genomes such as *M. truncatula* are sequenced and plant-specific SAGE databases become publicly available, the use of SAGE in understanding fundamental changes in gene expression should gain broad appeal in the plant research community.

PROTEOMICS

Two-dimensional polyacrylamide gel electrophoresis (2-D PAGE) has been established as the dominant technique for analysis of complex protein mixtures since its introduction in 1975.[23,24] The technique utilizes isoelectric focusing and polyacrylamide gel electrophoresis for first and second dimension separation, respectively. Currently, 2-D PAGE technology is capable of resolving some 10,000 proteins, with 2,000 proteins being typical experimental results.[25] A recent review describes the role of 2-D PAGE in proteomic and genetic studies of plant systems, including its use as a tool to investigate genetic diversity, phylogenetic relationships, mutant characterization, and drought tolerance.[26] 2-D PAGE has also been utilized in studies on plant defense-associated responses and responses to methyl jasmonate.[25-28]

Although 2-D PAGE analysis has been used for the last 20 years in protein profiling, it provides limited information on protein identification. Recent advances in mass spectrometry and the establishment of protein databases have substantially increased the ease and speed with which proteins can be identified.[29] The union of these technologies is the foundation for modern proteomic studies. The typical experiment begins with comparative digital imaging of the 2-D gels to detect variations in protein concentration or elution profile. These protein spots are excised, extracted, and identified by using a variety of mass spectrometry techniques.

Basically, two mass spectrometry (MS) techniques are used for protein identification. The first is peptide mass-mapping of proteolytic digest fragments.[29,30] The observed mass fragments can be searched against a theoretical list of proteolytic peptide maps predicted by a given database. Increased peptide mass accuracy has increased the success and selectivity of such searches.[31] If the database query is unsuccessful, the protein can be sequenced by using tandem mass spectrometry (MS/MS).[29] During the MS/MS experiment, only the peptide mass of interest is isolated or transmitted, thus discriminating against all other components of the mixture with different mass values. After isolation, the peptide is further fragmented by using a unimolecular or bimolecular (collision gas) strategy. Fragments observed in the isolated peptide MS/MS spectrum can then be rationalized to a sequence.

Initial proteome profiling in our program has been performed to generate representative 2-D PAGE protein profiles for stems, leaves, seedpods, roots, flowers, tissues, and suspension cell cultures. Proteins were systematically identified and cataloged by using peptide mass mapping and database searching. An interactive database of the results of these analyses can be found at the following web address: http://www.noble.org/2dpage/search.asp. Analytical and biological variances associated with the 2-D PAGE proteomics approach for *M. truncatula* have been determined and will function as baseline measurements for comparative protein profiling in elicitor-induced *M. truncatula* cell cultures. For a detailed description of approaches to protein profiling, see Sumner et al., this volume.

METABOLIC PROFILING

Metabolic profiling is the key to understanding how changes at the transcriptional and translational levels affect cellular function.[32] Unlike proteomics, a single analytical technique does not exist that is capable of profiling all the low molecular weight metabolites of the cell. Our approach is to profile metabolites of control and treatment tissues by using an assortment of analytical techniques including: high-performance liquid chromatography (HPLC), capillary electrophoresis (CE), gas chromatography (GC), mass spectrometry (MS), and various combinations of the above techniques such as GC/MS, LC/MS, and CE/MS.

Our initial experiments have focused on profiling isoflavonoids and related phenolics for the model system using ESI-MS.[33-35] As the program has progressed, we have continued to develop methods to extend our profiling range to include metabolite classes such as phenylpropanoids, lignins, terpenoids saponins, soluble sugars, sugar phosphates, complex carbohydrates, amino acids, and lipids. Method development also includes procedures for sequential extraction and parallel analysis. Sequential extraction segregates the metabolome into more manageable classes of chemical compounds with similar physical/chemical properties, thereby facilitating the use of parallel analytical profiling techniques. Profiling of elicitor-induced cell cultures and *M. truncatula* natural variants for flavonoids, lignins, other phenylpropanoids and triterpenoids, especially saponins, will be performed. For a detailed description of approaches to metabolite profiling, see Sumner et al., this volume.

MEDICAGO TRUNCATULA BIOLOGICAL MATERIALS

Insertional Mutagenesis

Activation tagging is an insertional mutagenesis approach that utilizes a transformation vector containing a multimeric series of plant virus transcriptional enhancers.[36] When inserted near a native gene, these enhancers have the ability to induce gene expression constitutively. A frequent consequence of this altered pattern of gene expression is a novel plant phenotype. Phenotypic screening of activation-tagged plant populations can result in the identification and characterization of new genes. For example, activation-tagging has been used to identify the FLOWERING LOCUS T (*FT*) in *Arabidopsis*.[37] In these experiments, an *Arabidopsis* mutant was identified which flowered early, independently of photoperiod length, and developed terminal flowers. When the gene adjacent to the activation-tag insertion site is expressed constitutively in wild-type plants, the mutant phenotype is reconstituted.[37] Since it is thought that the insertion of a transgene occurs at random sites within the plant genome, most gene or phenotypic targets can be addressed using this approach given a mutant population of sufficient size. We have proposed the generation of *M. truncatula* activation-tagged lines for use in gene expression, proteomic and metabolic profiling studies. In addition, promoter trap-tagged lines will be generated for the identification of a wide variety of tissue- and developmental-specific regulatory elements.

Pathway-specific transformation backgrounds can be generated to identify genes for specific biochemical pathways. In these plant backgrounds, regulatory elements for known genes in a preferred biochemical pathway are transcriptionally fused to a reporter gene. Under conditions in which the preferred pathway is induced, reporter gene activity will be observed. As an example, a plant background harboring a triterpene cyclase promoter/luciferase reporter gene construct could be used to identify plants in which the saponin pathway is up-regulated by screening for luciferase activity. Saponin pathway-specific transformation backgrounds will be generated and entered into the tagging program. These backgrounds will facilitate the rapid identification of genes involved in the saponin biosynthetic pathway.

The *M. truncatula* tagged lines will provide a valuable resource of materials to screen for traits of agronomic interest. While many of these traits are intriguing, we have neither the time nor established screens to perform the evaluation. Other groups with interests primarily in soybean have expressed interest in our mutant populations. These groups view *M. truncatula* as a potential model system for the study of legume secondary metabolite/nutraceutical pathways, a niche that *Arabidopsis* cannot fill in some instances.

Virus Induced Gene Silencing

The ability to suppress transiently mRNA accumulation of specific genes in a high-throughput fashion is a powerful tool in a genomics-scale approach to assign biological function to uncharacterized genes. Virus induced gene silencing, or VIGS, is a method to transiently interrupt gene function through RNA interference (Figure 11.3.). The exact mechanism by which VIGS operates is still unclear. It is known, however, that this approach harnesses the plant's natural ability to suppress the accumulation of foreign RNAs by an RNA-mediated defense mechanism against plant viruses.[38] The systemic signal by which this mechanism is induced is unknown, but it is thought to involve an RNA component.[39] Recently it was shown that inoculation of transcript from cloned viruses capable of expressing host sequences in plants led to silencing of the homologous host gene.[40] The infected plants displayed a phenotype representative of the loss of function of the host gene, and not of virus infection.

Should unique ESTs have no known role, it is possible to screen these sequences for function by using a virus vector to induce transient knockout of the gene in *M. truncatula*. To pursue this approach, we identified a virus capable of infecting *M. truncatula* in order to develop a vector for host gene expression. We are currently using portions of the alfalfa mosaic virus (AlMV) genome in order to construct VIGS-based gene knockout vectors. Putative function will be assigned to uncharacterized genes by screening suppressed plants at the visual, mRNA, and metabolite phenotypic levels. We anticipate that many of the *M. truncatula* uncharacterized genes are legume-specific, based on their absence from the genomes of mammalian, microbe, and other plant species. We hope that these unknown genes will give insight into the complex biochemical and physiological traits unique to the leguminous species.

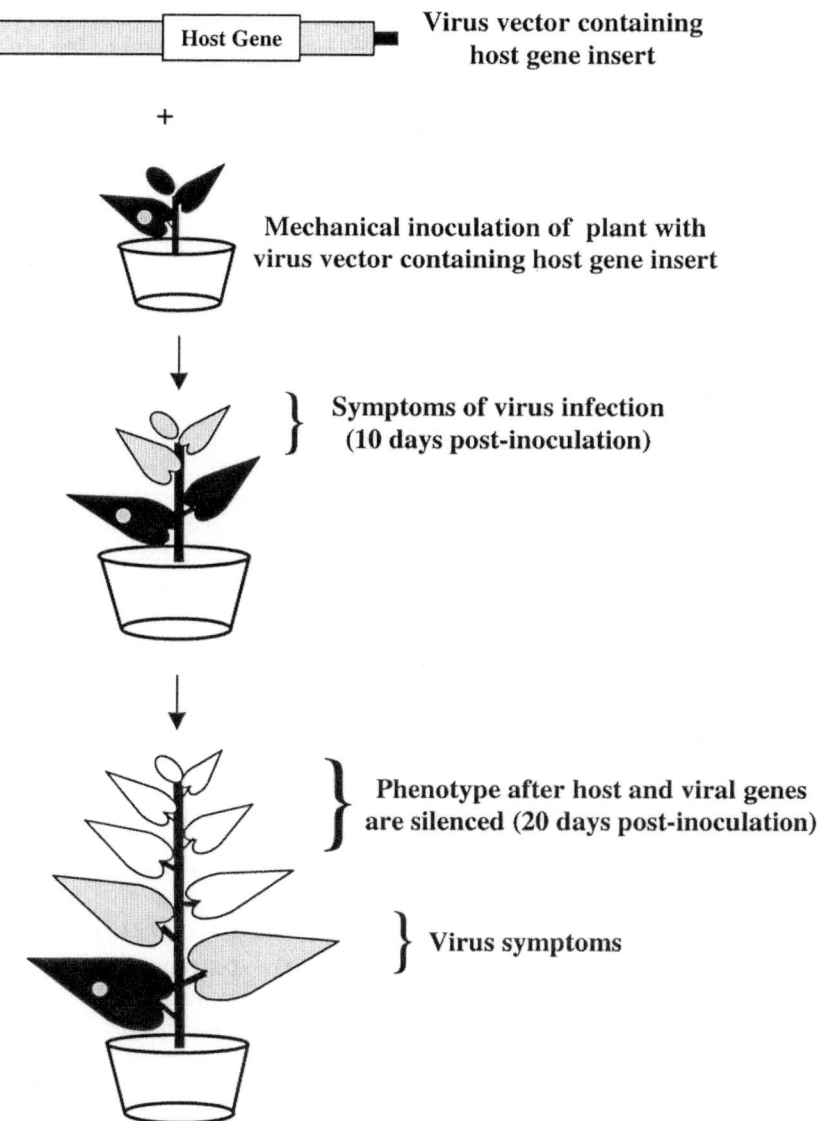

Figure 11.3: A schematic overview of virus induced gene silencing.

SUMMARY

We believe that *M. truncatula* is a good and highly developed model legume with a large research community and one that serves as an excellent model for developing new forage varieties. What is still necessary are bioinformatics tools for the processing, visualization, and integration of transcript, protein, and metabolite profiles and datasets (see Mendes et al., this volume). These tools will lead to a correlated view of gene expression and cellular response. The long-term impact will be the integration of transcript, protein, and metabolite data for plant mutants and natural variants that will advance all aspects of fundamental and applied legume research. This information will be used to develop agronomically important legume species, such as alfalfa and soybean, that (i) are more resistant to fungal and viral diseases, and drought, (ii) will provide higher crop yields with less need for chemical inputs, such as fertilizers and pesticides, and (iii) will produce natural chemicals that promote human and animal health (nutraceuticals). Higher yields and lower production costs will enhance the economy of rural agriculture, especially in developing nations, while a reduction in chemical usage will benefit the environment. Value-added traits such as increased levels of nutraceuticals will provide farmers with new crop alternatives and allow them to participate in the high value niche markets.

ACKNOWLEDGEMENTS

Richard A. Dixon, Robert A. Gonzales, Maria J. Harrison, Nancy L. Paiva, Lloyd W. Sumner, Liangjiang Wang, and members of their laboratory teams are acknowledged for their efforts. Callum Bell, Jennifer Weller, Peter Hraber, Bruno Sobral, Bill Beavis, and Mark Waugh are acknowledged for their contributions to MGI, and Pedro Mendes and Jennifer Weller for their contributions to the high-end bioinformatics portions of the project, as well as participation as joint collaborators on the NSF-funded program "An Integrated Approach to Functional Genomics and Bioinformatics in a Model Legume" (DBI-0109732). Kenneth Korth (University of Arkansas) contributed to EST analysis. Bruce Roe (University of Oklahoma) is directing the *M. truncatula* whole-genome sequencing program. The author acknowledges Lloyd W. Sumner and Richard A. Dixon for their contributions to this chapter and Shujun Yang for providing artwork. This project is supported by the National Science Foundation (DBI-0109732 and DBI-0110206), Forage Genetics International, and the Samuel Roberts Noble Foundation.

REFERENCES

1. COOK, D.R., *Medicago truncatula* - a model in the making!, *Curr. Opin. Plant Bio.*, 1999, **2**, 301-304.
2. TRIEU, A.T., BURLEIGH, S.H., KARDAILSKY, I.V., MALDONADO-MENDOZA, I.E., VERSAW, W.K., BLAYLOCK, L.A., SHIN, H., CHIOU, T.-J., KATAGI, H., DEWBRE, G.R., WEIGEL, D., HARRISON, M.J., Transformation of *Medicago truncatula* via infiltration of seedlings or flowering plants with *Agrobacterium*, *Plant J.*, 2000, **22**, 543-551.
3. HARRISON, M.J,. DIXON, R.A., Isoflavonoid accumulation and expression of defense gene transcripts during the establishment of vesicular arbuscular mycorrhizal associations in roots of *Medicago truncatula*, *Mol. Plant-Microbe Interact.*, 1993, **6**, 643-654.
4. PAIVA, N.L., OOMMEN, A., HARRISON, M.J, DIXON, R.A., Regulation of isoflavonoid metabolism in alfalfa, *Plant Cell, Tissue and Organ Cult.*, 1994, **38**, 213-220.
5. DIXON, R.A., HARRISON, M.J., PAIVA, N.L., The isoflavonoid phytoalexin pathway: from enzymes to genes to transcription factors, *Physiol. Plant.*, 1995, **93**, 385-392.
6. DIXON, R.A., Isoflavonoids: biochemistry, molecular biology, and biological functions. In: Comprehensive Natural Products Chemistry (U. Sankawa, ed.), Elsevier, Oxford. 1999, pp. 773-823
7. BELL, C.J., DIXON, R.A., FARMER, A.D., FLORES, R., INMAN, J., GONZALES, R.A., HARRISON, M.J., PAIVA, N.L., SCOTT, A.D., WELLER, J.W., MAY, G.D., The *Medicago* Genome Initiative: A model legume database, *Nuc. Acids Res.*, 2001, **29**, 114-117.
8. INMAN, J.T., FLORES, H.R., MAY, G.D., WELLER, J.W., BELL, C.J., A high-throughput distributed DNA sequence analysis and database system, *IBM Systems J.*, 2001, **40**, 464-486.
9. SCHENA, M., SHALON, D., DAVIS, R.W., BROWN, P.O., Quantitative monitoring of gene expression patterns with a complementary DNA microarray, *Science.*, 1995, **270**, 467-470.
10. DERISI, J.L., IYER, V.R., BROWN, P.O., Exploring the metabolic and genetic control of gene expression on a genomic scale, *Science.*, 1997, **278**, 680-686.
11. KEHOE, D.M., VILLAND, P., SOMERVILLE, S., DNA microarrays for studies of higher plants and other photosynthetic organisms, *Trends Plant Sci.*, 1999, **4**, 38-41.
12. DURRANT, W.E., ROWLAND, O., PIEDRAS, P., HAMMOND-KOSACK, K.E., JONES, J.D.G., cDNA-AFLP reveals a striking overlap in race-specific resistance and wound response gene expression profiles, *Plant Cell.*, 2000, **12**, 963-977.
13. MATSUMURA, H., NIRASAWA, S., TERAUCHI, R., Transcript profiling in rice (*Oryza sativa* L.) seedlings using serial analysis of gene expression (SAGE), *Plant J.*, 1999, **20**, 719-726.
14. ISHII, M., HASHIMOTO, S.I., TSUTSUMI, S., WADA, Y., MATSUSHIMA, K., KODAMA, T., ABURATANI, H., Direct comparison of genechip and SAGE on the quantitative accuracy in transcript profiling analysis, *Genome.*, 2000, **2**, 136-143.

15. LIU, Z., HALL, J.D., MOUNT, D.W., *Arabidopsis* uvh3 gene is a homolog of the *Saccharomyces cerevisiae* rad2 and human xpg DNA repair genes, *Plant J.*, 2001, **26**, 329-338.
16. JOHN, C.F., MORRIS, K., JORDAN, B.R., Ultraviolet-B exposure leads to up-regulation of senescence-associated genes in *Arabidopsis thaliana*, *J. Exp. Bot.*, 2001, **52**, 1367-1373.
17. COOLEY, N.M., HOLMES, M.G., ATTRIDGE, T.H., Growth and stomatal responses of temperate meadow species to enhanced levels of UV-a and UV-a+b radiation in the natural environment, *J. Photochem. Photobiol.*, 2000, **57**, 179-185.
18. BOCCALANDRO, H.E., MAZZA, C.A., MAZZELLA, M.A., CASAL, J.J., BALLARE, C.L., Ultraviolet B radiation enhances a phytochrome-b-mediated photomorphogenic response in *Arabidopsis*, *Plant Physiol.*, 2001, **2**, 780-788.
19. WADE, H.K., BIBIKOVA, T.N., VALENTINE, W.J., JENKINS, G.I., Interactions within a network of phytochrome, cryptochrome and UV-b phototransduction pathways regulate chalcone synthase gene expression in *Arabidopsis* leaf tissue, *Plant J.*, 2001, **6**, 675-685.
20. BIEZA, K., LOIS, R., An *Arabidopsis* mutant tolerant to lethal ultraviolet-b levels shows constitutively elevated accumulation of flavonoids and other phenolics, *Plant Physiol.*, 2001, **126**, 1105-1115.
21. VELCULESCU, V.E., ZHANG, L., ZHOU, W., VOGELSTEIN, J., BASRAI, M.A., BASSETT, D.E., HIETER, P., VOGELSTEIN, B., KINZLER, K.W., Characterization of the yeast transcriptome, *Cell*, 1997, **88**, 243-251.
22. HOLLOSY, F., Effects of ultraviolet radiation on plant cells, *Micron.*, 2002, **33**, 179-197.
23. O'FARRELL, P.H., High resolution two-dimensional electrophoresis, *J. Biol. Chem.*, 1975, **250**, 4007-4021.
24. BLACKSTOCK, W.P., WEIR, M.P., Proteomics: quantitative and physical mapping of cellular proteins, *Trends Biotech.*, 1999, **17**, 121-127.
25. KLOSE, J., KOBALZ, U., Two-dimensional electrophoresis of proteins: An updated protocol and implications for a functional analysis of the genome, *Electrophoresis*, 1995, **16**, 1034-1059.
26. THIELLEMENT, H., BAHRMAN, N., DAMERVAL, C., PLOMION, C., ROSSIGNOL, M., SANTONI, V., DeVIENNE, D., ZIVY, M., Proteomics for genetic and physiological studies in plants, *Electrophoresis.*, 1999, **20**, 2013-2026.
27. WAGONER, W., LOSCHKE, D.C., HADWIGER, L.A., Two-dimensional electrophoretic analysis of *in vivo* and *in vitro* synthesis of proteins in peas inoculated with compatible and incompatible *Fusarium solani*, *Physiological Plant Path.*, 1982, **20**, 99-107.
28. MUELLER-URI, F., PARTHIER, B., NOVER, L., Jasmonate-induced alteration of gene expression in barley leaf segments analyzed by *in vivo* and *in vitro* protein synthesis, *Planta*, 1988, **176**, 241-247.
29. YATES, J.R., Mass spectrometry and the age of the proteome, *J. Mass Spectrom.*, 1998, **33**, 1-19.

30. WOLF, B.P., SUMNER, L.W., SHIELDS, S.J., NIELSEN, K., GRAY, K.A., RUSSELL, D.H., Characterization of proteins utilized in the desulfurization of petroleum products by matrix-assisted laser desorption ionization time-of-flight mass spectrometry, *Anal. Biochem.*, 1998, **260**, 117-127.
31. JENSEN, O.N., PODTELEJNIKOV, A., MATTHIAS-MANN, M., Delayed extraction improves specificity in database searches by matrix-assisted laser desorption/ionization peptide maps, *Rapid Comm. Mass Spectrom.*, 1996, **10**, 1371-1378.
32. TRETHEWEY, R.N., KROTZKY, A.J., WILLMITZER, L., Metabolic profiling: A rosetta stone for genomics?, *Curr. Opin. Plant Biol.*, 1999, **2**, 83-85.
33. SUMNER, L., PAIVA, N.L., DIXON, R.A., GENO, P.W., High-performance liquid chromatography/continuous-flow liquid secondary ion mass spectrometry of flavonoid glucosides in leguminous plant extracts, *J. Mass Spectrom.*, 1996, **31**, 472-485.
34. BARNES, K.A., SMITH, R.A., WILLIAMS, K., DAMANT, A.P., SHEPHERD, M.J.A., Microbore high performance liquid chromatography/electrospray ionization mass spectrometry method for the determination of the phytoestrogens genistein and daidzein in comminuted baby foods and soya flour, *Rapid Comm. Mass Spectrom.*, 1998, **12**, 130-138.
35. WATSON, D.G., PITT, A.R., Analysis of flavonoids in tablets and urine by gas chromatography/mass spectrometry and liquid chromatography/mass spectrometry, *Rapid Comm. Mass Spectrom.*, 1998, **12**, 153-156.
36. WEIGEL, D., AHN, J.H., BLAZQUEZ, M.A., BOREVITZ, J.O., CHRISTENSEN, S.K., FANKHAUSER, C., FERRANDIZ, C., KARDIALSKY, I., MALANCHARUVIL, E.J., NEFF, M.M., NGUYEN, J.T., SATO, S., WANG, Z.-Y., XIA, Y., DIXON, R.A., HARRISON, M.J., LAMB, C.J., YANOFSKY, M.F., CHORY, J., Activation tagging in *Arabidopsis*, *Plant Physiol.*, 2000, **122**, 1003-1013.
37. KARDAILSKY, I., SHUKLS, V.K., AHN, J.H., DAGENAIS, N., CHRISTENSEN, S.K., NGUYEN, J.T., CHORY, J., HARRISON, M.J., WEIGEL, D., Activation tagging of the floral inducer *FT*, *Science*, 1999, **286**, 1962-1965.
38. DAWSON, W.O., Gene silencing and virus resistance: a common mechanism, *Trends Plant Sci.*, 1996, **1**, 107-108.
39. VOINNET, O., BAULCOMBE, D.C., Systemic signaling in gene silencing, *Nature*, 1997, **389**, 553.
40. BURTON, R.A., GIBEAUT, D.M., BACIC, A., FINDLAY, K., ROBERTS, K., HAMILTON, A., BAULCOMBE, D.C., FINCHER, G.B., Virus-induced silencing of a plant cellulose synthase gene, *Plant Cell.*, 2000, **12**, 691-705.

Chapter Twelve

STRUCTURALLY GUIDED ALTERATION OF BIOSYNTHESIS IN PLANT TYPE III POLYKETIDE SYNTHASES

Joseph P. Noel,[1,3,4] Joseph M. Jez,[1] Michael B. Austin,[1,3] Marianne E. Bowman,[1] and Jean-Luc Ferrer[2]

[1]*Structural Biology Laboratory, The Salk Institute for Biological Studies, 10010 North Torrey Pines Road, La Jolla, CA 92037, USA.*

[2]*IBS/LCCP, 41 rue Jules Horowitz, 38027 Grenoble cedex 1, France.*

[3]*Department of Chemistry and Biochemistry, University of California, San Diego, La Jolla, CA 92037-0634, USA.*

[4]*Author for correspondence, e-mail: noel@sbl.salk.edu*

Introduction	198
Chalcone Synthase / Stilbene Synthase Family of Plant Polyketide Synthases	199
General Catalytic Features	199
Chalcone Synthase, a Model Plant Polyketide Synthase	199
An Expanding Family of Type III Polyketide Synthases	203
Architecture and Mechanism of Chalcone Synthase	203
Scaffold for Successive Claisen Condensations	203
Structurally Derived Mechanism of Chalcone Formation	204
Manipulation of Plant Polyketide Biosynthesis	208
Chain Initiation and Starter Molecule Selection	208
Polyketide Chain Length Determination	209
Initiation and Chain Extension in 2-Pyrone Synthase	211
Cyclization by Aldol Condensation in Stilbene Synthase	214
Summary and Future Directions	218

INTRODUCTION

Polyketides comprise a large family of structurally diverse natural products that possess a wide range of biological activities in host organisms and, in purified form, constitute an important class of pharmacologically active compounds.[1] The structural diversity of these molecules results from varying the length and constituents of the polyketide mediated by enzymes that modify the final chemical scaffold. In plants, exploitation of polyketide biosynthetic pathways offers new avenues for altering disease resistance, raising micronutrient levels in crops, and enhancing production of known and novel nutraceuticals and pharmaceuticals in transgenic plants.[2,3] These agriculturally and therapeutically important natural products are synthesized by chain initiation with an acyl-CoA starter unit and successive Claisen condensations of extender units, derived from (methyl)malonyl-coenzyme A (CoA).

Three types of polyketide synthases (PKSs) are widely recognized and are classified based upon their unique architecture and gene structure.[4] In analogy to the fatty acid synthases (FASs), the distinction between the phylogenetically related type I and type II PKSs resides in the organization of their respective catalytic sites. Type I PKSs encompass biosynthetic machinery consisting of one or more multifunctional proteins that contain a different active site for each enzyme-catalyzed reaction used during polyketide chain assembly and modification. These lengthy polypeptide chains fall into two subgroups, the modular type I PKSs of bacteria [*i.e.*, 6-deoxyerythronolide B synthase (DEBS)][5] and the iterative type I PKSs of fungi [*i.e.*, 6-methylsalicylic acid synthase (MSAS)].[6] In contrast, the type II PKS system encompasses considerably smaller proteins. Each protein component includes a single active site that is used iteratively during the biosynthesis of polycyclic aromatic products (*i.e.*, actinorhodin and oxytetracycline).[7]

Recently, a new polyketide biosynthetic pathway in bacteria that parallels the well studied plant PKSs has been discovered that can assemble small aromatic metabolites.[8,9] These type III PKSs[10] are members of the chalcone synthase (CHS) and stilbene synthase (STS) family of PKSs previously thought to be restricted to plants.[11] The best studied type III PKS is CHS. Physiologically, CHS catalyzes the biosynthesis of 4,2',4',6'-tetrahydroxychalcone (chalcone). Moreover, in some organisms CHS works in concert with chalcone reductase (CHR) to produce 4,2',4'-trihydroxychalcone (deoxychalcone) (Fig. 12.1). Both natural products constitute plant secondary metabolites that are used as precursors for the biosynthesis of anthocyanin pigments, anti-microbial phytoalexins, and chemical inducers of *Rhizobium* nodulation genes.[12]

In this chapter, we describe the atomic resolution structural elucidation of several plant type III polyketide synthases, including chalcone synthase, 2-pyrone synthase, and stilbene synthase. Manipulation of the catalytic activity and specificity of these biosynthetic enzymes by using a structurally guided approach offers a novel means of predictably altering the natural chemical diversity that exists in important crop plants.

CHALCONE SYNTHASE / STILBENE SYNTHASE FAMILY OF PLANT POLYKETIDE SYNTHASES

General Catalytic Features

Members of the CHS/STS family of condensing enzymes are relatively modest-sized proteins of 40-47 kDa that function as homodimers. Each enzyme typically reacts with a cinnamoyl-CoA starter unit and catalyzes three successive chain extensions with reactive acetyl groups derived from enzyme catalyzed decarboxylation of malonyl-CoA.[11] Release of the resultant tetraketide together with or prior to polyketide chain cyclization and/or decarboxylation yields chalcone or resveratrol (a stilbene). Notably, CHS and STS catalyze identical reactions up to the formation of the intermediate tetraketide. Divergence occurs during the termination step of the biosynthetic cascade as each tetraketide intermediate undergoes a distinct cyclization reaction (Fig. 12.2).

CHS is ubiquitous in higher plants, while STS is more restricted and has likely evolved from CHS on at least three separate occasions.[13] Recently, the atomic resolution crystal structure of a CHS, CHS2 from the legume *Medicago sativa* (alfalfa), was determined by our group.[14] This structural study provides important regiochemical information concerning the reaction mechanism of plant polyketide synthases and serves as a necessary structural template for understanding the growing diversity of CHS-like enzymes involved in numerous secondary metabolic pathways.[15] Several of these new additions to the CHS/STS family that have emerged from plants deviate from the chalcone and stilbene biosynthetic model by utilizing non-phenylpropanoid starter units, varying the number of condensation reactions, and having different cyclization patterns (*e.g.*, acridone, 2-pyrone, and coumaroyltriacetic acid synthases).[16-18] Thus, plant enzymes in the CHS/STS family are growing in number and function and are not limited to cinnamoyl-CoA starter units.

Chalcone Synthase, a Model Plant Polyketide Synthase

The catalytic residues of *Medicago sativa* CHS2 sit at the intersection of the CoA binding tunnel and a large internal cavity that accommodates the growing polyketide chain (Fig. 12.3).[14] Mechanistic and crystallographic studies confirm the importance of Cys164 as the polyketide attachment site and indicate that His303 and Asn336 catalyze the decarboxylation of malonyl-CoA and facilitate extension of the

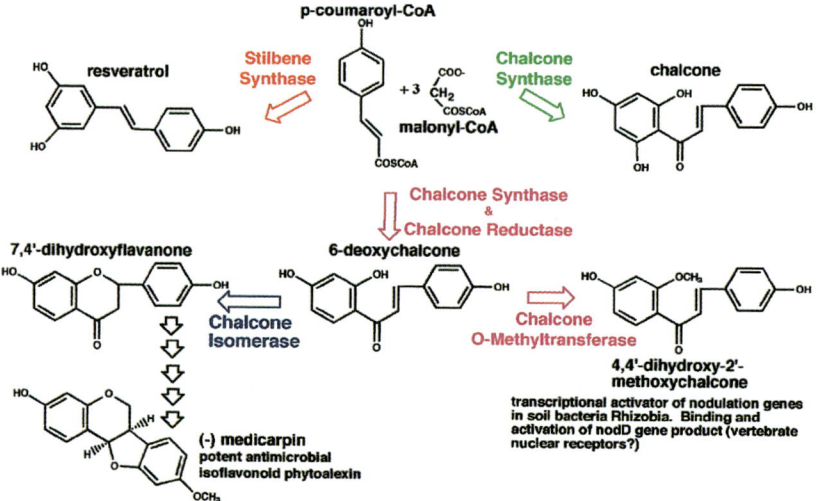

Figure 12.1: A brief summary of representative phenylpropanoid pathways.

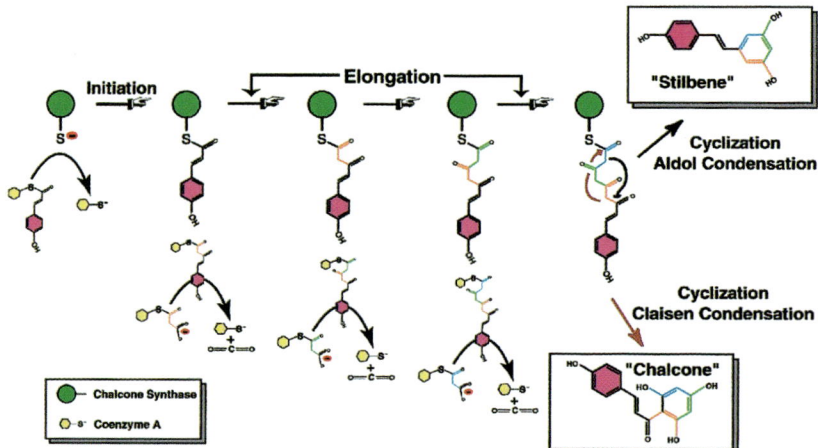

Figure 12.2: Schematic representation of the CHS and STS reactions. The reaction pathway highlights the initiation, elongation, and termination phases of the polyketide extension reaction.

SYNTHESIS – POLYKETIDE SYNTHASES

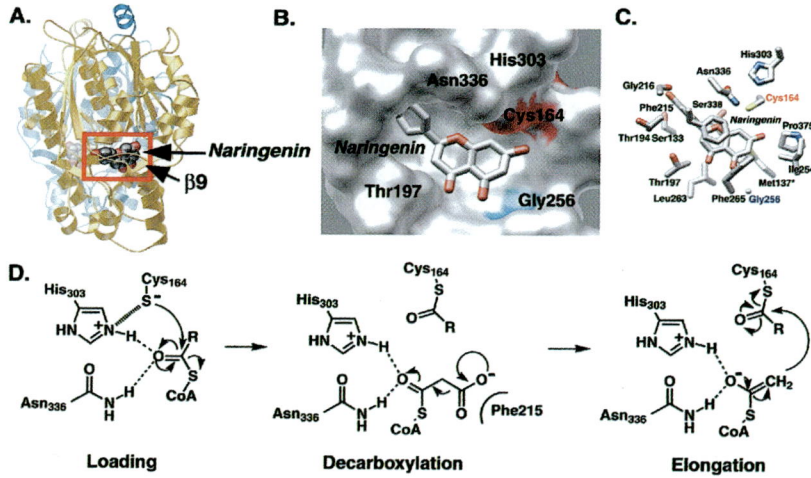

Figure 12.3: See following page.

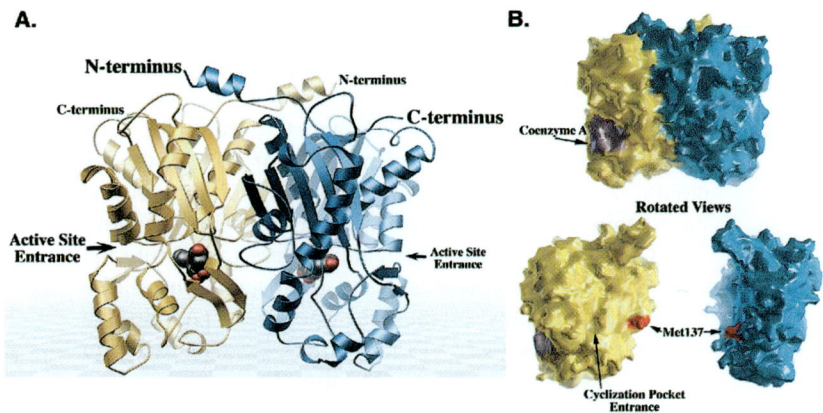

Figure 12.4: See following page.

Figure 12.3: Overall CHS-like fold and reaction mechanism. A. Ribbon diagram of CHS complexed with (2S)-naringenin (space-filling model). B. Surface representation of the active site cavity showing the complementary shape and size to naringenin. Surfaces corresponding to Phe215, Ile254, and Phe265 have been removed for clarity. The positions of Cys164 and Gly256 are indicated in red and blue, respectively. C. Close up view of residues forming the active site cavity. This orientation is slightly rotated from that shown in A and is positioned looking down the CoA binding tunnel into the active site. The coumaroyl-binding pocket is on the left side of the cavity. The right side of the pocket accommodates the growing polyketide chain and is the site where cyclization to chalcone occurs. D. Proposed reaction mechanism of CHS. The loading, decarboxylation, and elongation steps are shown. R is the coumaroyl moiety in the first reaction cycle, coumaroyl-acetyl group in the second cycle, and a coumaroyl-diacetyl group in the final cycle.

Figure 12.4: Overall architecture of the CHS dimer. A. Each monomer is colored in gold and blue, respectively, in this ribbon diagram. The N- and C-termini for each monomer are indicated. The position of the active site cavity is indicated by the position of the bound naringenin molecule (shown as a space-filling model). The CoA binding tunnel provides access to the internal cavity. B. Molecular surface representation of the CHS-CoA complex oriented as shown in (A). In the bottom panel, the two CHS monomers are separated and rotated slightly to highlight the flat dimerization interface along with the methionine side chain and dyad related hole in the backside of the CHS active site.

polyketide intermediates (Fig. 12.3D).[15,19] Conservation of these amino acid residues among all type III PKSs implies that these enzymes use a core set of reactions for starter molecule loading, malonyl-CoA decarboxylation, and polyketide chain elongation.[8,15]

SYNTHESIS – POLYKETIDE SYNTHASES

The CoA binding site juxtaposes activated CoA-linked thioesters with the bilobed initiation/elongation/cyclization cavity of CHS. One lobe of this cavity forms a coumaroyl-binding pocket, and the other accommodates the growing polyketide chain (Fig. 12.3A-B). The volume and shape of the initiation/elongation/cyclization cavity governs starter molecule selectivity, polyketide chain length, and the folding and cyclization pathway of the polyketide chain in different type III PKSs.

An Expanding Family of Type III Polyketide Synthases

Over the past several years, twelve CHS-like bacterial proteins have been identified, suggesting that this new polyketide biosynthetic pathway may also be widespread in bacteria. Genome sequence analyses of the model Gram-positive bacteria *Streptomyces coelicolor* (three genes), *Bacillus subtilis* (one gene), the human pathogen *Mycobacterium tuberculosis* (three genes), *Bacillus halodurans* (one gene), and *Deinococcus radiodurans* (one gene) revealed a number of CHS/STS homologues of unknown functions.[10]

Although none of the total genome-based type III PKSs has been biochemically characterized to date, three type III PKSs associated with secondary metabolic biosynthetic gene clusters have been functionally examined, and specific information exists as to the substrate and product specificity of each.[10] The structural and functional similarities of the bacterial type III PKSs with plant enzymes, such as CHS and STS, implies a common origin that may have evolved through horizontal gene transfer between plant-associated bacteria and higher plants. These suggestions are supported by distance matrix calculations and maximum likelihood analysis.[10]

ARCHITECTURE AND MECHANISM OF CHALCONE SYNTHASE

Scaffold for Successive Claisen Condensations

The 1.56 Å resolution crystal structure of alfalfa CHS2 reveals that the enzyme forms a symmetric dimer and provides a structural archetype for the type III PKS family.[14] The dimer interface is a flat surface delineated by two structural features. First, the N-terminal α-helix of monomer A entwines with the corresponding α-helix of monomer B. Second, a tight loop containing a *cis*-peptide bond between Met137 and Pro138 exposes the methionine side chain as a knob on the monomer surface. Across the interface, Met137 protrudes into a hole found in the surface of the adjoining monomer to form part of the active site cavity involved in polyketide chain cyclization (Fig. 12.4).

Structures of CHS complexed with different Coenzyme A (CoA) thioesters and product analogs (*i.e.*, naringenin and resveratrol) demonstrate that the active site is buried within an interior cavity located in the cleft between the upper and lower domains of each monomer (Fig. 12.3). Considering the complexity of the reaction mechanism leading to chalcone formation, there are remarkably few chemically reactive amino acids in the active site. In particular, four residues conserved in the known CHS-related enzymes (Cys164, Phe215, His303, and Asn336) define the catalytic machinery of CHS (described below). Access to the active site cavity is gained through a 16 Å long tunnel that forms the CoA binding site. Structures of CHS complexed with CoA, acetyl-CoA, and hexanoyl-CoA reveal that the pantetheine arm of each ligand extends through the tunnel to position the thioester-linked substrates near the active site cysteine.

Each CHS monomer consists of two structural domains (Fig. 12.5, left). The upper domain exhibits the α-β-α-β-α pseudo-symmetric motif observed in fatty acid β-ketoacyl synthases (KASs) (Fig. 12.5, right).[20] Both CHS and KAS use a cysteine as a nucleophile in the condensation reaction, and shuttle reaction intermediates via CoA thioester-linked molecules or ACPs, respectively. The conserved architecture of the upper domain maintains the three-dimensional position of the catalytic residues of each enzyme; Cys164, His303, and Asn336 in CHS correspond to a Cys, His, and His in KAS I and II.

These catalytic residues are also conserved in the sequences of the ketosynthase domains from modular and aromatic PKS (Fig. 12.5B). The structural differences in the lower domain of CHS create a larger active site cavity than that of KAS I and II, and this more spacious elongation cavity provides the necessary room for the intermediate tetraketide prior to formation of chalcone. In contrast, the active site of KAS II depicted in Figure 12.5 catalyzes the condensation reaction that elongates palmoleitic acid (C16:1) by a single acetate unit to *cis*-vaccenic acid (C18:1). The similar structural features and chemistry of these enzymes imply a common evolutionary origin for the CHS-like enzymes and the ketosynthases involved in fatty acid and polyketide biosynthesis. Most recently, KAS III (FabH), which provides ACP activated diketides to bacterial fatty acid synthases, has been structurally characterized.[21] Notably, this KAS III enzyme, unlike the previously characterized KAS I and II proteins, maintains a three dimensional architecture nearly identical to that of CHS. This conserved structural motif includes both the upper and lower domains depicted in the left panel of Figure 12.5A.

Structurally Derived Mechanism of Chalcone Formation

CHS orchestrates the condensation, cyclization, and aromatization of one *p*-coumaroyl-CoA and three malonyl-CoA molecules to produce chalcone (Fig. 12.2).[22] Transfer of the *p*-coumaroyl moiety from the CoA-linked starter molecule to Cys164 within the active site initiates the reaction sequence. Next, the sequential condensation of three acetate units, derived from malonyl-CoA, with the enzyme-bound coumaroyl moiety forms a tetraketide intermediate. Inherent in the condensation reaction is decarboxylation of malonyl-CoA to an acetyl-CoA carbanion that serves as a nucleophile during the successive chain elongation reactions. Four amino acids (Cys164, Phe215, His303, and Asn336), situated at the intersection of the CoA-binding tunnel and the active site cavity, play essential and distinct roles during malonyl-CoA decarboxylation and chalcone formation (Fig. 12.3D). A series of functional studies that focused on the properties of site-directed mutants of these residues established roles for each amino acid side chain during polyketide formation in the CHS/STS family of PKSs.[15,19,23,24]

During the initial loading reaction, the nucleophilic thiolate anion of Cys164 attacks the thioester carbonyl, resulting in transfer of the coumaroyl moiety to the cysteine side chain. The thiolate anion of Cys164 is maintained by an ionic interaction with the imidazolium cation of His303. In turn, His303 and Asn336 form hydrogen bonds with the thioester carbonyl oxygen, further enhancing the formation of the tetrahedral transition state for reaction initiation. Collapse of the transition state expels CoA, which dissociates from the enzyme, leaving a coumaroyl-thioester linked to Cys164 (Fig. 12.3D).

Next, malonyl-CoA binds, and its bridging methylene carbon is positioned near the carbonyl carbon of the enzyme-bound coumaroyl-thioester. Through its side chain amide group, Asn336 forms hydrogen bonds that orient the thioester carbonyl of malonyl-CoA near His303, while Phe215 provides a non-polar environment for the terminal carboxylate. This latter interaction likely facilitates decarboxylation by shifting the negative charge centered on the carboxylate moiety to the thioester carbonyl oxygen bounded by His303 and Asn336. In addition, the phenyl ring of Phe215 likely assists in the necessary reorientation of the sigma bond of malonyl-CoA that undergoes cleavage as decarboxylation proceeds. Ideally, the bridging methylene is positioned perpendicular to the thioester carbonyl group sequestered by His303 and Asn336.

During the elongation step, attack of the newly formed carbanion on the carbonyl carbon of the enzyme-bound coumaroyl thioester releases the thiolate anion of Cys164, while the coumaroyl group is transferred to the acetyl moiety of the CoA thioester. As for the initial loading stage of polyketide formation, hydrogen bonds from His303 and Asn336 stabilize the tetrahedral transition state during this transfer reaction. Recapture of the elongated coumaroyl-acetyl-diketide-CoA by Cys164 and the subsequent dissociation of CoA set the stage for two additional rounds of malonyl-CoA decarboxylation and polyketide elongation, resulting in formation of the final tetraketide reaction intermediate (Fig. 12.2).

The final step in chalcone formation involves an intramolecular Claisen condensation that encompasses the three acetate units derived from malonyl-CoA. During cyclization, the nucleophilic methylene group nearest the coumaroyl moiety attacks the carbonyl carbon of the thioester linked to Cys164. Ring closure is proposed to proceed through an internal proton transfer from the nucleophilic carbon to the carbonyl oxygen.[14] Breakdown of this tetrahedral intermediate expels the newly formed ring system from Cys164. Subsequent aromatization of the trione ring

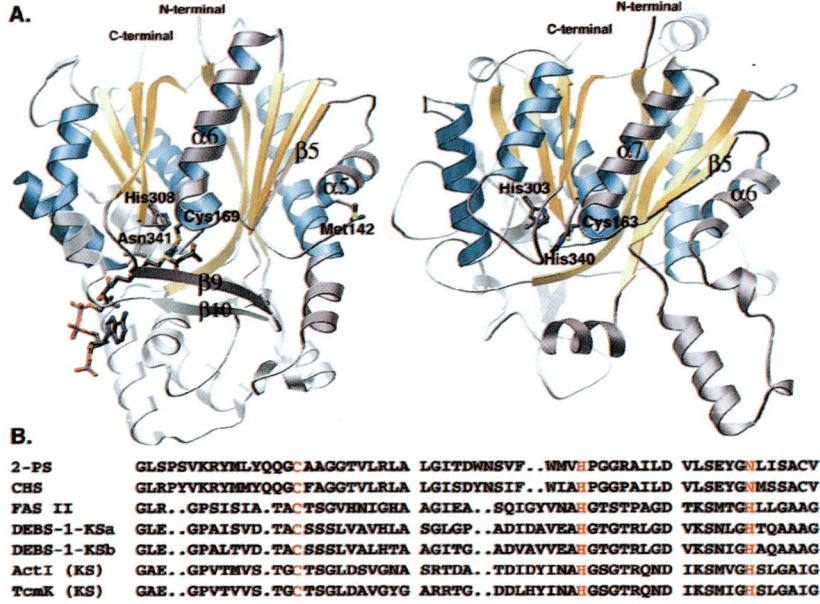

```
2-PS          GLSPSVKRYMLYQQGCAAGGTVLRLA  LGITDWNSVF..WMVHPGGRAILD  VLSEYGNLISACV
CHS           GLRPYVKRYMMYQQGCFAGGTVLRLA  LGISDYNSIF..WIAHPGGPAILD  VLSEYGNMSSACV
FAS II        GLR..GPSISIA.TACTSGVHNIGHA  AGIEA..SQIGYVNAHGTSTPAGD  TKSMTGHLLGAAG
DEBS-1-KSa    GLE..GPAISVD.TACSSSLVAVHLA  SGLGP..ADIDAVEAHGTGTRLGD  VKSNLGHTQAAAG
DEBS-1-KSb    GLE..GPALTVD.TACSSSLVALHTA  AGITG..ADVAVVEAHGTGTRLGD  VKSNIGHAQAAAG
ActI (KS)     GAE..GPVTMVS.TGCTSGLDSVGNA  SRTDA..TDIDYINAHGSGTRQND  IKSMVGHSLGAIG
TcmK (KS)     GAE..GPVTVVS.TGCTSGLDAVGYG  ARRTG..DDLHYINAHGSGTRQND  IKSMIGHSLGAIG
```

Figure 12.5: A. Comparison of the CHS monomer (left) and β-ketoacyl synthase monomer (right). The structurally conserved secondary structure of each monomer's upper domain is colored in blue (α-helix) and gold (β-strand). Portions of each protein monomer forming the dimer interface are colored purple. The side-chains of the catalytic residues of CHS (Cys164, His303, Asn336) and β-ketoacyl synthase (Cys163, His303, His340) are shown. B. Sequence conservation of the catalytic residues of CHS, 2-PS, β-ketoacyl synthase (FAS II), and the ketosynthase modules of 6-deoxyerythronolide B synthase (DEBS), actinorhodin synthase (ActI) and tetracenomycin synthase (TcmK). The catalytic residues are in red.

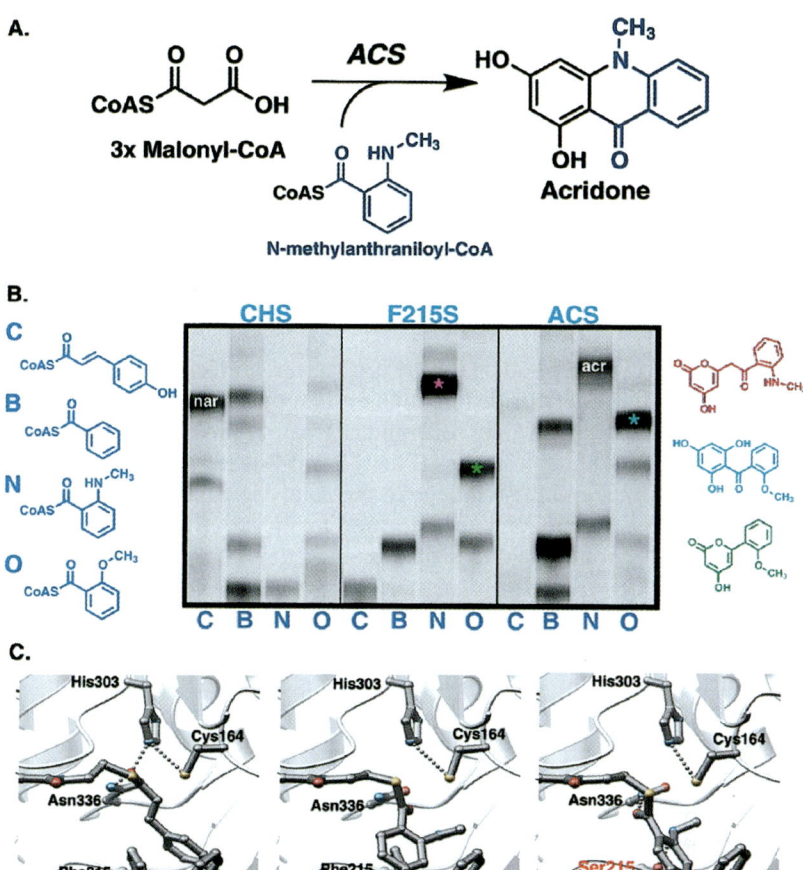

Figure 12.6: Starter molecule engineering. A. Reaction catalyzed by ACS. B. Thin layer chromatography screening for enzymatic activity with different starter molecules. C. Views illustrate the active site of the F215S mutant (right), wild-type CHS with N-methylanthraniloyl-CoA (center), and wild-type CHS with *p*-coumaroyl-CoA (left) modeled at the active site entrances. The catalytic residues, Cys 164, His 303, and Asn 336, and Phe 265 are shown. In wild-type CHS, N-methylanthraniloyl-CoA clashes with Phe 215 to prevent the CoA thioester from adopting the conformation depicted in (A). The wild-type - *p*-coumaroyl-CoA model emphasizes the ability of the propanoid linker to extend the phenolic ring deeper into the active site cavity.

through a second series of facile internal proton transfers yields chalcone. The type III PKSs generate molecular diversity in their products by selection of different starter molecules, by varying the length of the polyketide chain, and through regiospecific control of the polyketide cyclization reaction.[10,11] Elucidating how type III PKSs achieve specificity in polyketide formation is essential for manipulating the activities of these enzymes both *in vitro* and *in vivo*.

MANIPULATION OF PLANT POLYKETIDE BIOSYNTHESIS

Chain Initiation and Starter Molecule Selection

In vitro, CHS accepts disparate CoA starter molecules, including aromatic and aliphatic CoA-thioesters of different chain lengths.[25-28] The CoA binding tunnel extending from the exterior of CHS to an interior cavity provides access to the buried catalytic residues. Two phenylalanines (Phe215 and Phe265) reside at the juncture of the CoA binding tunnel and the entrance to the catalytic cavity. Phe215 is conserved in all type III PKSs except for benzalacetone synthase (BAS), which maintains a leucine at this position and catalyzes the condensation of a single acetate unit to *p*-coumaroyl-CoA.[29] Given the different starter molecule preferences of other type III PKSs, the identity of residue 215 in natural PKS sequences would not appear at first glance to influence starter molecule selection. On the other hand, residue 265 is a phenylalanine in CHS and almost all type III PKSs, but is a valine in both ACS isoforms from *Ruta graveolens*.[30,31] This difference suggests that Phe265 may play a role in determining starter molecule specificity since ACS initiates its chain extension reactions by using N-methylanthraniloyl-CoA (Fig. 12.6A).

We examined the effect of mutating these two phenylalanines (F215S and F265V) on starter molecule selectivity within the CHS active site. Contrary to our expectations, the F265V mutant does not influence either the broad substrate tolerance of CHS or its selectivity for a sub-class of CoA thioester starter molecules. However, the F215S mutant preferentially uses N-methylanthraniloyl-CoA as a starter molecule to yield a tetraketide lactone (Fig. 12.6B). To date, this activity of the CHS F215S mutant has not been observed in nature. Structural analysis of the F215S mutant suggests that widening the active site entrance allows binding of the shorter and bulkier N-methylanthraniloyl-CoA starter molecule in a catalytically productive conformation that is situated near Cys164, His303, and Asn336. By using the experimentally determined structure of the F215S mutant as a starting point, modeling of the loading stage of the PKS reaction suggests that the productive orientation of the N-methylanthraniloyl moiety is assisted by hydrogen bond formation between the serine side chain at position 215 and the methylamine group of the anthraniloyl moiety (Fig. 12.6C) (Jez and Noel, unpublished observations).

Our results demonstrate that type III PKSs use multiple regiochemical mechanisms for achieving specificity during the loading of a given CoA-thioester starter molecule on the active site cysteine. Furthermore, the success of this structurally guided approach indicates that specific point mutations in the PKS active site can expand the biosynthetic repertoire of these type III PKSs. Further modification of polyketide assembly may be achieved by additional rounds of site specific mutagenesis and targeted randomization of the catalytic surface.

Polyketide Chain Length Determination

Given Gly256's position on the catalytic surface surrounding the elongation pocket of CHS, site-directed mutagenesis was used to replace Gly256 in the CHS initiation-elongation-cyclization cavity with alanine, valine, leucine, or phenylalanine residues (Fig. 12.3B). These mutations were used to establish the relationship between the size of the active site cavity and terminal chain lengths of resultant polyketide products. X-ray crystallographic and catalytic characterization of the G256A, G256V, G256L, and G256F mutants demonstrate that structural alterations in the active site cavity correlate with functional changes in the kinetic and specificity properties of each CHS mutant.[32]

Following formation of a thioester-linked tetraketide from sequential condensation of *p*-coumaroyl-CoA and three malonyl-CoAs, wild type CHS catalyzes an intramolecular Claisen condensation yielding chalcone (Fig. 12.7). In this step, the acidic methylene group nearest the coumaroyl moiety loses a proton, with the resultant carbanion attacking the thioester carbonyl carbon. Acridone synthase,[16] homoeriodictyol/eriodictyol synthase,[33] benzophenone synthase,[34] valerophenone synthase,[35] and 2,4-diacetylphloroglucinol synthase[9] catalyze identical regiospecific Claisen condensation reactions of the malonyl-derived portion of a tetraketide intermediate. Formation of these physiological products requires a defined conformation of the linear tetraketide intermediate for the cyclization reaction to proceed efficiently.

Based on the three-dimensional structure of CHS, we proposed that the initiation/elongation/cyclization cavity serves as a structural template that selectively stabilizes a particular folded conformation of the linear tetraketide, allowing the Claisen condensation to proceed from C6 to C1 of the reaction intermediate.[14] In contrast, CTAL formation can occur either in solution or alternatively while sequestered in the enzyme active site. In either case, enolization of the C5 ketone followed by nucleophilic attack on the C1 ketone with either a hydroxyl group (in solution) or the cysteine thiolate (enzyme bound) as the leaving group results in CTAL. Similar lactones are commonly formed as by-products of *in vitro* reactions in other PKS systems.[36-38]

Structural changes within the elongation/cyclization pocket of CHS affect the ratio of Claisen-derived products versus lactone products, with both physiological, *i.e.*, *p*-coumaroyl-CoA, and non-physiological starter molecules. The product profiles of the G256A and G256V mutants with *p*-coumaroyl-CoA show an increase in the amount of CTAL produced versus the amount of naringenin formed, although the overall catalytic efficiency of each mutant is not dramatically altered. The three-dimensional structures of the G256A and G256V mutants exhibit small variations in the surface topology of the elongation cavity.[32] While at first glance these changes appear insignificant, they most likely alter the conformation of the tetraketide intermediate enough to interfere with chalcone formation. Previous mutagenesis studies of CHS that demonstrate increased CTAL production also suggest that structural differences at other positions in the initiation/elongation/cyclization cavity alter the shape of the tetraketide intermediate.[39,40] For example, mutation of Thr197 to a leucine in CHS causes a complete derailment of the normal tetraketide cyclization reaction resulting in CTAL as the sole product.[40] The current study establishes a direct link between structural changes in the active site cavity and functional differences in the cyclization reactions leading to chalcone and CTAL formation, respectively.

Triketide styrylpyrones are commonly found in fungi and occur as secondary metabolites in *Pinus strobus*, *Equisetum arvense*, and *Piper methylsticum*.[11] Although there is no reported purification or cloning of an authentic styrylpyrone synthase, under certain *in vitro* reaction conditions wild-type CHS forms styrylpyrone products, suggesting this may be a vestigial activity in CHS family members. Originally, formation of *bis*-noryangonin from *p*-coumaroyl-CoA and two malonyl-CoAs by CHS was reported as an artifact of high reductant concentrations in enzyme assays.[41] Later experiments with aromatic starter molecules larger than *p*-coumaroyl-CoA, like feruloyl-CoA, showed that CHS yields triketide styrylpyrones in preference to chalcone-like products.[42] Likewise, when *p*-coumaroyl analogs bearing a halogen in place of the hydroxyl group are used as the starter molecule, the corresponding styrylpyrones are formed.[27] Also, use of phenylacetyl-CoA, hexanoyl-CoA, isovaleryl-CoA, and isobutyryl-CoA as starter molecules for *Pinus strobus* CHS produces triketide lactones in significant amounts.[26]

These results demonstrate that replacement of Gly 256 with larger amino acid residues causes a similar derailment. Since the CHS G256L and G256F mutants produce *bis*-noryangonin, our experiments demonstrate that a smaller active site cavity reduces the number of acetate additions made to the coumaroyl-starter unit in a predictive manner from three to two, resulting in formation of a triketide product.

The overall backbone architecture of the CHS initiation/elongation/cyclization cavity, which is also structurally conserved in *Gerbera hybrida* 2-PS,[40] is a versatile scaffold that through natural selection has generated a robust array of secondary metabolites in various plants and microorganisms. The sequence databases include nearly four hundred CHS-related sequences from plants and microbes; however, the substrate and product specificities of many of the proteins encoded by these sequences remain undetermined.

The cloned and characterized CHS-like enzymes that form tetraketide products all have a glycine at position 256. In comparison, the G256L mutation occurs naturally in *Gerbera hybrida* 2-PS, CHS-A and -B from various *Ipomoea* (morning glory) species, and *Petunia hybrida* CHS-B.[43,44] The metabolic role of the CHS-variants in *Ipomoea* and *Petunia* remains to be established. CHS-G from *Petunia hybrida* presents an alanine side chain at position 256; however, the function of this enzyme is also unresolved.[43] Comparison of evolutionary rate variations among CHS-like enzymes suggests that *Ipomoea* CHS-A and CHS-B and *Petunia* CHS-B have acquired or are evolving new functions.[43-45] Biochemical characterization of the *Ipomoea* and *Petunia* enzymes will likely demonstrate that these enzymes do not produce chalcones, but possibly function in the biosynthesis of other secondary metabolites, such as styrylpyrones.

Notably, natural variation in the type III PKS active site cavity, like that observed in *Ipomoea* and *Petunia*, does not result in functionally impaired enzymes, but in fact, generates catalytically active enzymes that display both altered substrate and product specificities. Sequential increases in the side chain volume of position 256 in alfalfa CHS2 result in decreases in polyketide chain length and predictable shifts in the ratio of tetraketide to triketide reaction products.[32] These results functionally link the volume of the elongation/cyclization lobe in type III PKS to chain length determination.

Initiation and Chain Extension in 2-Pyrone Synthase

The structure of 2-PS complexed with acetoacetyl-CoA was determined by molecular replacement using CHS as a search model and refined to 2.05 Å resolution. The overall fold of 2-PS contains the α-β-α-β-α motif found in CHS.[40] In addition, the positions of the catalytic residues of 2-PS (Cys169, His308, and Asn341), CHS (Cys164, His303, Asn336), and KAS II (Cys163, His303, and His340) are structurally analogous. As expected from sequence homology, the structures of 2-PS and CHS are nearly identical, and they superimpose with an root mean square deviation of 0.64 Å for the two proteins' C_α-atoms. Similar to CHS, the 2-PS dimerization surface buries 1805 Å^2 of surface area per monomer, and a loop containing a *cis*-peptide bond between Met142 and Pro143 allows the methionine of one monomer to protrude into the adjoining monomer's active site.

2-PS and CHS maintain identical catalytic residues and highly conserved CoA binding sites, but form structurally distinct reaction products. Comparison of the initiation/elongation cavities of 2-PS and CHS reveals four amino acid differences between these two proteins that may account for each protein's individual chemistry. In 2-PS, Leu202, Met259, Leu261, and Ile343 replace Thr197, Ile254, Gly256, and Ser338, respectively, of CHS (Fig. 12.8). The triketide methylpyrone was modeled into the 2-PS initiation/elongation cavity, based on the position of acetoacetyl-CoA. When viewed next to the active site cavity of the CHS-naringenin complex structure, the changes in 2-PS are notable and significant. Clearly, the 2-PS

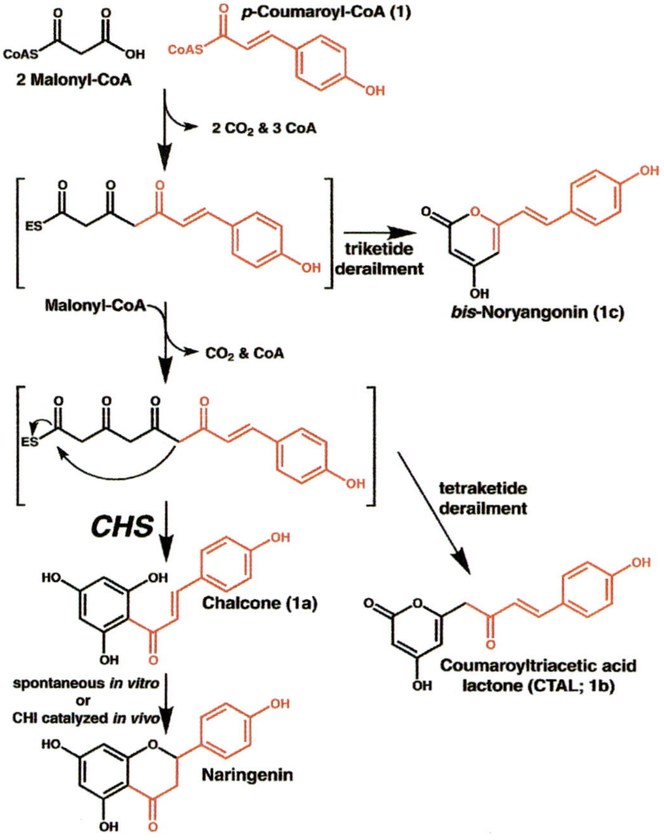

Figure 12.7: Derailment reactions leading to tri- and tetraketide products.

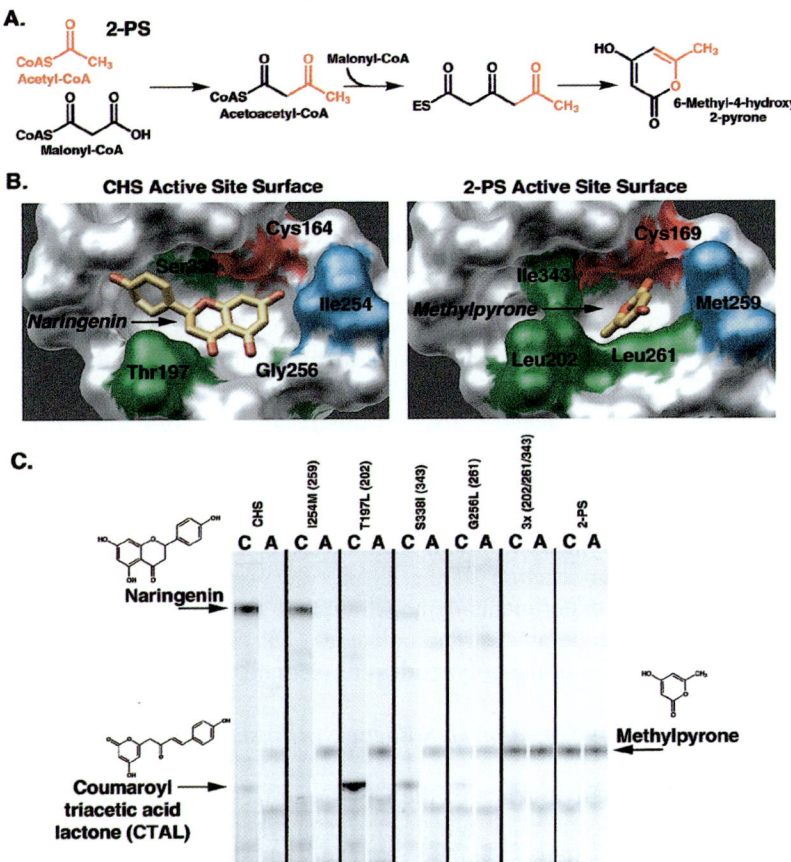

Figure 12.8: A. 2-PS reaction. B. Surface representations of the CHS (left) and 2-PS (right) active site cavities are shown. The catalytic cysteines (red), the three positions that convert CHS into 2-PS (green), and the substitution that does not affect product formation (blue) are highlighted. C. TLC analysis of CHS, 2-PS, and CHS mutant enzymes. The radiogram shows the radiolabeled products produced by incubation of each protein with [^{14}C]malonyl-CoA and either *p*-coumaroyl-CoA (C) or acetyl-CoA (A). Numbering of mutants corresponds to CHS with 2-PS numbering in parenthesis. Positions of reaction products and their identities are indicated.

active site cavity cannot accommodate the larger chalcone product. The four amino acid substitutions reduce cavity volume in CHS to 274 Å^3 in 2-PS. Leu202 and Ile343 occlude the region of the 2-PS cavity corresponding to the coumaroyl-binding site of CHS. Replacement of Gly256 in CHS by Leu261 in 2-PS severely reduces the size of the elongation cavity. Substitution of Met259 in 2-PS for Ile254 in CHS produces only a modest alteration in cavity volume.[40] The x-ray crystal structures of 2-PS and CHS unambiguously identify the structural differences between these two PKSs and imply that the size of the active site cavity controls starter molecule selectivity and limits the chain length in polyketide products.

To demonstrate this principle, the active site cavity of CHS was modified to resemble that of 2-PS by site-directed mutagenesis. Kinetic characterization and identification of reaction products confirmed that a combination of three amino acid substitutions (T197L/G256L/S338I) in CHS changed starter molecule preference from *p*-coumaroyl-CoA to acetyl-CoA, and this resulted in formation of a triketide instead of a tetraketide product (Fig. 12.8C). Surprisingly, introduction of each substitution as a single mutation prevented chalcone formation but did not interfere with generation of the tetraketide intermediate, since the reaction product was identified as coumaroyltriacetic acid lactone (CTAL). In these mutants, the regiospecific cyclization reaction of the tetraketide was derailed. As noted in the previous section, these experiments demonstrate that substitutions of residues lining the active site cavity can derail the intramolecular Claisen condensation reaction to a non-specific lactonization of the polyketide intermediate.

Cyclization by Aldol Condensation in Stilbene Synthase

Duplication and divergence of the CHS gene has given rise to an expanding superfamily of homologous plant and bacterial polyketide synthases (PKSs). One emergent family is the stilbene synthases (STSs), which share 70-90% amino acid identity with CHSs, and produce the medicinally relevant anti-fungal compound resveratrol. CHS and STS both catalyze the sequential addition of three acetate units to a *p*-coumaroyl-CoA starter molecule.[11] Cyclization of the resulting linear tetraketide intermediate in the active site of CHS occurs via an intramolecular Claisen condensation, resulting in covalent linkage of carbon C6 to C1 (numbering from attachment to the active site cysteine). However, in the STS active site this same linear intermediate undergoes an intramolecular aldol condensation, covalently joining carbons C2 and C7 (Fig. 12.2).

The prevailing model of type III PKS divergence predicts that STS will have diverged by introducing changes in the steric bulk of residues lining the active site, thus achieving a different productive conformation of the linear intermediate. However, homology modeling of STS by using the structure of CHS predicts no significant structural differences whatsoever. In fact, no evidence supporting this (or any other) theoretical explanation of STS's aldol condensation mechanism is apparent. To further complicate matters, STS has apparently evolved from CHS on at least three independent occasions, as exemplified by the well-characterized pine, grapevine, and peanut STS subfamilies, each of which exhibit more homology with the CHS sequences of their own species than they do with each other.[13] Comparative amino acid analyses of these three subfamilies has failed to reveal a STS consensus sequence, or provide any useful insight into the structural determinants of cyclization specificity.

To resolve the issue of cyclization specificity, the x-ray crystal structure of the stilbene synthase from pine was determined to atomic resolution. This information allowed the mutagenic conversion of alfalfa CHS to a functional STS, and crystal structures of this engineered STS were solved, in the apo form and with resveratrol bound in the active site (Austin and Noel, unpublished). These experiments support a mechanistic proposal, which prompted further mutagenic and modeling experiments. This work has allowed the elucidation of the structural and mechanistic basis for cyclization specificity (aldol versus Claisen condensation) in the CHS family of type III PKSs.

The first STS structure solved was that of a pinosylvin-forming STS from *Pinus sylvestris*. Pine trees can by-pass the C4H reaction and directly produce the CoA thioester of cinnamate that allows this STS to utilize a non-substituted cinnamoyl-CoA starter *in vivo*.[11] However, when presented *in vitro* with *p*-coumaroyl-CoA, the enzyme proves to be comparable in activity to STS enzymes from organisms that utilize the para-substituted cinnamoyl starter.

This crystal structure, solved to 2.1 Å by molecular replacement with a homology model based on the structure of alfalfa CHS, confirms that no major fold rearrangement has taken place relative to CHS. Aside from a presumably irrelevant one-residue insertion in a remote solvent-exposed loop, overlays reveal four regions of pine STS with backbone movements relative to CHS (Fig. 12.9). One region exhibiting conformational changes resides near the dimer interface along the back wall of the active site. This region spans areas 1-3 shown in Figure 12.9. The observed changes are centered on a conformational kink (area 2) in a seven-residue segment of the ordered polypeptide strand located along the dimer interface. Only minor repositioning of the exposed first and last residues is evident, with the largest displacement occurring near a proline residue introduced into the buried portion of the strand. In pine STS, compensatory changes (relative to CHS) both above (area 3) and below (area 1) this strand are necessary to accommodate the new conformation of area 2. The second region of pine STS that exhibits a modification of the polypeptide backbone position encompasses a four-residue sequence defining one outside edge of the opening to the CoA-binding tunnel (area 4).

The result of the conformational changes in area 2 that require compensatory changes in the strand above (area 3) and the helix below (area 1) creates a slight concavity at the back of the active site. Upon closer examination, this new opening

A.

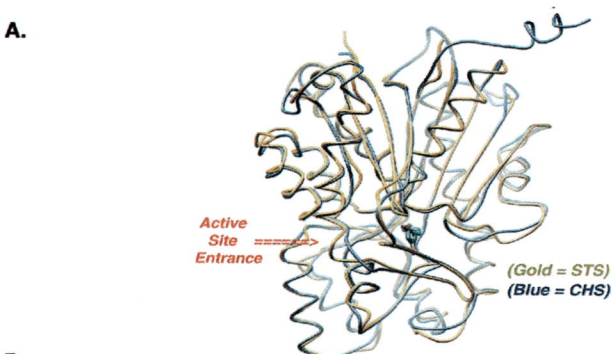

B.
18 mutations in 4 areas converts *M. sativa* CHS to a stilbene synthase

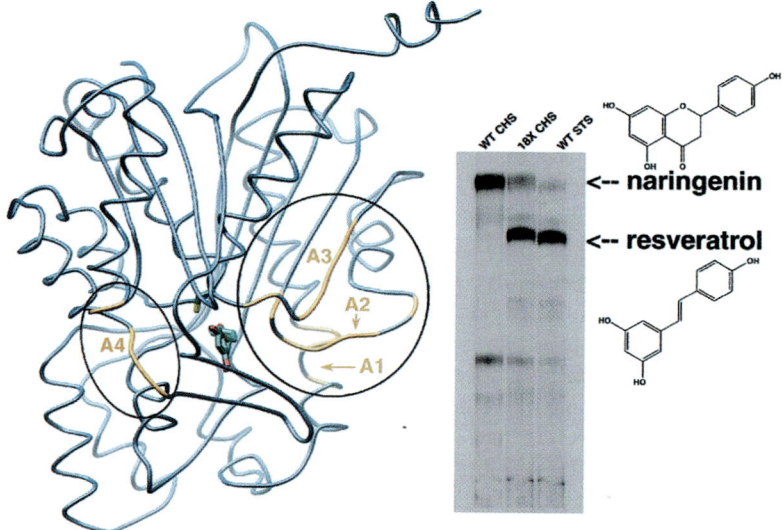

Figure 12.9: Structural comparisons of CHS and STS and functional conversion of CHS into a STS. A. Alpha carbon representation of the CHS and STS monomers. B. Expanded view of the STS monomer with areas 1 (D96A, V98L, V99A, V100M), 2 (T131S, S133T, G134T, V135P, M137L), 3 (Y157V, M158G, M159V, Y160F, Q162H), and 4 (L268K, K269G, D270A, G273D) highlighted. The 18 residues and designated mutants encompassing areas 1-4 were changed in CHS resulting in functional conversion to an efficient STS. The TLC analysis of wild-type CHS, STS, and the 18XCHS mutant illustrate the functional conversion.

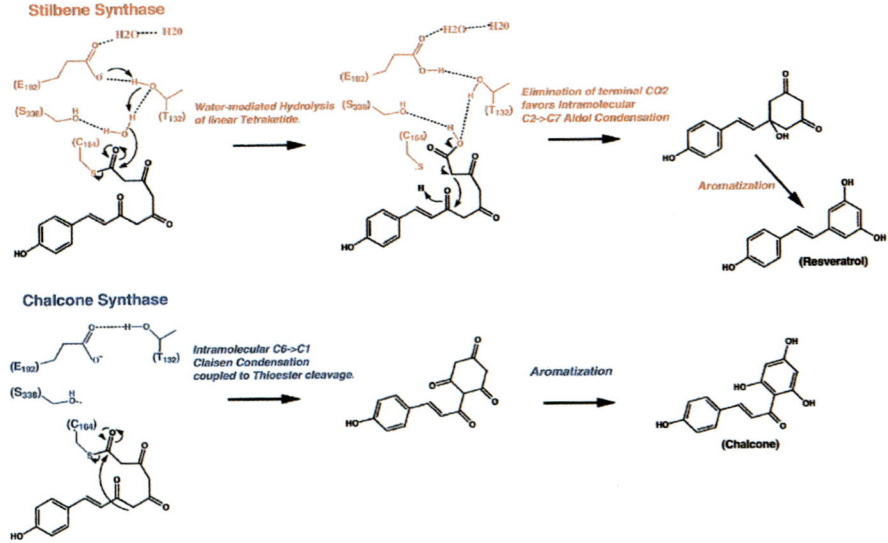

Figure 12.10: Proposed cyclization mechanism in STS and CHS.

in the back wall of the active site serves as the mouth of a back-door tunnel to the outside, and is just large enough to accommodate a line of hydrogen-bonded waters that extend from the surrounding solvent to the backside of the catalytic machinery of STS.

The changes that occur in area 4 are similar to conformational changes that occur in 2-PS. In 2-PS, this change was shown to enhance the binding of CoA, by positioning an arginine over the bound adenine ring and bonded to one of the phosphates of the CoA molecule.[40] In the case of pine STS, a glycine residue replaces this arginine, thus eliminating the possibility of enhanced CoA binding through electrostatic interactions

An extensive mutagenesis protocol was used to probe the contribution of these observed changes in the STS fold towards cyclization specificity in CHS. Initially, an 18X mutant of alfalfa CHS was created to probe the mechanistic relevance of the structural differences observed in STS. Introduction of the pine STS primary sequence, consisting of 18 amino acid changes in areas 1-3, into alfalfa CHS results in an enzyme with similar kinetic efficiency to wild-type CHS.[15] However, the mutant now produces resveratrol as the major product by using coumaroyl-CoA and malonyl-CoA in assays (Fig. 12.9B). Further deconvolution of the necessary mutations in areas 1-3 will ultimately elucidate the precise nature accompanying the structural and mechanistic differences between CHS and STS.

Nevertheless, based upon the observed changes, it appears that the conformational changes seen in STS allow for the positioning of a water molecule near the back wall of the active site abutting the catalytic cysteine. In turn, this water molecule is connected through a conserved threonine residue to the side chain carboxylate of a buried glutamate residue. This arrangement supports a catalytic mechanism that involves prior hydrolysis of the tetraketide intermediate and subsequent ring closure facilitated by decarboxylative condensation (Fig. 12.10). Future structural studies of a number of CHS and STS mutants, combined with mechanistic experiments, will be used to firmly establish the detailed catalytic mechanism accompanying formation of resveratrol in the STS family of plant type III polyketide synthases.

SUMMARY AND FUTURE DIRECTIONS

Polyketide synthases (PKS) produce an array of natural products with different biological activities and pharmacological properties by varying the starter and extender molecules that form the final polyketide. Recent studies of the simplest PKS, the chalcone synthase-like enzymes involved in the biosynthesis of flavonoids, anthocyanin pigments, and anti-microbial phytoalexins, have yielded insight on the molecular basis of this biosynthetic versatility. Understanding the structure-function relationship in these PKSs provides a foundation for manipulating polyketide formation and suggests strategies for further increasing the scope of polyketide biosynthetic diversity.

With the structural and mechanistic information currently available, approaches towards re-engineering polyketide formation in these PKSs can be exploited by using a structurally directed approach. While type III PKSs are architecturally simple, they arguably represent the most sophisticated PKSs mechanistically. Embodied within their homodimeric architecture is the catalytic machinery necessary for starter molecule recognition and loading, for malonyl-CoA decarboxylation and polyketide chain extension, and for multiple pathways accompanying chain termination. Their simple gene and protein architecture makes them amenable to study by using a variety of sophisticated approaches that includes heterologous biosynthesis, *in vitro* and *in vivo* biochemical analysis, directed and random approaches towards enzyme engineering, and atomic resolution protein x-ray crystallography.

ACKNOWLEDGMENTS

I acknowledge the contributions of my dear friends and colleagues including Joachim Schröder of the Institut für Biologie II, Biochemie der Pflanzen, Universitat Freiburg, Richard A. Dixon, Director of the Plant Biology Division, Samuel Roberts Noble Foundation, and Bradley Moore in the Division of Medicinal Chemistry, College of Pharmacy, University of Arizona, who have contributed much to our understanding of type III polyketide synthases. Work in the Noel laboratory has been supported by a grant from the National Science Foundation (MCB9982586) and an innovation grant provided by the Salk Institute for Biological Studies.

REFERENCES

1. O'HAGAN, D., The Polyketide Metabolites, Horwood, Chichester, United Kingdom, 1991.
2. HAIN R., REIF, H.J., KRAUSE, E., LANGEBARTELS, R., KINDL, H., VORNAM, B., WIESE, W., SCHMELZER, E., SCHREIER, P.H., STOCKER, R.H., STENZEL, K., Disease resistance results from foreign phytoalexin expression in a novel plant, *Nature*, 1993, **361**, 153-156.
3. DIXON, R.A., ARNTZEN, C.J., Transgenic plant technology is entering the era of metabolic engineering, *Trends Biotechnol.*, 1997, **15**, 441-444.
4. HOPWOOD, D.A., Genetic contributions to understanding polyketide synthases, *Chem. Rev.*, 1997, **97**, 2465-2497.
5. KHOSLA, C. Harnessing the biosynthetic potential of modular polyketide synthases, *Chem. Rev.*, 1997, **97**, 2577-2590.
6. FUJII, I., Polyketide biosynthesis in filamentous fungi. In: Comprehensive Natural Products Chemistry, Vol. 1, Polyketides and Other Secondary Metabolites Including Fatty Acids and Their Derivatives (U. Sankawa ed.), Elsevier, Amersterdam, 1999, pp. 409-441.
7. RAWLINGS, B.J., Biosynthesis of polyketides (other than actinomycete macrolides), *Nat. Prod. Rep.*, 1999, **16**, 425-484.
8. FUNA, N., OHNISHI, Y., FUJII, I., SHIBUYA, M., EBIZUKA, Y., HORINOUCHI, S., A new pathway for polyketide synthesis in microorganisms, *Nature*, 1999, **400**, 897-899.
9. BANGERA, M.G., THOMASHOW, L.S., Identification and characterization of a gene cluster for the synthesis of the polyketide antibiotic 2,4-diacetylphloroglucinol from *Pseudomonas fluorescens* Q2-87, *J. Bacteriol.*, 1999, **181**, 3155-3163.
10. MOORE, B.S., HOPKE, J.N., Discovery of a new bacterial polyketide biosynthetic pathway, *Chem. Bio. Chem.*, 2001, **2**, 35-38.
11. SCHRODER, J., The chalcone/stilbene synthase-type family of condensing enzymes. In: Comprehensive Natural Products Chemistry, vol. 1, Polyketides and Other Secondary Metabolites Including Fatty Acids and Their Derivatives (U. Sankawa ed.), Elsevier, Amersterdam, 1999, pp. 749-771.

12. DIXON, R.A., PAIVA, N.L., Stress-induced phenylpropanoid metabolism, *Plant Cell*, 1995, **7**, 1085-1097.
13. TROPF, S., LANZ, T., RENSING, S.A., SCHRODER, J., SCHRODER, G., Evidence that stilbene synthases have developed from chalcone synthases several times in the course of evolution, *J. Mol. Evol.*, 1994, **38**, 610-618.
14. FERRER, J.-L., JEZ, J.M., BOWMAN, M.E., DIXON, R.A., NOEL, J.P., Structure of chalcone synthase and the molecular basis of plant polyketide synthesis, *Nature Struct. Biol.*, 1999, **6**, 775-784.
15. JEZ, J.M., FERRER, J.-L., BOWMAN, M.E., DIXON, R.A., NOEL, J.P., Dissection of malonyl-coenzyme A decarboxylation from polyketide formation in the reaction mechanism of a plant polyketide synthase, *Biochemistry*, 2000, **39**, 890-902.
16. JUNGHANNS, K.T., KNEUSEL, R.E., BAUMERT, A., MAIER, W., GROGER, D., MATERN, U., Molecular cloning and heterologous expression of acridone synthase from elicited *Ruta graveolens* L cell suspension cultures, *Plant Molec. Biol.*, 1995, **27**, 681-692.
17. ECKERMANN, S., SCHRODER, G., SCHMIDT, J., STRACK, D., EDRADA, R.A., HELARIUTTA, Y., ELOMAA, P., KOTILAINEN, M., KILPELAINEN, I., PROKSCH, P., TEERI, T.H., SCHRODER, J., New pathway to polyketides in plants, *Nature*, 1998, **396**, 387-390.
18. AKIYAMA, T., SHIBUYA, M., LIU, H.M., EBIZUKA, Y. p-Coumaroyltriacetic acid synthase, a new homologue of chalcone synthase, from *Hydrangea macrophylla* var. *thunbergii*, *Eur. J. Biochem.*, 1999, **263**, 834-839.
19. JEZ, J.M., NOEL, J.P., Mechanism of chalcone synthase: pKa of the catalytic cysteine and the role of the conserved histidine in a plant polyketide synthase, *J. Biol. Chem.*, 2000, **275**, 39640-39646.
20. HUANG W, JIA, J., EDWARDS, P., DEHESH, K., SCHNEIDER, G., LINDQVIST, Y., Crystal structure of β-ketoacyl-acyl carrier protein synthase II from *E. coli* reveals the molecular architecture of condensing enzymes, *EMBO J.*, 1998, **17**, 1183-1191.
21. DAVIES, C., HEATH, R.J., WHITE, S.W., ROCK, C.O., The 1.8 Å crystal structure and active-site architecture of β-ketoacyl-acyl carrier protein synthase III (FabH) from *Escherichia coli, Structure*, 2000, **8**, 185-195.
22. KREUZALER, F., HAHLBROCK, K., Enzymic synthesis of an aromatic ring from acetate units: partial purification and some properties of flavanone synthase from cell-suspension cultures of *Petroselinum hortense*, *Eur. J. Biochem.*, 1975, **56**, 205-213.
23. LANZ T., TROPF, S., MARNER, F.J., SCHRODER, J., SCHRODER, G., The role of cysteines in polyketide synthases: site-directed mutagenesis of resveratrol and chalcone synthases, two key enzymes in different plant-specific pathways, *J. Biol. Chem.*, 1991, **266**, 9971-9976.
24. SUH, D.Y., KAGAMI, J., FUKUMA, K., SANKAWA, U., Evidence for catalytic cysteine-histidine dyad in chalcone synthase, *Biochem. Biophy. Res. Comm.*, 2000, **275**, 725-730.

25. SCHUZ, R., HELLER, W., HAHLBROCK, K., Substrate specificity of chalcone synthase from *Petroselinum hortense*: formation of phloroglucinol derivatives from aliphatic substrates, *J. Biol. Chem.*, 1983, **258**, 6730-6734.
26. ZUURBIER, K.W.M., LESER, J., BERGER, T., HOFTE, A.J.P., SCHRODER, G., VERPOORTE, R., SCHRODER, J., Hydroxy-2-pyrone formation by chalcone and stilbene synthase with nonphysiological substrates, *Phytochemistry*, 1998, **49**, 1945-1951.
27. ABE, I., MORITA, H., NOMURA, A., NOGUCHI, H., Substrate specificity of chalcone synthase: enzymatic formation of unnatural polyketides from synthetic cinnamoyl-CoA analogues, *J. Am. Chem. Soc.*, 2000, **122**, 11242-11243.
28. MORITA, H., TAKAHASHI, Y., NOGUCHI, H., ABE, I., Enzymatic formation of unnatural aromatic polyketides by chalcone synthase, *Biochem. Biophys. Res. Comm.*, 2000, **279**, 190-195.
29. ABE, I., TAKAHASHI, Y., MORITA, H., NOGUCHI, H., Benzalacetone synthase: a novel polyketide synthase that plays a crucial role in the biosynthesis of phenylbutanones in *Rheum palmatum*, *Eur. J. Biochem.*, 2001, **268**, 3354-3359.
30. LUKACIN, R., SPRINGOB, K., URBANKE, C., ERNWEIN, C., SCHRODER, G., SCHRODER, J., MATERN, U., Native acridone synthases I and II from *Ruta graveolens* L. form homodimers, *FEBS Lett.*, 1999, **448**, 135-140.
31. SPRINGOB, K., LUKACIN, R., ERNWEIN, C., GRONING, I., MATERN, U., Specificities of functionally expressed chalcone and acridone synthases from *Ruta graveolens*, *Eur. J. Biochem.*, 2000, **267**, 6552-6559.
32. JEZ, J.M., BOWMAN, M.E., NOEL, J.P., Structure-guided programming of polyketide chain-length determination in chalcone synthase, *Biochemistry*, 2001, **40**, (in press).
33. CHRISTENSEN, A.B., GREGERSEN, P.L., SCHRODER, J., COLLINGE, D.B., A chalcone synthase with an unusual substrate preference is expressed in barley leaves in response to UV light and pathogen attack, *Plant Mol. Biol.*, 1998, **37**, 849-857.
34. BEERHUES, L., Benzophenone synthase from cultured cells of *Centaurium erythraea*, *FEBS Lett.*, 1996, **383**, 264-266.
35. PANIEGO, N.B., ZUURBIER, K.W.M., FUNG, S.-Y., VAN DER HEIJDEN, R., SCHEFFER, J.C.C., VERPOORTE, R., Phlorisovalerophenone synthase, a novel polyketide synthase from hop (*Humulus lupulus* L.) cones, *Eur. J. Biochem.*, 1999, **262**, 612-616.
36. DIMROTH, P., WALTER, H., LYNEN, F., Biosynthesis of 6-methylsalicylic acid, *Eur. J. Biochem.*, 1970, **13**, 98-110.
37. SPENCER, J.B., JORDAN, P.M., Purification and properties of 6-methylsalicylic acid synthase from *Penicillium patulum*, *Biochem. J.*, 1992, **288**, 839-846.
38. KUROSAKI, F., KIZAWA, Y., NISHI, A., Derailment product in NADPH-dependent synthesis of a dihydroisocoumarin 6-hydroxymellein by elicitor-treated carrot cell extracts, *Eur. J. Biochem.*, 1989, **185**, 85-89.
39. SUH, D.-Y., FUKUMA, K., KAGAMI, J., YAMAZAKI, Y., SHIBUYA, M., EBIZUKA, Y., SANKAWA, U., Identification of amino acid residues important in the cyclization reactions of chalcone and stilbene synthases, *Biochem. J.*, 2000, **350**, 229-235.

40. JEZ, J.M., AUSTIN, M.B., FERRER, J.-L., BOWMAN, M.E., SCHRODER, J., NOEL, J.P., Structural control of polyketide formation in plant-specific polyketide synthases, *Chem. Biol.*, 2000, **7**, 919-930.
41. KREUZALER, F., HAHLBROCK, F., Enzymatic synthesis of aromatic compounds in higher plants. Formation of *bis*-noryangonin (4-hydroxy-6[4-hydroxystyryl]2-pyrone) from p-coumaroyl-CoA and malonyl-CoA, *Arch. Biochem. Biophys.*, 1975, **169**, 84-90.
42. HRAZDINA, G., KREUZALER, F., HAHLBROCK, K., GRISEBACH, H., Substrate specificity of flavanone synthase from cell suspension cultures of parsley and structure of release products in vitro, *Arch. Biochem. Biophys.*, 1976, **175**, 392-399.
43. KOES, R.E., SPELT, C.E., VAN DEN ELZEN, P.J.M., MOL, J.N.M., Cloning and molecular characterization of the chalcone synthase multigene family of *Petunia hybrida*, *Gene*, 1989, **81**, 245-257.
44. DURBIN, M.L., LEARN, G.H., HUTTLEY, G.A., CLEGG, M.T., Evolution of the chalcone synthase gene family in the genus *Ipomoea*, *Proc. Natl. Acad. Sci. USA*, 1995, **92**, 3338-3343.
45. RAUSHER, M.D., MILLER, R.E., TIFFIN, P., Patterns of evolutionary rate variation among genes of the anthocyanin biosynthetic pathway, *Mol. Biol. Evol.*, 1999, **16**, 266-274.

Chapter Thirteen

THE ROLE OF CYTOCHROMES P450 IN BIOSYNTHESIS AND EVOLUTION OF GLUCOSINOLATES

Barbara Ann Halkier,[1,2*] Carsten Hørslev Hansen,[1,2,3] Michael Dalgaard Mikkelsen,[1,2] Peter Naur,[1,2] and Ute Wittstock.[1,2,4]

[1]*Plant Biochemistry Laboratory, Dept. of Plant Biology, The Royal Veterinary and Agricultural University, Thorvaldsensvej 40, DK-1871 Frederiksberg C, Copenhagen, Denmark*

[2]*Center for Molecular Plant Physiology (PlaCe), The Royal Veterinary and Agricultural University*

[3]*Max-Planck Institute of Molecular Plant Physiology, Am Mühlenberg 1, 14476 Potsdam, Germany*

[4]*Max-Plank Institute of Chemical Ecology, Karl-Marx Str., Jena, Germany*

[*]*Author for correspondence, e-mail: bah@kvl.dk*

Introduction	224
The Biosynthetic Pathway of Glucosinolates	226
Cytochromes P450 of the CYP79 Family in Glucosinolate Biosynthesis	227
Lessons from the Biosynthesis of Cyanogenic Glucosides	227
Characterization of CYP79A2 from *Arabidopsis*	227
Characterization of CYP79B2 and CYP79B3 from *Arabidopsis*	231
Characterization of CYP79F1 and CYP79F2 from *Arabidopsis*	232
Evolution	233
CYP79s in the Biosynthesis of Cyanogenic Glucosides and Glucosinolates	233
The Oxime-Metabolizing Enzyme as Branch Point between the Cyanogenic Glucoside and the Glucosinolate Pathway	235
Biochemical Characterization of the Oxime-Metabolizing Enzyme in Glucosinolate Biosynthesis	235
Molecular Identification of Oxime-Metabolizing Enzyme in Glucosinolate Biosynthesis	236
Heterologous Expression of CYP79 Homologues in *E.coli*	238
Metabolic Engineering of Glucosinolate Profiles in Transgenic Plants	240
Summary and Future Perspectives	242

INTRODUCTION

Glucosinolates are naturally occurring S-glucosides of thiohydroximate-O-sulfonates, with variable side chains derived from amino acids. To date, more than 100 different glucosinolates are known. They are only found in the order Capparales and in the genus *Drypetes* (Euphorbiaceae, Euphorbiales). Glucosinolates co-occur with endogenous thioglucosidases called myrosinases that upon tissue damage hydrolyze glucosinolates into unstable aglucones. Depending on the nature of the side chain, pH of the medium, the presence of certain ions (*e.g.,* ferrous ions), and the presence of epithiospecifier proteins, aglucones rearrange into a variety of degradation products, such as isothiocyanates, nitriles, thiocyanates, and oxazolidine-2-thiones (Fig. 13.1). The degradation products are responsible for the characteristic flavour of *Brassica* vegetables and have different biological effects ranging from antimicrobial and inflammatory to goitrogenic activities. Generally, glucosinolates or rather their degradation products are believed to function as defensive compounds against pathogens and generalist herbivores and as attractants to insects that are specialized feeders on Brassicaceae.[1-5] In the last decade, glucosinolates have received significant attention, as it has been discovered that certain degradation products have anticarcinogenic properties.[6] Particularly, sulphoraphane (4-methylsulfinylbutylisothiocyanate) derived from 4-methylsulfinylbutylglucosinolate has been found to be a potent cancer-preventive phytochemical.[7] Sulphoraphane has been shown to induce a quinone reductase, which is a phase II enzyme in the detoxification process in mammals. In contrast, indole carbinol derived from indole glucosinolates is a compound with carcinogenic potential, as it induces phase I enzymes that might convert procarcinogens into carcinogens.[5,8] The wide range of biological effects of glucosinolates and their degradation products has resulted in a strong interest in elucidating their role in plant-insect-interactions and in controlling the level of individual glucosinolates to improve nutritional quality and disease resistance of agriculturally important *Brassica* crops. This has prompted a desire to understand the biosynthetic pathway. The present review will focus on recent advances in identification of cytochrome P450-dependent monooxygenases (cytochromes P450) catalyzing the conversion of amino acids to oximes and the oxidation of oximes in the biosynthetic pathway of glucosinolates. The data provides implications for the evolution of glucosinolate biosynthesis.

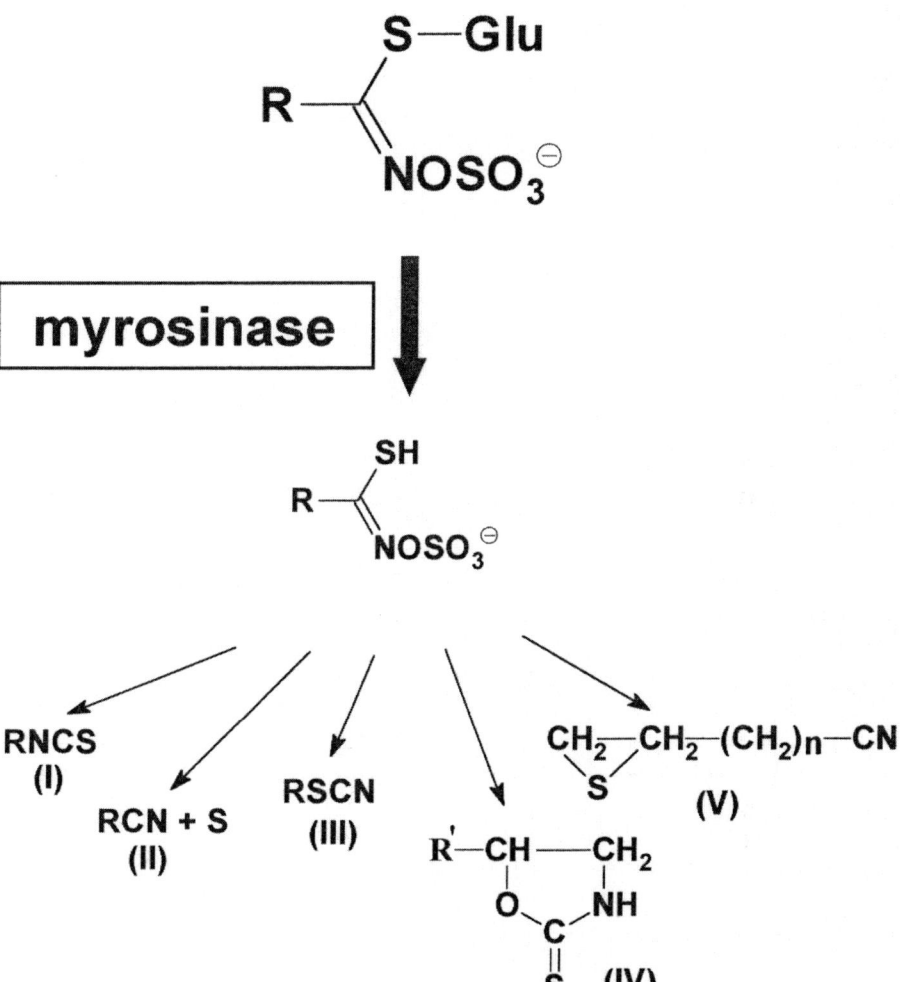

Figure 13.1: Degradation of glucosinolates. Hydrolysis is catalyzed by myrosinases and gives rise to different degradation products dependent on the structure of the glucosinolate side chain and the hydrolysis conditions. (I) isothiocyanates, the major product at pH >7; (II) nitriles, the major product at pH <4; (III) thiocyanates, produced from 2-propenyl-, benzyl-, and 4-methylthiobutylglucosinolates; (IV) oxazolidine-2-thiones, produced from glucosinolates with β-hydroxylated side chains, (V) epithionitriles, produced in the presence of epithiospecifier proteins.

THE BIOSYNTHETIC PATHWAY OF GLUCOSINOLATES

Glucosinolates are derived from amino acids. The precursor amino acids comprise seven protein amino acids (alanine, valine, leucine, isoleucine, phenylalanine, tyrosine, and tryptophan), a number of chain-elongated methionines $(CH_3\text{-}S\text{-}(CH_2)_2\text{-}(CH_2)_n\text{-}CH(NH_2)\text{-}COOH$, n = 1-9), and chain-elongated phenylalanine and tyrosine.[5] The high chemical diversity is obtained by secondary modifications of the side chains and/or the glucose moiety. The biosynthesis comprises three independent stages.[2,9] First, chain-elongated protein amino acids are synthesized, if needed. Secondly, the precursor amino acids are converted to the core structure of the parent glucosinolates. Finally, the parent glucosinolates may undergo secondary modifications such as thiol oxidation, desaturation, and esterification.

Most of the intermediates in the conversion of the precursor amino acid to the parent glucosinolate are known. *In vivo* biosynthetic studies have shown that *N*-hydroxy amino acids, oximes, thiohydroximates, and desulfoglucosinolates are precursors of glucosinolates. The conversion of amino acids to oximes is a key step, and the nature of the enzymes that catalyze this step have been the subject of much controversy. Independent biochemical studies have indicated that three different enzymes are involved, namely cytochrome P450-dependent monooxygenases, flavin-containing monooxygenases, and peroxidases.[9] From these studies, it appeared that the nature of the oxime-forming enzymes may depend on the nature of the amino acid to be converted, implicating that three different types of enzymes have evolved to catalyze the formation of oximes. Recently, this has been disproved by the identification of cytochromes P450 of the CYP79 family that catalyze oxime formation from aliphatic and aromatic amino acids as well as from tryptophan (see below).[10-14]

The conversion of oximes to the parent glucosinolates is less well understood. Ettlinger and Kjær[15] have proposed that the oxime is oxidized to an *aci*-nitro compound that is then conjugated with a sulfur donor to produce an *S*-alkyl thiohydroximate. It has been suggested that the conjugation is catalyzed by a glutathione-*S*-transferase-type of enzyme, although no evidence is available.[16] Biochemical studies have indicated that a C-S lyase hydrolyzes the *S*-alkyl thiohydroximate to a thiohydroximic acid, which is subsequently glucosylated to a desulfoglucosinolate by a soluble UDPG-thiohydroximic acid glucosyl transferase (S-GT). Several S-GTs have been characterized biochemically.[17,18] Finally, the desulfoglucosinolate is sulfated by a soluble 3'-phosphoadenosine 5'-phosphate PAPS-desulfoglucosinolate sulfotransferase.[19-21] It is common for the enzymes that metabolize the oxime to the parent glucosinolates to have a high substrate-specificity with respect to the functional groups, but a low substrate-specificity with respect to the side chain.[2]

CYTOCHROMES P450 OF THE CYP79 FAMILY IN GLUCOSINOLATE BIOSYNTHESIS

Lessons from the Biosynthesis of Cyanogenic Glucosides

Our laboratory has a long tradition of studying the biosynthetic pathways of cyanogenic glucosides and glucosinolates (Fig. 13.2). A working hypothesis has been that glucosinolates are evolutionarily related to cyanogenic glucosides, as both groups of compounds are derived from amino acids and have oximes as intermediates. The biosynthetic pathway of the tyrosine-derived cyanogenic glucoside dhurrin from *Sorghum bicolor* (L.) Moench has been fully elucidated.[22] The first enzyme in the pathway is a cytochrome P450, P450tyr (designated CYP79A1), which catalyzes the conversion of tyrosine to the corresponding oxime p-hydroxyphenylacetaldoxime.[23,24] The mechanism of this reaction has been studied extensively. The biosynthetic studies included stoichiometric measurements of oxygen consumption in combination with product formation, $^{18}O_2$ labeling experiments, as well as feeding studies using isotopically labeled tyrosine.[25-27] The studies showed that CYP79A1 is an N-hydroxylase that catalyzes two consecutive N-hydroxylation reactions that are followed by a dehydration and a decarboxylation reaction. The oxime is subsequently converted by a cytochrome P450, P450ox (designated CYP71E1), to the aglucone p-hydroxymandelonitrile,[28,29] which is glucosylated by UDPG:p-hydroxymandelonitrile glucosyl transferase.[30] Surprisingly, the pathway is catalyzed by only three enzymes, two multifunctional cytochromes P450 and a soluble UDPG-glucosyltransferase. The cDNAs encoding the three enzymes have been isolated by using a classical biochemical approach of purifying the enzymes, obtaining amino acid sequences, and using degenerate oligonucleotides to design probes to screen a cDNA library. To date, PCR approaches have been used to isolate CYP79s from other cyanogenic plants. These include CYP79E1 and E2 from *Triglochin maritima* L. with tyrosine as substrate,[31] and CYP79D1 and D2 from cassava (*Manihot esculenta* CRANTZ) with valine and isoleucine as substrates[32] (Fig. 13.3).

Characterization of CYP79A2 from Arabidopsis

In microsomes from *Sinapis alba* L.,[33,34] *Tropaeolum majus* L.,[35,36] and *Carica papaya* L.,[37] the aromatic amino acids (tyrosine and phenylalanine) have been shown to be converted to the corresponding oximes by cytochrome P450-dependent monooxygenases. The conversion of tyrosine to the corresponding oxime in microsomes from *S. alba* was approximately 1000 fold lower than in microsomes from the cyanogenic sorghum.[33] This made a biochemical approach for the isolation

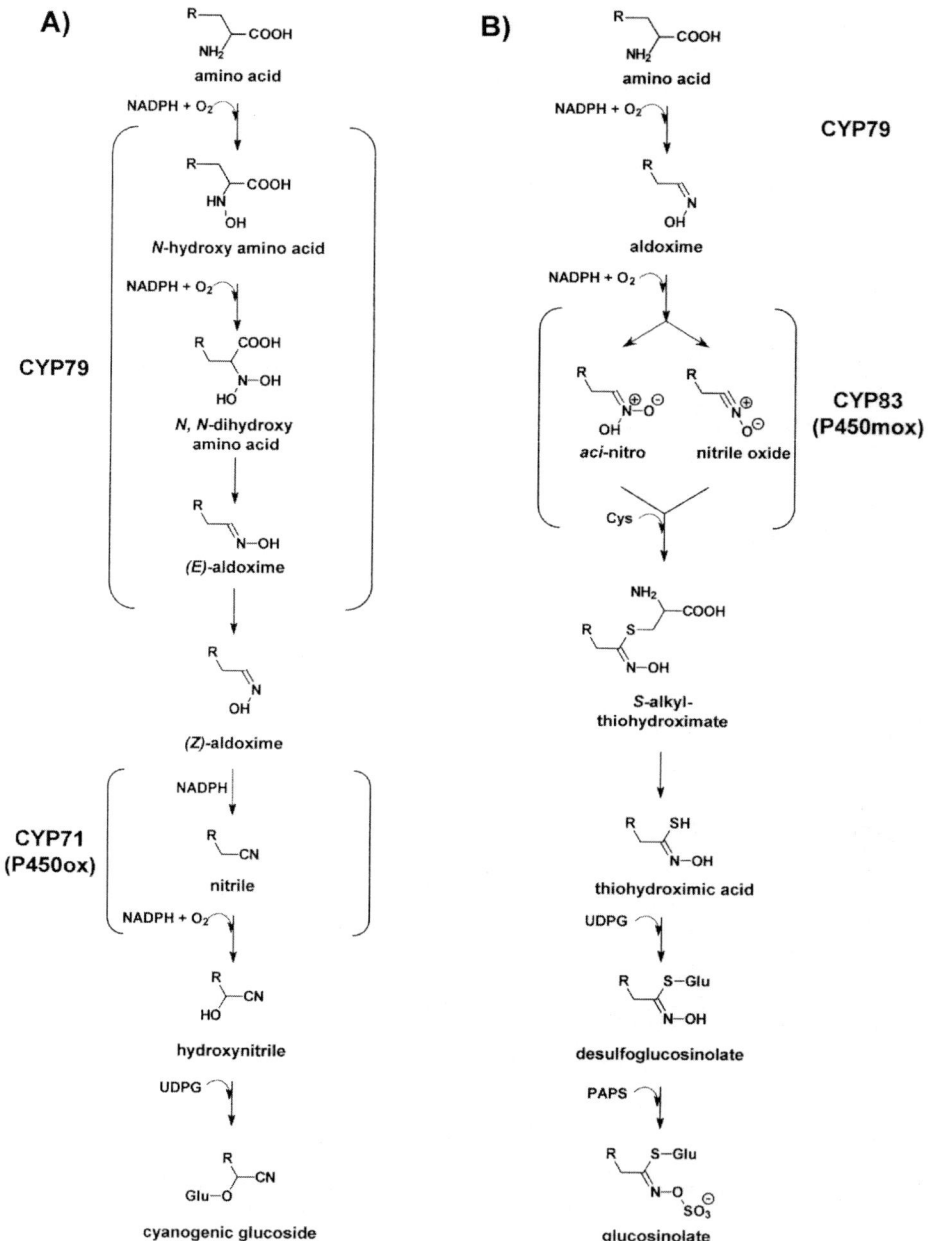

Figure 13.2: Biosynthetic pathways of (A) cyanogenic glucosides and (B) glucosinolates. The CYP79s are assumed to catalyze the same reaction in both pathways. It is not known whether the oxime is oxidized to an *aci*-nitro compound or a nitrile oxide in the glucosinolate pathway.

of the genes in the glucosinolate pathway less attractive. As sequence information from the *Arabidopsis* Genome Initiative (AGI) became available, it was used to identify *Arabidopsis* homologues of the sorghum CYP79A1. With the completion of the genome, seven CYP79 genes were identified in *Arabidopsis* (Fig. 13.3). The first CYP79 from *Arabidopsis* to be cloned and characterized was CYP79A2.[12] The full-length cDNA of *CYP79A2* was isolated from an *Arabidopsis* cDNA library by extensive PCR amplification. Heterologous expression of the recombinant protein in *Escherichia coli* showed that CYP79A2 catalyzes the conversion of phenylalanine to phenylacetaldoxime. This provided the first evidence that cytochromes P450 belonging to the CYP79 family were involved in the biosynthesis of glucosinolates. The substrate specificity of CYP79A2 is narrow, as neither tyrosine, homophenylalanine, tryptophan nor methionine were metabolized by the recombinant enzyme. The kinetic data showed that CYP79A2 has high affinity to phenylalanine (K_m 6.7 µmol liter^{-1}), which, however, is metabolized at a low rate (turnover number 0.24 min^{-1}) compared with the conversion of tyrosine by recombinant CYP79A1 (turnover number 350 min^{-1}).[12]

Further evidence for the involvement of CYP79A2 in glucosinolate biosynthesis was provided by accumulation of the phenylalanine-derived benzylglucosinolate in transgenic *Arabidopsis* plants expressing CYP79A2 under the control of the CaMV35S (35S) promoter.[12] Benzylglucosinolate is only sporadically observed in roots and cauline leaves of wild-type plants (*A. thaliana* ecotype Columbia). However, the seeds are known to contain the homophenylalanine-derived 2-phenylethylglucosinolate. This raises the question whether homophenylalanine is a substrate for CYP79A2 *in vivo*. One would expect that increased levels of 2-phenylethylglucosinolate would accumulate in the seeds of the transgenic plants, assuming that the chain-elongation is not ratelimiting. HPLC analysis showed that the content of 2-phenylethylglucosinolate was unchanged in seeds of the transgenic plants compared with seeds of wild-type plants, whereas seeds of the transgenic plants accumulated high levels of benzylglucosinolate.[12] This supported the data obtained with the recombinant protein, and indicated that CYP79A2 also *in vivo* converts phenylalanine and not homophenylalanine to the corresponding oxime. Furthermore, the data showed that the formation of the oxime is the rate-limiting step in the biosynthesis of benzylglucosinolate in *Arabidopsis*.

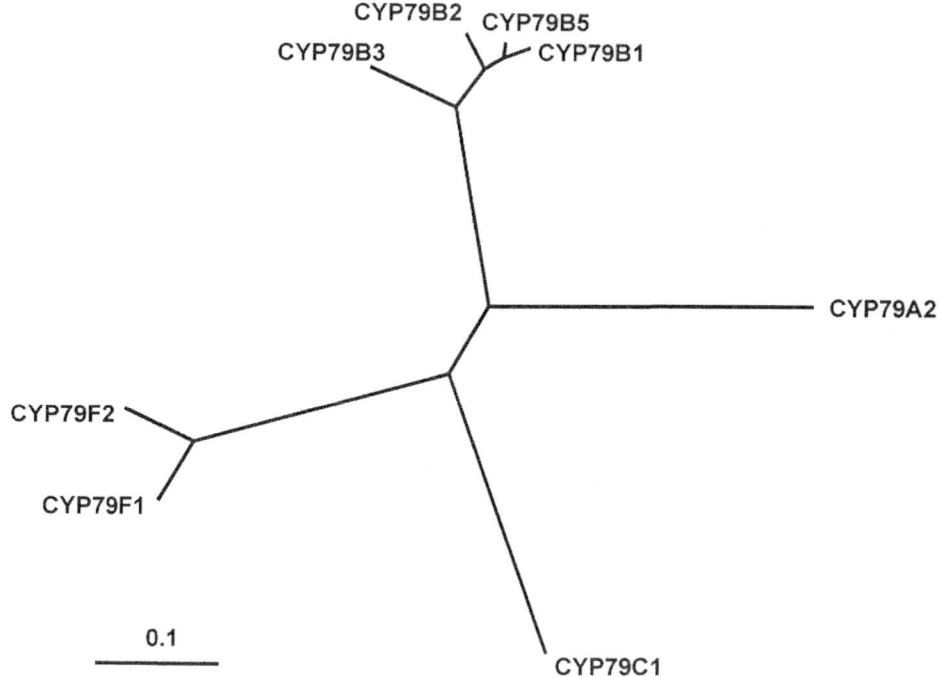

Figure 13.3: Phylogenetic tree of the CYP79s from glucosinolate-producing plants. All the CYP79s are from *Arabidopsis thaliana* (Columbia), except for CYP79B5, which is from *Brassica napus*.

Further evidence for the involvement of CYP79A2 in glucosinolate biosynthesis was provided by accumulation of the phenylalanine-derived benzylglucosinolate in transgenic *Arabidopsis* plants expressing CYP79A2 under the control of the CaMV35S (35S) promoter.[12] Benzylglucosinolate is only sporadically observed in roots and cauline leaves of wild-type plants (*A. thaliana* ecotype Columbia). However, the seeds are known to contain the homophenylalanine-derived 2-phenylethylglucosinolate. This raises the question whether homophenylalanine is a substrate for CYP79A2 *in vivo*. One would expect that increased levels of 2-phenylethylglucosinolate would accumulate in the seeds of the transgenic plants, assuming that the chain-elongation is not ratelimiting. HPLC analysis showed that the content of 2-phenylethylglucosinolate was unchanged in seeds of the transgenic plants compared with seeds of wild-type plants, whereas seeds of the transgenic plants accumulated high levels of benzylglucosinolate.[12] This supported the data obtained with the recombinant protein, and indicated that CYP79A2 also *in vivo* converts phenylalanine and not homophenylalanine to the corresponding oxime. Furthermore, the data showed that the formation of the oxime is the rate-limiting step in the biosynthesis of benzylglucosinolate in *Arabidopsis*.

The sporadic occurrence of benzylglucosinolate corresponded well with the observation that the *CYP79A2* mRNA was a low abundant transcript that was not detectable by Northern blotting or RT-PCR.[12] In *Arabidopsis* plants transformed with a CYP79A2 promoter-GUS construct, blue staining was restricted to the hydathodes (Hansen and Halkier unpubl. results). The specific expression pattern together with the known antimicrobial effects of the degradation products of benzylglucosinolate suggests that CYP79A2 produces a glucosinolate that may function as a 'guard' at the hydathodes, the gate to the vascular system.

Characterization of CYP79B2 and CYP79B3 from Arabidopsis

The second CYP79 homologue from *Arabidopsis* that was identified and characterized was CYP79B2.[13,14] Using the 5'-end truncated EST T42902 as probe, a full-length cDNA of *CYP79B2* was isolated. In parallel, *CYP79B2* was isolated in a yeast screen designed to identify *Arabidopsis* cDNAs that were able to confer resistance to the toxic 5-fluorotryptophan.[14] Heterologous expression in *E. coli* showed that the recombinant protein catalyzes the conversion of tryptophan to indole-3-acetaldoxime (IAOX), a precursor of indole glucosinolates and of the plant hormone indole-3-acetic acid (IAA).[13,14] The homologue CYP79B3, that is 85.2% identical at the amino acid level to CYP79B2, has been shown to have the same enzymatic activity as CYP79B2.[14]

The conversion of tryptophan to IAOX has previously been shown to be catalyzed by peroxidase-like enzymes (TrpOxE) isolated from plasmamembranes of Chinese cabbage, maize, sunflower, tobacco, pea, and *Arabidopsis*.[38,39] The activity of recombinant CYP79B2 was clearly distinct from the TrpOxE activity, as it was severely inhibited by $MnCl_2$, H_2O_2, and 2,4-dichlorophenol.[13] Thus, it appears that different enzymes catalyze the conversion of tryptophan to IAOX. *Arabidopsis* expressing CYP79B2 under control of the 35S promoter showed a significant increase in the two most abundant indole glucosinolates, indol-3-ylmethylglucosinolate and 4-methoxyindol-3-ylmethylglucosinolate.[13] No significant increase was seen for N-methoxyindol-3-ylmethylglucosinolate, probably due to low activity of the downstream enzymes. The data are consistent with the involvement of CYP79B2 in the biosynthesis of indole glucosinolates.

Besides its role as an intermediate in indole glucosinolate biosynthesis, it has been suggested that IAOX is involved in IAA biosynthesis.[40,41] This has been supported by the biochemical characterization of IAOX-metabolizing enzymes that convert IAOX to either indole-3-acetaldehyde,[42,43] indole-3-acetonitrile,[42,44] or directly to IAA.[45] The previously described peroxidase-like activity (TrpOxE) is found in many different plant species that do not produce glucosinolates. It may, therefore, represent a general biosynthetic pathway for IAA via IAOX. In addition, a flavin monooxygenase, identified in an activation tagged IAA-overexpression mutant with increased IAA level, has been shown to be able to hydroxylate tryptamine to N-hydroxytryptamine, which suggests that it may be involved in the conversion of tryptophan to IAOX in the IAA pathway.[47] Furthermore, an IAA-synthase complex of 160-180 kD has been isolated from *Arabidopsis*, suggesting a new way for plants to synthesize IAA.[46] Based on the evolutionary relationship between the biosynthesis of cyanogenic glucosides and glucosinolates, it is likely that CYP79B2 is involved in biosynthesis of indole glucosinolates. However, it cannot be excluded that CYP79B2 (and/or CYP79B3) is involved in the biosynthesis of IAA or that there is a crosstalk between the biosynthetic pathways of indole glucosinolates and IAA at the IAOX branch point. It is expected that the phenotype of a double knock-out mutant of CYP79B2 and CYP79B3 (if viable) will answer the question whether the CYP79Bs are involved in the IAA biosynthesis. We have not been able to identify members of the CYP79B subfamily in tobacco, suggesting that if CYP79B2 or CYP79B3 is involved in IAA biosynthesis, this pathway might be specific to cruciferous plants and not generally found in the plant kingdom.

Characterization of CYP79F1 and CYP79F2 from Arabidopsis

The next CYP79s that appeared in the *Arabidopsis* sequence database were *CYP79F1* and *CYP79F2*. CYP79F1 is 88% identical at the amino acid level with CYP79F2, and the two genes are located on the same chromosome, only separated by 1638 bp.[10,11] This suggests that the two genes were formed by gene duplication and that the respective proteins might catalyze similar reactions. To determine the substrate-specificity of CYP79F1, an EST (ATTS5112) containing the full length *CYP79F1* cDNA was used to express CYP79F1 in *E. coli*. Both dihomomethionine and trihomomethionine were metabolized by CYP79F1, resulting in the formation of 5-methylthiopentanaldoxime and 6-methylthiohexanaldoxime, proven by GC-MS analysis and comparison with authentic standards.[10] More recently, we have found that the short-chain methionine derivatives homo- to tetrahomomethionine (n=1-4), as well as the long-chain methionine derivatives penta- and hexahomomethionine (n=5-6) are converted to the corresponding oximes by CYP79F1 (Hansen, Glawischnig and Halkier, unpubl. data). Neither methionine nor other protein amino acids tested were substrates for CYP79F1. The turnover numbers for the conversion of dihomomethionine and trihomomethionine by recombinant CYP79F1 are 0.23 min^{-1} and 0.15 min^{-1}, respectively. Characterization of recombinant CYP79F2 indicates that this enzyme primarily catalyzes the conversion of methionine derivatives with n=5-6 (Chen, Glawischnig, and Halkier, unpubl. results).

Several *CYP79F1* knockout mutants have been isolated.[11,48] Analysis of the glucosinolate profiles showed that glucosinolates derived from chain-elongated methionines with n=1-4 were completely abolished, whereas glucosinolates derived from chain-elongated methionines with n=5-6 were slightly upregulated in the mutants.[11] This suggested that CYP79F1 does not metabolize the chain-elongated methionines with more than four methylene groups. However, the recent investigations of the substrate specificity of CYP79F1 (see above) disprove this conclusion. This shows that although the identification and characterization of mutants has greatly facilitated progress in the field, the data obtained might not always be sufficient to explain biochemical pathways. For correct interpretation of mutant phenotypes, the biochemical characterization of the respective proteins is a necessity.

Transgenic *Arabidopsis* plants expressing the *CYP79F1* cDNA under control of the CaMV35S promoter had either reduced or increased levels of aliphatic glucosinolates.[10] In the plants with reduced content of total aliphatic glucosinolates, both *CYP79F1* and *CYP79F2* were downregulated, as demonstrated by RT-PCR. In addition, the CYP79F1 substrates dihomomethionine and trihomomethionine accumulated 50 and 16 fold, respectively.[10] The high level of the chain-elongated methionines indicates that the enzymes catalyzing the chain-elongation are not subject to feedback inhibition. Furthermore, it suggests that the enzymes that catalyze additional chain-elongation cycles may be rate-limiting in the biosynthesis of longer-chain methionine homologues. In addition to the altered expression levels of *CYP79F1* and *CYP79F2*, *CYP79B2* was upregulated approximately twofold in the 35S::CYP79F1 plants with reduced levels of aliphatic glucosinolates.[10] This resulted in an increased content of indole glucosinolates.

The transgenic *Arabidopsis* plants with altered content of aliphatic glucosinolates possess a characteristic morphological phenotype characterized by production of multiple axillary shoots.[11,48] Since *Arabidopsis* is able to tolerate overexpression of cytochromes P450 of the CYP79 family leading to a two to five fold increase in glucosinolate content without similar changes in the appearance of the plants,[49] it seems unlikely that the morphological changes resulted from the presence or absence of specific glucosinolates. The accumulation of very high levels of chain-elongated methionine homologues in the transgenic plants suggests that the morphological phenotype may be a pleiotropic effect caused by disturbance of the plant's sulfur metabolism, in which methionine plays a central role.

EVOLUTION

CYP79s in the Biosynthesis of Cyanogenic Glucosides and Glucosinolates

Cyanogenic glucosides and glucosinolates are related groups of natural plant products, as both are derived from amino acids and have oximes as intermediates.

Cyanogenic glucosides occur widely throughout the plant kingdom, where they are present in angiosperms, gymnosperms, and ferns. This indicates that cyanogenesis arose as an early evolutionary event.[50] In contrast, glucosinolates are found exclusively among dicotyledoneous plants in the order Capparales and in the genus *Drypetes* in the evolutionarily distant order Euphorbiales.[51] This has led to speculation that the biosynthesis of glucosinolates evolved from the cyanogenic pathway, and that homologous enzymes might catalyze the conversion of amino acids to oximes in both pathways. CYP79A1 from sorghum was the first oxime-forming enzyme to be characterized (see above). This prompted a search for homologues of this cytochrome P450 in *Arabidopsis* and in other glucosinolate-producing plants, as well as in plants that do not contain cyanogenic glucosides or glucosinolates. Before the characterization of the *Arabidopsis* CYP79s had been accomplished, a PCR study showed that *CYP79* homologues are found in distantly related families within the order Capparales as well as in the glucosinolate-producing genus *Drypetes* (syn. Puntranjiva) in the order Euphorbiales.[52] The PCR study was performed with degenerate *CYP79*-specific primers that were designed based on the sequence alignment of CYP79A1 from sorghum, CYP79B1 from *S. alba*, and CYP79B2 from *Arabidopsis* (the three CYP79s of which full-length clones were available at that time). Using these primers, it was possible to obtain PCR products from genomic DNA from six glucosinolate-producing species (*S. alba, B. napus, Arabidopsis, T. majus, C. papaya, Puntranjiva roxburghii*), but not from tobacco.[52] Traditionally, Tropaeolaceae and Caricaceae have not been regarded as belonging to the order Capparales, although the presence of glucosinolates and myrosinase in these families suggests a close relation to this order. Evidence for the existence of an expanded order Capparales including Tropaeolaceae and Caricaceae, as initially suggested by Dahlgren several decades ago,[53,54] was provided by a combination of morphological and phytochemical data as well as molecular data based on the DNA sequences of the chloroplast gene *rbcL* and the nuclear ribosomal 18S RNA gene.[51,55] The genus *Drypetes*, however, is phylogenetically distant from the major glucosinolate clade[52] and represents a second unrelated glucosinolate clade in the order Euphorbiales. Identification of CYP79 homologues in both Capparales and Euphorbiales strongly indicates that in both clades, glucosinolates evolved based on a cyanogenic predisposition, suggesting convergent evolution.

The recent finding that cytochromes P450 belonging to the CYP79 family catalyze the conversion of aliphatic and aromatic amino acids as well as tryptophan to the corresponding oximes in the biosynthesis of glucosinolates in *Arabidopsis* is consistent with this hypothesis.[10-14] Furthermore, the closest relative to the glucosinolate-hydrolyzing myrosinases are the β-glucosidases that hydrolyze cyanogenic glucosides.[56] This suggests that myrosinases (which are the only described β-thioglucosidases in nature) have been recruited from the cyanogenic β-glucosidases.

Generally, glucosinolates and cyanogenic glucosides are not found in the same species, and it has been suggested that their presence is mutually exclusive. *C. papaya* represents an interesting exception, as it is the only species known to have both a glucosinolate and a cyanogenic glucoside, in this case both derived from phenylalanine.[37] This suggests that the previously identified CYP79 in *C. papaya*[52] may serve both pathways.

Interestingly, no tryptophan-derived cyanogenic glucoside has yet been identified. Apparently, CYP79s of the glucosinolate pathway have acquired new substrates after having diverged from the 'cyanogenic' CYP79s. The biosynthesis of tryptophan-derived glucosinolates seems to be a recent evolutionary event, as indole glucosinolates are present only in five families in the order Capparales, namely the Brassicaceae, Tropaeolaceae, Bataceae, Resedaceae, and Capparaceae.[5,57] The CYP79 family forms a distinct phylogenetic group within the cytochromes P450 of the A clan, which represents cytochromes P450 primarily involved in secondary metabolism. Consequently, it is expected that as yet uncharacterized cytochromes P450 of the CYP79 family in cyanogenic and glucosinolate-producing plants catalyze oxime production from all precursor amino acids of cyanogenic glucosides and glucosinolates.

THE OXIME-METABOLIZING ENZYME AS BRANCH POINT BETWEEN THE CYANOGENIC GLUCOSIDE AND THE GLUCOSINOLATE PATHWAY

If glucosinolates have evolved from cyanogenic glucosides, the next question to ask is where is the branch point between the two pathways and how did it arise? In the pathway of the cyanogenic glucoside dhurrin in sorghum, the oxime-metabolizing enzyme is a cytochrome P450 named P450ox (CYP71E1). P450ox converts the oxime to a hydroxynitrile, which is subsequently glucosylated to the final product dhurrin by a soluble glucosyltransferase.[22] Our working hypothesis has been that a mutated homologue of P450ox (called P450*mox*) catalyzes the oxime-metabolizing step in the pathway of glucosinolates. According to this hypothesis, P450*mox* would not oxidize the oxime to a hydroxynitrile as in the cyanogenic glucoside pathway, but instead would oxidize the oxime to a toxic or reactive compound, which would be detoxified by general detoxification reactions into glucosinolates.

Biochemical Characterization of the Oxime-Metabolizing Enzyme in Glucosinolate Biosynthesis

S. alba has been used as a model plant for biochemical studies of the oxime-metabolizing enzyme in the biosynthetic pathway of glucosinolates.[33] The major glucosinolate in *S. alba* is the tyrosine-derived *p*-hydroxybenzylglucosinolate. This glucosinolate is synthesized via the same oxime intermediate, *p*-hydroxyphenylacetaldoxime, as the tyrosine-derived cyanogenic glucoside dhurrin. The microsomal fraction isolated from jasmonate-treated etiolated seedlings of *S. alba* was used as source of enzyme, as it has previously been shown that treatment with jasmonate resulted in approximately an eightfold increase in the incorporation of radiolabeled tyrosine into *p*-hydroxybenzylglucosinolate by seedlings of *S. alba*.[33] After administration of radiolabeled *p*-hydroxyphenylacetaldoxime to microsomes of *S. alba*, small amounts of several radiolabeled compounds were detected by TLC.[58] However, when radiolabeled *p*-hydroxyphenylacetaldoxime was administered to microsomes from *S. alba* in the presence of NADPH and a sulfur donor cysteine, a new water-soluble radiolabeled compound accumulated in high amounts in the reaction mixture. The compound was identified by mass spectrometry as *S*-(*p*-hydroxyphenylacetohydroximoyl)-L-cysteine,[58] the proposed *S*-alkyl thiohydroximate intermediate in glucosinolate biosynthesis.[15] The S-atom in the thiohydroximate intermediate is likely to be derived from cysteine, as *in vivo* feeding studies have shown that cysteine is the most efficient sulfur donor.[59] Furthermore, C-S lyases in higher plants capable of hydrolyzing *S*-substituted derivatives of cysteine such as *S*-cysteine thiohydroximates have a requirement for the presence of the α-hydrogen atom and an unsubstituted amino group.[60,61] Identification of the *S*-(*p*-hydroxyphenylacetohydroximoyl)-L-cysteine compound was hampered by the fact that it undergoes internal cyclization to produce the corresponding 2-substituted thiazoline-4-carboxylic acid and release hydroxylamine *in vitro*.[58] When *N*-acetylcysteine was used as sulfur donor, the corresponding product *S*-(*p*-hydroxyphenylacetohydroximoyl)-*N*-acetylcysteine did not undergo cyclization.[58] This product, however, is not a substrate for a C-S lyase, as it does not have an unsubstituted amino group. This indicates that the expected C-S lyase in the glucosinolate pathway is tightly coupled to the sulfur conjugating enzyme. The production of *S*-(*p*-hydroxyphenylacetohydroximoyl)-L-cysteine and *S*-(*p*-hydroxyphenylacetohydroximoyl)-*N*-acetylcysteine was inhibited by the cytochrome P450 inhibitor tetcyclasis, indicating that the reaction was dependent on a cytochrome P450.[58]

Molecular Identification of Oxime-Metabolizing Enzyme in Glucosinolate Biosynthesis

The enzymatic activity of the oxime-metabolizing enzyme in *S. alba* was too low to pursue a classical biochemical approach to isolate the genes involved. Because of the assumed evolutionary relationship between glucosinolates and cyanogenic glucosides, a cytochrome P450 related to the oxime-metabolizing enzyme in the cyanogenic pathway was anticipated. Based on homology to P450ox (CYP71E1), several candidate genes were identified in the *Arabidopsis* genome and subjected to phylogenetic analysis. In contrast to the homogenous CYP79 family, which only comprises cytochromes P450 that are *N*-hydroxylases catalyzing the conversion of amino acids to the corresponding oximes, the CYP71 family with 52 members is the largest cytochrome P450 family in *Arabidopsis*. The many CYP71s have very different catalytic functions and cluster into different subgroups. In addition to the cytochromes P450 designated as CYP71s, members of the CYP83 family belong to the CYP71 family, based on sequence similarity (http://www.drnelson.utmem.edu). Among the candidate genes, *CYP71B7*, *CYP71B6*, and *CYP83B1* exist as full-length ESTs and were expressed in *E. coli*.[58] When radiolabeled *p*-hydroxyphenylacetaldoxime and cysteine were administered to a reaction mixture containing spheroplasts of *E. coli* expressing CYP83B1 and purified NADPH:cytochrome P450-reductase in the presence of NADPH, exactly the same biochemistry was observed as in the reaction mixtures containing microsomes from *S. alba*.[58] The expression pattern of CYP83B1 and phenotypes of *CYP83B1* mutants (S. Bak, pers. comm.) were in agreement with CYP83B1 being an oxime-metabolizing enzyme in the glucosinolate biosynthetic pathway (Fig. 13.4).

In the conversion of oxime to S-alkyl thiohydroximate, the oxime has to be oxidized to a nitro or *aci*-nitro compound or a nitrile oxide prior to conjugation with the sulfur donor. *Aci*-nitro compounds are tautomers of nitro compounds and are formed from nitro compounds under alkaline conditions. As no nitro compounds were detected in reactions performed with recombinant CYP83B1, there is no indication that nitro compounds are intermediates in the glucosinolate pathway.[58] CYP83B1 may produce a nitrile oxide, which is a very reactive compound and likely to be subject of nucleophilic attack by, for example, cysteine. In the absence of nucleophiles, nitrile oxides dimerize to form furoxans (1,2,5-oxadiazole 2-oxide). Based on the strong reactivity of nitrile oxides, it is likely that conjugation with the sulfur donor is under strict control, partly to prevent dimerization of the nitrile oxide and partly to ensure that only a cysteine conjugate is formed that is hydrolyzable by a C-S lyase. The cysteine conjugation might be carried out by a tightly coupled glutathione-*S*-transferase, as previously proposed.[58] Alternatively, CYP83B1 may have a binding site not only for the oxime, but also for cysteine, although this would be highly unusual.

Characterization of recombinant CYP83B1 heterologously expressed in yeast showed that IAOX is a high affinity substrate for CYP83B1, whereas aromatic amino acids are low affinity substrates for CYP83B1.[62,63] Another member of the CYP83 family in *Arabidopsis*, CYP83A1 with 65% amino acid identity to CYP83B1, was shown to have low affinity for IAOX, but high affinity for the aromatic [64] and aliphatic oximes (Naur and Halkier, unpublished results). This indicates that CYP83A1 and CYP83B1 are not redundant in the plant.

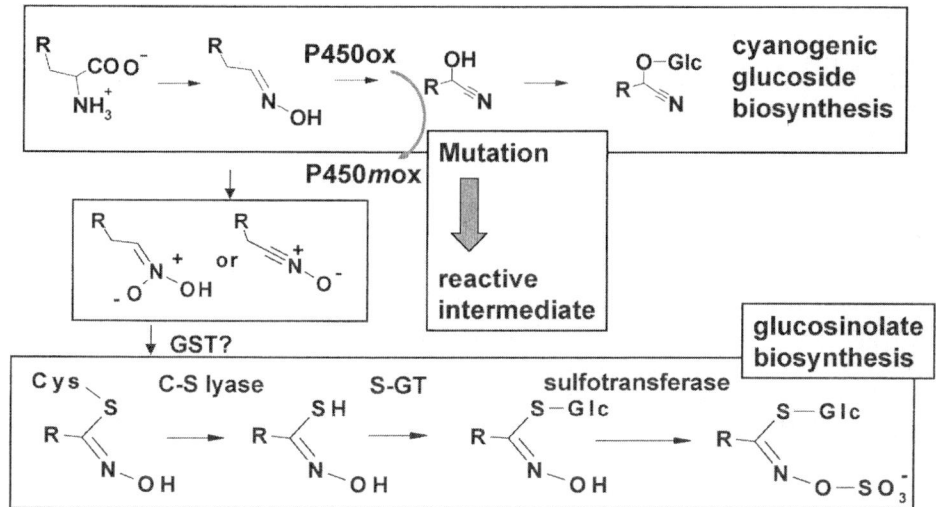

Figure 13.4: The proposed evolutionary relationship between cyanogenic glucosides and glucosinolates. The oxime-metabolizing enzyme in the glucosinolate pathway (CYP83B1/ P450*m*ox) is related to the oxime-metabolizing enzyme in the cyanogenic pathway (CYP71E1/P450ox). This is consistent with the hypothesis that P450ox was mutated into a P450*m*ox that converted the oxime into a toxic compound that the plant detoxified into a glucosinolate. The amino acids shown to be substrates for CYP79s are given.

Several knockout mutants of *CYP83B1* have been identified due to their strong IAA overexpression phenotype.[62,64] The expected result of a blockage of CYP83B1 in glucosinolate-producing plants is the accumulation of IAOX. As IAOX is a precursor of IAA (Fig. 13.4), an increased flux into the synthesis of IAA in the seedling stage would be expected and might be responsible for the observed phenotype. The *sur2* knockout mutant of *CYP83B1* has increased levels of IAA.[64] Accordingly, CYP83B1 has been described as a regulator of auxin production by controlling the flux of IAOX into IAA and indole glucosinolates.[62]

HETEROLOGOUS EXPRESSION OF CYP79 HOMOLOGUES IN E. COLI

E. coli has proved to be a good host organism for heterologous expression of several cytochromes P450 of the CYP79 family. The first CYP79 to be expressed in

E. coli was CYP79A1 from S. bicolor.[65] CYP79A1 was expressed in the strain JM109 by use of the expression vector pSP19g10L.[66] N-terminal modifications were designed into four different expression constructs to apply previously obtained experience for high level expression of eukaryotic cytochromes P450 in E. coli.[67] These included constructs encoding the native protein and three proteins that were N-terminally 'bovine-modified' and N-terminally truncated to different degrees.[65] The term 'bovine-modification' indicates that the first eight amino acids were substituted for the eight N-terminal amino acids of the bovine cytochrome P450 CYP17A.[67] All the recombinant proteins were functionally expressed, as evidenced by the ability of the transformed E. coli cells to produce p-hydroxyphenylacetaldoxime upon administration of radiolabeled tyrosine to the whole E. coli cells. This showed that the endogenous electron-donating enzyme system of E. coli, flavodoxin and NADPH:flavodoxin-reductase, was able to donate electrons to the heterologous cytochrome P450, and that tyrosine was taken up by the cells. Only one of the constructs, TYRΔ(1-25)$_{bov}$, produced sufficient levels of functional protein (up to 900 nmol (1 culture)$^{-1}$) to provide a CO difference spectrum.

We applied the same principles to heterologous expression of the CYP79 homologues from *Arabidopsis*, but surprisingly the *Arabidopsis* CYP79s behaved differently from CYP79A1 with respect to their ability to being produced in E. coli and with respect to their interaction with the electron-donating system of E. coli. For CYP79B2, the construct expressing the native protein gave the highest level of expression.[13] However, the expression level was not high enough to obtain a CO difference spectrum. In the case of CYP79F1, only one construct encoding the 'native' protein was made. With this construct, an expression level of 110 nmol (1 culture)$^{-1}$ was obtained, as evidenced by CO difference spectroscopy.[10] However, in the third case,[12] the 'native' protein had catalytic activity, but only a chimeric protein with the N-terminus of CYP79A1 (TYRΔ(1-25)$_{bov}$) and the catalytic domain of CYP79A2 was expressed at high enough levels (50 nmol (1 culture)$^{-1}$) to enable CO difference spectroscopy.[12] For CYP79B2 and CYP79F1, the expression level of functional protein was increased (up to 200 fold in the case of CYP79B2), when the E. coli strain C43(DE3) was used instead of strain JM109. In contrast, the level of functional CYP79A2 was only slightly increased in strain C43(DE3) compared with strain JM109. C43(DE3) is a mutant E. coli strain that has been selected for its ability to accommodate large amounts of heterologous membrane proteins.[68] For all three CYP79s, activity was only obtained when spheroplasts of E. coli transformed with the expression construct were reconstituted with sorghum NADPH:cytochrome P450 reductase in the presence of NADPH. When we reconstituted spheroplasts of E. coli expressing CYP79A1 with NADPH:cytochrome P450 reductase, a higher level of activity was obtained than when only E. coli's own electron-donating system was used (Chen and Halkier, unpubl. results). This shows that although E. coli's electron-donating system is able to support CYP79A1, it is a less efficient redox partner than the NADPH:cytochrome P450 reductase from sorghum. Possibly, this may apply to other cytochromes P450.

From the experiences with heterologous expression of CYP79s in *E. coli*, it is concluded that the strategy for success is unpredictable. In some cases, the construct expressing the 'native' protein was the most efficient one, in other cases, the construct for the N-terminally modified protein provided the highest level of functional protein. Generally, the strain C43(DE3) seems to be more suitable for expression of cytochromes P450 than JM109. It is common for all CYP79s that the enzyme activity of the recombinant proteins is stimulated by addition of 3 mM glutathione. The stimulatory effect of glutathione has been observed previously,[65,69] but its mechanism is unknown.

METABOLIC ENGINEERING OF GLUCOSINOLATE PROFILES IN TRANSGENIC PLANTS

The identification of the enzymes catalyzing the conversion of amino acids to oximes in both the cyanogenic glucoside and the glucosinolate pathway provides molecular tools for genetic engineering of the glucosinolate profile of *Arabidopsis*, with respect to both endogenous glucosinolates and introduction of new glucosinolates. Alterations of the profile of endogenous glucosinolates have been obtained in transgenic plants with overexpression or downregulation of different endogenous CYP79s (Fig. 13.5). The outcome of the engineering process is dependent on the availability of the precursor amino acid and its role in other pathways of primary and secondary metabolism. Overexpression of CYP79s that have protein amino acids as substrates resulted in an increased content of the corresponding glucosinolates. This was seen in *Arabidopsis* plants overexpressing CYP79A2, which have phenylalanine as substrate. In 35S::CYP79A2 plants, benzylglucosinolate constituted up to 50% of the total glucosinolates in some of the transgenic lines[12] (Wittstock and Halkier unpubl. results). The 35S::CYP79A2 plants had no morphological phenotype.

When the expression of CYP79B2 was driven by the 35S promoter, a 4-fold increase in indole glucosinolates was observed for approximately 20% of the transgenic lines, whose appearance also resembled that of wild-type plants.[13] The majority (approximately 80%) of the 35S::CYP79B2 lines exhibited dwarfism, did not develop inflorescences, and, therefore, did not produce seeds. A possible explanation for this phenotype could be that the IAOX produced by CYP79B2 was partly channeled into production of the plant hormone IAA, which would disturb the growth and development of the plants.

When the expression of CYP79B2 was driven by the 35S promoter, a 4-fold increase in indole glucosinolates was observed for approximately 20% of the transgenic lines, whose appearance also resembled that of wild-type plants.[13] The

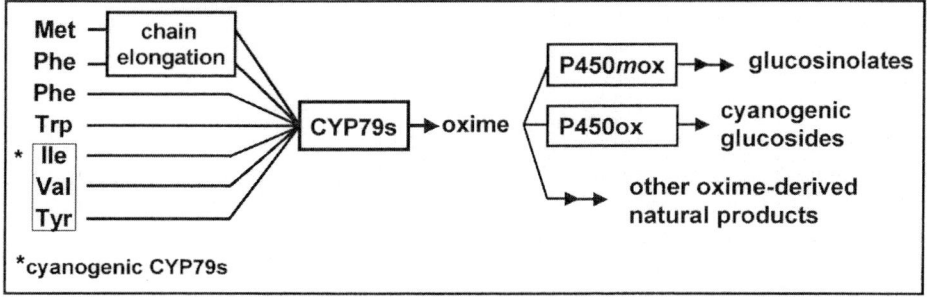

Figure 13.5: Metabolic engineering of oxime-derived natural products in *Arabidopsis* plants. Up- and downregulation of endogenous as well as exogenous CYP79s provides a powerful tool for alteration of the glucosinolate profiles and introduction of novel glucosinolates, as well as for engineering oxime-derived natural products.

majority (approximately 80%) of the 35S::CYP79B2 lines exhibited dwarfism, did not develop inflorescences, and, therefore, did not produce seeds. A possible explanation for this phenotype could be that the IAOX produced by CYP79B2 was partly channeled into production of the plant hormone IAA, which would disturb the growth and development of the plants.

When the substrate for a CYP79 is a chain-elongated amino acid, the chain-elongation step might be rate-limiting. In this case, upregulation of the corresponding CYP79 would have little effect. Downregulation of such a CYP79 might result in metabolic stress, due to accumulation of the chain-elongated precursor amino acids in the plant. This is illustrated by the transgenic plants, which are up- or downregulated in CYP79F1 or which do not express CYP79F1 at all due to a knock-out mutation.[10,11,48] Upregulation of the level of CYP79F1 resulted only in a slight increase in the level of glucosinolates derived from chain-elongated methionines, indicating that oxime formation is not the only rate-limiting step.[10,11] The co-suppressed and knockout mutants of CYP79F1 had a pronounced bushy phenotype with curly leaves. This is probably due to metabolic stress caused by the increased levels of dihomomethionine and trihomomethionine in the transgenic lines.[10]

Identification of CYP79s from other plant species, including cyanogenic plants, has provided means for alteration of glucosinolate profiles by introduction of exogenous glucosinolates. We have introduced the sorghum CYP79A1 into *Arabidopsis*.[49] This resulted in the production of high levels of the tyrosine-derived *p*-hydroxybenzylglucosinolate, not known to accumulate in *Arabidopsis*. Only small amounts of other compounds derived from *p*-hydroxyphenylacetaldoxime accumulated in the plants. This showed that the *p*-hydroxyphenylacetaldoxime produced by the 'cyanogenic' CYP79A1 was efficiently channeled into the biosynthetic pathway of glucosinolates leading to the production of high amounts of *p*-hydroxybenzylglucosinolate. The result is in agreement with previous biochemical data, which showed that the downstream enzymes in the glucosinolate pathway have a low specificity for the side chain of the oxime. CYP79D1 and CYP79D2 from cassava catalyze the conversion of valine and isoleucine to the corresponding oximes in the biosynthesis of the cyanogenic glucosides linamarin and lotaustralin.[32] Introduction of the cyanogenic CYP79D2 into *Arabidopsis* resulted in the production of the valine- and isoleucine-derived glucosinolates, not known to be natural constituents of *Arabidopsis* (Fig. 13.5) (Mikkelsen and Halkier unpubl. results).

It appears from the described examples from *Arabidopsis* that CYP79s are good targets for metabolic engineering of glucosinolate profiles. It seems possible to engineer 'custom-designed' glucosinolate profiles in *Arabidopsis* by combination of overexpression and downregulation of various CYP79s. The level of the accumulated glucosinolate derived directly from protein amino acids reflects the efficiency of the individual CYP79s, whereas the chain-elongation step might be rate-limiting for the accumulation of glucosinolates derived from chain-elongated amino acids.

SUMMARY AND FUTURE PERSPECTIVES

Within the last few years, considerable advances in our understanding of glucosinolate biosynthesis have been achieved. It has been demonstrated that cytochromes P450 belonging to the CYP79 family catalyze the conversion of the precursor amino acids to the corresponding oximes in the biosynthesis of glucosinolates and cyanogenic glucosides. This supports the idea that glucosinolates have evolved based on a 'cyanogenic' predisposition. Identification of the CYP79s provides molecular tools that enable metabolic engineering of plants with new glucosinolate profiles by altering the level of endogenous CYP79s and by introducing new CYP79s to a plant. The oxime-metabolizing enzyme (CYP83B1/P450*mox*) in *Arabidopsis* has been shown to be related to the oxime-metabolizing enzyme (CYP71E1, P450ox) of the cyanogenic glucoside pathway. This is consistent with the hypothesis that a mutation of P450ox led to the production of toxic compounds that were detoxified into glucosinolates.

While the elucidation of the glucosinolate pathway from the precursor amino acid to the glucosinolate core structure can be regarded as being close to completion, we still have a lot to learn about the genes and enzymes involved in the biosynthesis of chain-elongated precursor amino acids and secondary modifications of the side chain. The wide range of biological activities of glucosinolate degradation products and the dual role of glucosinolates as deterrents and attractants in plant-insect-interactions suggest specific functions for individual glucosinolates and the existence of a complex regulatory network. Little is known about the mechanisms for regulation of the biosynthesis of individual glucosinolates in response to specific signaling molecules. In addition, it is largely unknown how glucosinolates are turned over and translocated in a physiologically safe manner within the plant.

Identification of the CYP79 provides molecular tools that enable metabolic engineering of plants with new glucosinolate profiles by altering the level of endogenous CYP79s and by introducing new CYP79s to a plant. Metabolic engineering at the level of side chain modifications may allow specific alterations of glucosinolate profiles, which may facilitate further investigations of the physiological and ecological role of glucosinolates. This approach, in combinations with introduction and/or knockouts of various oxime-metabolizing enzymes, has great potential for the design of 'biotech crops' with novel natural products and with improved pest resistance and increased nutritional value.

ACKNOWLEDGEMENTS

Karina Peitersen is thanked for technical assistance in writing the manuscript. Our work on glucosinolates has been financially supported by grants from the Danish Governmental Biotechnology Programme, the Danish National Research Foundation, the Danish Agricultural and Veterinary Research Council and Director Ib Henriksens Foundation.

REFERENCES

1. CHEW, F.S., Biological effects of glucosinolates. In: Biologically Active Natural Products Potential Use in Agriculture (H.G. Cutler, ed.), American Chemical Society, Washington DC. 1988, pp. 155-181.
2. HALKIER, B.A., Glucosinolates. In: Naturally occurring Glycosides (R. Ikan, ed.), John Wiley & Sons, London. 1999, pp. 193-223.
3. RASK, L., ANDRÉASSON, E., EKBOM, B., ERIKSSON, S., PONTOPPIDAN, B., MEIJER, J., Myrosinase gene family evolution and herbivore defense in Brassicaceae, *Plant Mol. Biol.*, 2000, **42**, 93-113.
4. MITHEN R., DEKKER M., VERKERK R., RABOT S., JOHNSON I., The nutritional significance, biosynthesis and bioavailability of glucosinolates in human foods, *J. Sci. Food and Agricult.*, 2000, **80**, 967-984.
5. FAHEY, J.W., ZALEMANN, A.T., TALALAY, P., The chemical diversity and distribution of glucosinolates and isothiocyanates among plants, *Phytochemistry*, 2001, **56**, 5-51.
6. TALALAY, P., ZHANG, Y., Chemoprotection against cancer by isothio-cyanates and glucosinolates, *Biochem. Soc. Trans.*, 1996, **24**, 806-810.
7. ZHANG, Y., TALALAY, P., CHO, C-G., POSNER, G.H., A major inducer of anticarcinogenic protective enzymes from broccoli: Isolation and elucidation of structure, *Proc. Natl. Acad. Sci. USA*, 1992, **89**, 2399-2403.

8. VERHOEVEN, D.T.H., VERHAGEN, H., GOLDBOHM, R.A., VAN DER BRANDT, P.A., VAN POPPEL, G., A review of mechanisms underlying anticarcinogenicity by *Brassica* vegetables, *Chemico Biological Interactions*, 1997, **103**, 79-129.
9. HALKIER, B.A., DU, L., The biosynthesis of glucosinolates, *Trend Plant Sci.* 1997, **2**, 425-431.
10. HANSEN, C.H., WITTSTOCK, U., OLSEN, C.E., HICK, A.J., PICKETT, J.A., HALKIER, B.A., Cytochrome P450 CYP79F1 from *Arabidopsis* catalyzes the conversion of dihomomethionine and trihomomethionine to the corresponding aldoximes in the biosynthesis of aliphatic glucosinolates, *J. Biol. Chem.*, 2001, **276**, 11078-11085.
11. REINTANZ, B., LEHNEN, M., REICHELT, M., GERSHENZON, J., KOWALCZYK, M., SANDBERG, G., GODDE, M., UHL, R., PALME, K., *bus*, a bushy *Arabidopsis CYP79F1* knockout mutant with abolished synthesis of short chain aliphatic glucosinolates, *Plant Cell*, 2000, **13**, 351-367.
12. WITTSTOCK, U., HALKIER, B.A., Cytochrome P450 CYP79A2 from *Arabidopsis thaliana* L. catalyzes the conversion of L-phenylalanine to phenylacetaldoxime in the biosynthesis of benzylglucosinolate, *J. Biol. Chem.*, 2000, **275**, 14659-14666.
13. MIKKELSEN, M.D., HANSEN, C.H., WITTSTOCK, U., HALKIER, B.A., Cytochrome P450 CYP79B2 from *Arabidopsis* catalyzes the conversion of tryptophan to indole-3-acetaldoxime, a precursor of indole glucosinolates and indole-3-acetic acid, *J. Biol. Chem.*, 2000, **275**, 33712-33717.
14. HULL, A.K., VIJ, R., CELENZA, J.L., *Arabidopsis* cytochromes P450 that catalyze the first step of tryptophan-dependent indole-3-acetic acid biosynthesis, *Proc. Natl. Acad. Sci. USA*, 2000, **97**, 2379-2384.
15. ETTLINGER, M.G., KJÆR, A., Sulfur compounds in plants, *Rec. Adv. Phytochemistry*, 1968, **1**, 59-144.
16. WALLSGROVE, R.M., BENNETT, R.N., The biosynthesis of glucosinolates in Brassicas. In: Amino Acids and Their Derivatives in Higher Plants, (R.M. Wallsgrove, ed.), Cambridge University Press, Cambridge. 1995, pp. 243-259.
17. REED, D.W., DAVIN, L., JAIN, J.C., DELUCA, V., NELSON, L., UNDERHILL, E.W., Purification and properties of UDP-glucose: thiohydroximate glucosyltransferase from *Brassica napus* L. seedlings, *Arch. Biochem. Biophys.*, 1993, **305**, 526-532.
18. GUO, L., POULTON, J.E., Partial purification and characterization of *Arabidopsis thaliana* UDPG: Thiohydroximate glucosyltransferase, *Phytochemistry*, 1994, **36**, 1133-1138.
19. GLENDENING, T.M., POULTON, J.E., Glucosinolate biosynthesis. Sulfation of desulfoglucosinolate by cell-free extracts of cress (*Lepidium sativum* L.) seedlings, *Plant Physiol.*, 1988, **86**, 319-321.
20. JAIN, J.C., GROOTWASSINK, J.W.D., KOLENOVSKY, A.D., UNDERHILL, E.W., Purification and properties of 3-phosphoadenosine-5-phosphosulphate: desulphoglucosinolate sulphotransferase from *Brassica juncea* cell cultures, *Phytochemistry*, 1990, **29**, 1425-1428.

21. MARSOLAIS, F., GIDDA, S.K., BOYD, J., VARIN, L., Plant soluble sulfotransferases: structural and functional similarity with mammalian enzymes. In: Evolution of Metabolic Pathways (J.T. Romeo, R. Ibrahim, L. Varin, V. de Luca, eds.), Elsevier Science Ltd., Amsterdam. 2000, pp. 433-456.
22. JONES, P.R., ANDERSEN, M.D., NIELSEN, J.S., HØJ, P.B., MØLLER, B.L., The biosynthesis, degradation, transport and possible function of cyanogenic glucosides. In: Evolution of Metabolic Pathways, Recent Advances in Phytochemistry (J.T. Romeo, R. Ibrahim, L. Varin, V. de Luca, eds.), Elsevier, New York. 2000, **34**, pp. 191-247.
23. SIBBESEN, O., KOCH, B., HALKIER, B.A., MØLLER, B.L., Cytochrome P450$_{TYR}$ is a multifunctional heme-thiolate enzyme catalyzing the conversion of l-tyrosine to p-hydroxyphenylacetaldehyde oxime in the biosynthesis of the cyanogene glucoside dhurrin in Sorghum bicolor (L.) Moench, J. Biol. Chem., 1995, **270**, 3506-3511.
24. KOCH, B., SIBBESEN, O., SVENDSEN, I., MØLLER, B.L., The primary sequence of cytochrome P-450tyr, the multifunctional N-hydroxylase catalyzing the conversion of L-tyrosine to p-hydroxy-phenylacetaldehyde oxime in the biosynthesis of the cyanogenic glucoside dhurrin in Sorghum bicolor (L.) Moench, Arch. Biochem. Biophys., 1995, **323**, 177-186.
25. HALKIER, B.A., OLSEN, C.E., MØLLER, B.L., The biosynthesis of cyanogenic glucosides in higher plants. The (E)- and (Z)-isomers of p-hydroxyphenylacetaldehyde oxime as intermediates in the biosynthesis of dhurrin in Sorghum bicolor (L.) Moench, J. Biol. Chem., 1989, **264**, 19487-19494.
26. HALKIER, B.A., MØLLER, B.L., The biosynthesis of cyanogenic glucosides in higher plants. Identification of three hydroxylation steps in the biosynthesis of dhurrin in Sorghum bicolor (L.) Moench and the involvement of 1-aci-nitro-2-(p-hydroxyphenyl)ethane as an intermediate, J. Biol. Chem., 1990, **265**, 21114-21121.
27. HALKIER, B.A., LYKKESFELDT, J., MØLLER, B.L., 2-Nitro-3-(p-hydroxyphenyl)propionate and aci-1-nitro-2-(p-hydroxyphenyl)ethane, two novel intermediates in the biosynthesis of the cyanogenic glucoside dhurrin in Sorghum bicolor (L.) Moench, Proc. Natl. Acad. Sci. USA, 1991, **88**, 487-491.
28. BAK, S., KAHN, R.A., NIELSEN, H.L., MØLLER, B.L., HALKIER, B.A., Cloning of three A-type cytochromes P450, CYP71E1, CYP98, and CYP99 from Sorghum bicolor (L.) Moench by a PCR approach and identification by expression in Escherichia coli of CYP71E1 as a multifunctional cytochrome P450 in the biosynthesis of the cyanogenic glucoside dhurrin, Plant Mol. Biol., 1998, **36**, 393-405.
29. KAHN, R.A., BAK, S., SVENDSEN, I., HALKIER, B.A., MØLLER, B.L., Isolation and reconstitution of cytochrome P450ox and in vitro reconstitution of the entire biosynthetic pathway of the cyanogenic glucoside dhurrin from sorghum, Plant Physiol., 1997, **115**, 1661-1670.
30. JONES, P.R., MØLLER, B.L., HØJ, P.B. The UDP-glucose:p-Hydroxymandelonitrile-O-glucosyltransferase that catalyzes the last step in synthesis of the cyanogenic glucoside dhurrin in Sorghum bicolor, J. Biol. Chem., 1999, **274**, 35483-35491.

31. NIELSEN, J.S., MØLLER, B.L., Cloning and expression of cytochrome P450 enzymes catalysing the conversion of tyrosine to p-hydroxyphenyl-acetaldoxime in the biosynthesis of cyanogenic glucosides in *Triglochin maritime*, *Plant Physiol.*, 2000, **122**, 1311-1321.
32. ANDERSEN, M.D., MØLLER, B.L., Cytochromes P450 from Cassava (*Manihot esculenta* Crantz) catalyzing the first steps in the biosynthesis of the cyanogenic glucosides linamarin and lotaustralin cloning, functional expression in *Pichia pastoris* and substrate specificity of the isolated recombinant enzymes, *J. Biol. Chem.*, 2000, **275**, 1966-1975.
33. DU, L., LYKKESFELDT, J., OLSEN, C.E., HALKIER, B.A., Involvement of cytochrome P450 in oxime production in glucosinolate biosynthesis as demonstrated by an in vitro microsomal enzyme system isolated from jasmonic acid-induced seedlings of *Sinapis alba* L, *Proc. Natl. Acad. Sci. USA*, 1995, **92**, 12505-12509.
34. BENNETT, R.N., KIDDLE, G., WALLSGROVE, R.M., Involvement of cytochrome P450 in glucosinolate biosynthesis in white mustard, *Plant Physiol.*, 1997, **114**, 1283-1291.
35. DU, L., HALKIER, B.A., Isolation of a microsomal enzyme system involved in glucosinolate biosynthesis from seedlings of *Tropaeolum majus* L, *Plant Physiol.*, 1996, **111**, 831-837.
36. BENNETT, R.N., DAWSON, G.W., HICK, A.J., WALLSGROVE, R.M., Glucosinolate biosynthesis: further characterization of the aldoxime-forming microsomal monoxygenases in oilseed rape leaves, *Plant Physiol.*, 1995, **109**, 299-305.
37. BENNETT, R.N., KIDDLE, G., WALLSGROVE, R.M., Biosynthesis of benzylglucosinolate, cyanogenic glucosides and phenylpropanoids in *Carica Papaya, Phytochemistry*, 1997, **45**, 59-66.
38. LUDWIG-MÜLLER, J., HILGENBERG, W., A plasma membrane-bound enzyme oxidases L-tryptophan to indole-3-acetaldoxime, *Physiol. Plant.*, 1988, **74**, 240-250.
39. LUDWIG-MÜLLER, J., RAUSCH, T., LANG, S., HILGENBERG, W., Plasma membrane-bound high plant isoenzymes convert tryptophan to indole-3-acetaldoxime, *Phytochemistry*, 1990, **29**, 1397-1400.
40. NORMANLY, J., BARTEL, B., Redundancy as a way of life – IAA metabolism, *Current Opinion in Plant Biology*, 1999, **2**, 207-213.
41. CELENZA, J.L., Metabolism of tyrosine and tryptophan – new genes for old pathways, *Current Opinion in Plant Biology*, 2001, **4**, 234-240.
42. RAJAGOPAL, R., LARSEN, P., Metabolism of indole-3-acetaldoxime in plants, *Planta*, 1972, **103**, 45-54.
43. HELMLINGER, J., RAUCH, T., HILGENBERG, W., Metabolism of ^{14}C-indole-3-acetaldoxime by hypocotyls of chinese cabbage, *Phytochemistry*, 1985, **24**, 2497-2502.
44. LUDWIG-MÜLLER, J., HILGENBERG, W., Conversion of indole-3-acetaldoxime to indole-3-acetonitrile by plasma membranes from Chinese cabbage, *Physiol. Plant.*, 1990, **79**, 311-318.

45. HELMINGER, J., RAUSCH, T., HILGENBERG, W., A soluble protein factor from Chinese cabbage converts indole-3-acetaldoxime to IAA, *Phytochemistry*, 1987, **26**, 615-618.
46. ZHAO, Y., CHRISTENSEN, S.K, FANKHAUSER, C., CASHMAN, J.R., COHEN, J.D., WEIGEL, D., CHORY, J., A role for flavin monoxygenase-like enzymes in auxin biosynthesis, *Science*, 2001, **291**, 306-309.
47. MÜLLER, A., WEILER, E.W., IAA-synthase, an enzyme complex from *Arabidopsis thaliana* catalyzing the formation of indole-3-acetic acid from (S)-tryptophan, *Biol. Chem.*, 2000, **381**, 679-686.
48. TANTIKANJANA, T., YONG, J.W.H., LETHAM, D.S., GRIFFITH, M., HUSSAIN, M., LJUNG, K., SANDBERG, G., SUNDARESAN, V., Control of axillary bud initiation and shoot architecture in *Arabidopsis* through the SUPERSHOOT gene, *Genes and Development*, 2001, **15**, 1577-1588.
49. BAK, S., OLSEN, C.E., PETERSEN, B.L., MØLLER, B.L., HALKIER, B.A., Metabolic engineering of *p*-hydroxybenzylglucosinolate in *Arabidopsis* by expression of the cyanogenic CYP79A1 from *Sorghum bicolor*, *Plant J.*, 1999, **20**, 663-671.
50. SAUPE, S.G., Cyanogenic compounds and angiosperm phylogeny. In: Phytochemistry and Angiosperm Phylogeny (D.A. Young and D.S. Seigler, eds.), Praeger Scientific, New York, 1981, pp. 81-116.
51. RODMAN, J.E., SOLTIS, P.S., SOLTIS, D.E., SYTSMA, K.J., KAROL, K.G., Parallel evolution of glucosinolate biosynthesis inferred from congruent nuclear and plastid gene phylogenies, *Amer. J. Bot.*, 1998, **85**, 997-1006.
52. BAK, S., NIELSEN, H.L., HALKIER, B.A., The presence of CYP79 homologues in glucosinolate-producing plants show evolutionary conservation of the enzymes in the conversion of amino acid to aldoxime in the biosynthesis of cyanogenic glucosides and glucosinolates, *Plant Mol. Biol.*, 1998, **38**, 725-734.
53. DAHLGREN, R., A system of classification of the angiosperms to be used to demonstrate the distribution of characters, *Botaniska Notiser*, 1975, **128**, 119-147.
54. DAHLGREN, R., A commentary on a diagrammatic presentation of the angiosperms in relation to the distribution of character states, *Plant Syst. Evol.*, 1977, Supplement **1**, 253-283.
55. RODMAN, J.E., KAROL, K.G., PRICE, R.A., SYTSMA, K.J. 1996. Molecules, morphology, and Dahlgrens expanded order Capparales, *Systematic Bot.*, 1996, **21**, 289-307.
56. BURMEISTER, W.P., COTLAZ, S., DRIGUEZ, H., LORI, R., PALMIERI, S., HENRISSAT, B., The crystal structures of *Sinapis alba* myrosinase and a covalent glycosyl-enzyme intermediate provide insights into the substrate recognition and active-site machinery of an *S*-glycosidase, *Structure*, 1997, **5**, 663-675.
57. GRIFFITHS, D.W., DEIGHTON, N., BIRCH. A.N.E., PATRIAN, B., BAUR, R., STÄDTLER, E. Identification of glucosinolates on the leaf surface of plants from the Cruciferae and other closely related species, *Phytochemistry*, 2001, **57**, 693-700.
58. HANSEN, C.H., DU. L., NAUR, P., OLSEN, C.E., AXELSEN, K.B., HICK, A.J., PICKETT, J.A., HALKIER, B.A., CYP83B1 is the oxime-metabolizing enzyme in the glucosinolate pathway in *Arabidopsis, J. Biol. Chem.*, 2001, **276**, 24790-24796.

59. WETTER, L.R., CHISHOLM, M.D., Sources of sulfur in the thioglucosides of various higher plants, *Can. J. Biochem.*, 1968, **46**, 931-935.
60. KIDDLE, G.A., BENNETT, R.N., HICK, A.J., WALLSGROVE, R.M., C-S lyase activities in leaves of crucifers and non-crucifers, and the characterzation of three classes of C-S lyase activities from oilseed rape (*Brassica napus* L.), *Plant, Cell and Environment*, 1999, **22**, 433-445.
61. SCHWIMMER, S., KJAER, A., Purification and specificity of the C-S lyase of *Albizzia lophanta*, *Biochim. Biophys. Acta*, 1960, **42**, 316-324.
62. BAK, S., TAX, F.E., FELDMANN, K.A., GALBRAITH, D.A., FEYEREISEN, R., CYP83B1, a cytochrome P450 at the metabolic branchpoint in auxin and indole glucosinolate biosynthesis in *Arabidopsis thaliana*, *Plant Cell*, 2001, **13**, 101-111.
63. BAK, S., FEYEREISEN, R., The involvement of two P450 enzymes, CYP83B1 and CYP83A1, in auxin homeostasis and glucosinolate biosynthesis, *Plant Physiol.*, 2001, **127**, 108-118.
64. BARLIER, I., KOWALCZYK, M., MARCHANT, A., LJUNG, K., BHALERAO, R., BENNETT, M., SANDBERG, G., BELLINI, C., The SUR2 gene of *Arabidopsis thaliana* encodes the cytochrome P450 CYP83B1, a modulator of auxin homeostasis, *Proc. Natl. Acad. Sci. USA*, 2000, **97**, 14819-14824.
65. HALKIER, B.A., NIELSEN, H.L., KOCH, B., MØLLER, B.L., Purification and characterization of recombinant cytochrome P450$_{TYR}$ expressed at high levels in *Escherichia coli*, *Arch. Biochem. Biophys.*, 1995, **322**, 369-377.
66. BARNES, H.J., Maximizing expression of eukaryotic cytochromes P450 in *Escherichia coli*, *Meth. Enzymol.*, 1996, **272**, 3-14.
67. BARNES, H.J., ARLOTTO, M.P., WATERMAN, M.R., Expression and enzymatic activity of recombinant cytochrome P450 17á-hydroxylase in *Escherichia coli*, *Proc. Natl. Acad. Sci. USA*, 1991, **88**, 5597-5601.
68. MIROUX, B., WALKER, J.E., Over-production of proteins in *Escherichia coli*: Mutant hosts that allow synthesis of some membrane proteins and globular proteins at high levels, *J. Mol. Biol.* 1996, **260**, 289-298.
69. GILLAM, E.M.J., BABA, T., KIM, B-R., OHMORI, S., GUENGERICH, F.P., Expression of modified human cytochrome P450 3A4 in *Escherichia coli* and purification and reconstitution of the enzyme, *Arch. Biochem. Biophys.*, 1993, **305**, 123-131.

INDEX

α-/β-amyrin synthase, 84
Acacia victoriae, 44
Acridone synthase, 209
Actinorhodin, 198
Activation tagging, 32, 112, 114, 119-120, 135; *see also T-DNA tagging*
Active site, 198, 203-204, 208-211, 214-215, 217-218
Agrobacterium, 112, 119
  A. rhizogenes, 119
Aldol condensation, 214
Alfalfa, 32, 44, 47, 104, 172, 180, 190, 192, 199, 203, 211, 215, 217
Alfalfa mosaic virus (AlMV), 190
Alkaloids, 119, 164-165, 167, 169-173, 175-176
  benzo[*c*]phenanthridine, 165
  biosynthesis, 164-165, 170-172, 176
  bisbenzylisoquinoline, 167
  indole, 119, 164, 173
  isoquinoline, 171, 176
  phenanthrene, 173
  steroidal, 82, 89
Allelopathy, 44
Amino acid levels, 67, 180
Analytical variance, 39
Anesthetic molecules, 126
Anthocyanidins, 137
Anthocyanins, 98-99, 114-116, 118-119, 126, 137, 198, 218
Antibody, 105, 164
  scFv antibodies, 96, 105
Anticancer, 44, 126, 180, 224
Antifungal, 44, 84-85, 214

Anti-inflammatory, 44
Antimicrobial, 170, 198, 218, 224, 231
Antinutritional effects, 89
Apoptosis, 44
Arabidopsis, 7, 19, 22-26, 63-64, 68-70, 72-75, 84, 86, 95-96, 98-99, 101, 103-106, 112, 114, 116, 118-120, 126-127, 129-132, 135, 137-141, 152, 165, 180, 184-187, 189-190, 223, 227, 229-234, 236-237, 239-242
  Arabidopsis genome, 22-25, 98, 118, 127, 129, 141, 186-187, 236
  Arabidopsis mutants, 63, 73
  Arabidopsis thaliana, 7, 68-69, 84, 86, 152, 229, 230
Atomic resolution protein x-ray crystallography, 218
Attractants, 224, 242
Auxin, 96, 101, 103-104, 106, 119, 135, 238; *see also Indole-3-acetic acid*
  biosynthesis, 125, 135
  transport, 96, 101, 103-104, 106
Avena, 81, 84-85
  A. strigosa, 81, 85, 88
Avenacins, 84-85, 89
  biosynthesis, 84

β-amyrin, 84-86, 88-89
β-amyrin synthase (AsbAS1), 81, 85-89
β-ketoacyl synthases (KASs), 204, 211

INDEX

Bacillus halodurans, 203
Bacillus megaterium, 68
Bacillus subtilis, 203
Bacteria, 23-24, 68, 136, 198, 203
Bacterial artificial chromosome (BAC) clones, 184, 186
Basil, 146
Benzalacetone synthase (BAS), 208
Benzo[*c*]phenanthridine alkaloids, 165
Berbamunine, 167, 169
  synthase, 167, 169
Berberine, 165, 167, 169-170, 172
Berberine bridge enzyme, 165, 167, 169, 172
Berberis, 167, 169
  B. stolonifera, 169
Bioinformatics, 2, 3, 181, 192
Biological variance, 36, 188
Bioplastics, 64, 68
Biotech crops, 243
Biotic stress, 115
Bisbenzylisoquinoline alkaloids, 167
Brassica, 224
Brassicaceae, 224, 235
Brassinolide (BL), 132-133, 135
Brassinosteroids (BR), 126, 132
  biosynthetic pathway, 132

C_1 enzymes, 15-16, 18, 22-23, 26, 116, 118-119, 209, 214
C_1 metabolism, 15-16, 18-19, 26
CHO-THF cycloligase (5-FCL), 22, 26
Caenorhabditis elegans, 126, 165
Capillary electrophoresis (CE), 4, 37, 50, 52, 71, 188
Capparales, 224, 234-235
Carbohydrates, 35, 50-51, 154-155, 158, 188
Carica papaya, 227, 234-235
Caricaceae, 234
Carotenoids, 119

Catharanthus, 119, 173
  C. roseus, 173-174
Cathasterone, 132
Cauliflower, 112, 139
cDNA-based microarrays, 185; *see also Microarray*
Cell cultures, 36, 119, 165, 185, 188-189
Cereals, 85, 88
Chain-elongation, 229-230, 233, 241-242
Chalcone synthase / stilbene synthase family (CHS/STS), 199, 205
Chalcones, 98, 172, 198-199, 204-205, 208-211, 214, 218
  chalcone isomerase (CHI), 98, 102, 104-105, 118
  chalcone reductase (CHR), 172, 198
  chalcone synthase (CHS), 98, 101-102, 104, 115, 118, 198-199, 203-204, 208-211, 214-215, 217-218
  tetrahydroxychalcone, 198
Chemical complexity, 35, 37
Chemical defenses, 81-82, 84-85, 136, 180, 187, 224, 242; *see also Protection*
Chickpea, 180
Chloroplast genome, 184
Cholesterol lowering, 44
Cinnamoyl-CoA, 199, 215
Cloning, 26, 112, 119-120, 135, 146, 169, 172, 210
Clustering, 8, 127, 181, 183, 185
CoA thioesters, 204-205, 207-208, 215
Codeine, 164, 172
Codeinone, 172-173
Codeinone reductase, 172-173
Computational biology, 2, 4, 8
Condensed tannins, 98-99
Coumaric acid, 115

Coumaroyltriacetic acid lactone (CTAL), 214
Crop alternatives, 192
Crop plants, 32, 65, 88, 99, 180, 192, 198-199, 224
Crosstalk, 99, 232
Cruciferous, 135, 232
Crystallographic studies, 199
Cyanogenic, 227, 234-236, 240-242
Cyanogenic glucosides, 227, 234-236, 242
  biosynthesis, 232
  pathway, 235, 242
Cycloartenol, 82, 84-86, 88-89
  cycloartenol synthases (CS), 84, 86, 88-89
CYP71s, 129, 237
CYP78A family, 130, 139-140
CYP79 family, 135, 223, 226-227, 229, 232-235, 237-243
CYP79A1, 227, 229, 234, 238-239, 241
CYP83s, 130, 134-135
CYP83B1, 130, 135, 237, 238, 242
CYP90 family, 132-133
Cytochromes P450 (P450s), 125-127, 129-133, 135, 137-138, 140, 167, 169-170, 226-227, 229, 233-235, 237-240, 242
  A-type P450s, 127
  non-A-type P450s, 127
  P450 reductases, 126
Cytochrome P450-dependent monooxygenases, 85, 104, 133, 151, 167, 224, 226-227, 229, 233-236, 238-240, 242

Daidzein, 136
Dammarane, 82, 84
Defensive compounds against pathogens, 136, 224
Deinococcus radiodurans, 203
Deoxy-D-xylulose 5-phosphate (DXP), 146, 151, 153, 156
Deoxy-D-xylulose 5-phosphate reductoisomerase (DXR), 146, 156, 158
Deoxy-D-xylulose 5-phosphate synthase (DXPS), 152, 156
Deoxocathasterone, 132
Deoxoteasterone, 132
Deoxyerythronolide B synthase (DEBS), 198
Deterrents, 242
Dhurrin, 227, 235-236
Dihydroflavonol reductase (DFR), 98-99, 102, 104-105, 115, 118
Dimethylallyl alcohol (DMAPP), 149, 151, 154-156
Disease resistance, 82, 85, 88-89, 180, 198, 224
Diterpenes, 149
DNA chips, 4
Drosophila melanogaster, 126
Drug production, 82, 89
Drypetes, 224, 234
Dwarfism, 132, 240-241
Dynamic range, 36-37, 52

Ectopic misexpression, 125, 130
Electrospray ionization mass spectometry (ESI/MS), 36
Enstoma russelianum, 137
Enzyme engineering, 218; *see also* Metabolic engineering
Equisetum arvense, 210
Escherichia coli, 19, 22, 25, 150-154, 156, 165, 229, 231-232, 237-240
Eschscholzia californica, 165, 169-170
Essential oils, 146-147, 150
  biosynthesis, 146, 151, 158
  peppermint, 149, 151
Euphorbiaceae, 224

Euphorbiales, 224, 234
Evolution, 87-88, 101, 180, 224, 234
Evolutionary relationship, 232, 236; see also Phylogenetically related
Expressed sequence tags (EST), 4, 16, 18, 22, 25-26, 48, 76, 85, 101, 145-147, 151, 154, 164, 176, 181, 183, 185-186, 190, 192, 231-232, 237
Expression; see also Gene expression
   analysis, 72, 140, 182
   arrays, 68, 75
   heterologous expression, 84, 152, 154, 176, 238-240
   serial analysis of gene expression (SAGE), 34, 185-187
   studies, 130

Fatty acid synthases (FASs), 198, 204
Flavanone 3-hydroxylase (F3H), 98, 102, 104
Flavonoids, 36, 40, 42, 96, 98-99, 101, 103-106, 115, 118, 137, 171, 187, 189, 218
   biosynthesis, 98-99, 103, 105, 106, 118, 171
   enzyme complex, 104
   flavonoid 3'-hydroxylase (F3'H), 98, 104
   glycosides, 40
   metabolism, 96, 99, 105, 106
   methylated, 146
   mutants, 98, 101, 103
   pathway, 96, 98, 101, 105-106
Flavonols, 98, 106
   flavonol synthase (FLS), 98, 106
Folates, 16
Forage, 99, 180-181, 192
Formamidase, 22-23, 26
Formyl-THF deformylase, 22-23, 26
Formyltetrahydrofolate cycloligase, 22
Forward genetics, 129
Fosmidomycin, 146, 156

Functional genomics, 2-4, 7-10, 32, 34, 55, 64, 75-76, 114, 120, 158
Functional redundancy, 120, 140; see also Multifunctional
Fungal pathogens, 85
Fusarium, 85

Gaeumannomyces graminis, 85
Gain-of-function mutations, 112, 119
Gas chromatography – mass spectrometry (GC/MS), 36-40, 48-49, 52-53, 64, 70, 72, 74, 76, 188
GenBank, 23, 173, 183
Gene expression, 4, 6, 10, 32, 34, 55, 68, 71, 76, 96, 101, 181, 185-187, 189, 190, 192; see also Expression
Gene knock-outs, 131, 141, 181, 190, 233, 238, 241
Gene redundancy, 140
Genetic approaches, 32, 96, 130; see also Mutational approaches
Genetic perturbation, 32, 52
Genistein, 136-137
Genome, 2-4, 16, 18, 22, 26, 32, 35, 75, 112, 119, 126, 129-130, 164, 180-181, 183, 189-190, 203, 229
   Arabidopsis, 22-25, 98, 118, 127, 129, 141, 186-187, 236
   Avena, 86, 181, 183, 192
   chloroplast, 184
   Medicago truncatula, 32, 181, 183, 192
   yeast, 187
Genome sequencing, 18, 75, 164, 181, 192
Genomics, 2-4, 7-10, 16, 18-19, 22-23, 26, 32, 55, 67, 75, 190
Gepasi, 9
Geranyl diphosphate synthase, 151
Gerbera hybrida, 210-211
Gibberellins, 126
Glandular trichomes, 146

INDEX

Global profiling, 35, 38
Glucokinase, 67, 73
Glucosinolates, 126, 135, 224, 226-227, 229, 231-236, 238, 240-243
  biosynthesis, 224, 229-231, 236, 242
  pathway, 229, 235-237, 240, 242
Glutathione, 23, 101, 186, 226, 237, 240
Glutathione S-transferases (GST), 101, 115
Glycerrhiza, 172
Glycosyltransferases, 85
Glycyrrhiza echinata, 135
Glycyrrhiza glabra, 84
Grasses, 85, 180
Green alga, 151
Gypsophila paniculata, 84

Health promoting, 136, 180
Heme-thiolate monoxygenases, 126
Herbicide, 72, 114
Heterologous biosynthesis, 218
Heterologous expression, 84, 152, 154, 176, 238-240
Hexose phosphates, 65, 67
Hierarchical cluster analysis (HCA), 31, 53, 54
High performance liquid chromatography (HPLC), 38, 41, 44, 47-48, 50, 52, 115, 146, 155, 188, 229-230
Homocysteine S-methyltransferase (HMT), 18, 22, 24-25
Homology-based approaches, 99
Host-symbiont interactions, 185
HPLC/PDA/MS, 41, 44, 47
Hydroxynitrile, 235

Improved pest resistance, 243
Indole, 103, 119, 135, 164, 173, 224, 231, 233, 235, 238, 240
Indole alkaloids, 119, 164, 173

Indole-3-acetaldoxime (IAOX), 231
Indole-3-acetic acid (IAA), 103, 125, 135, 231, 238, 240-241; *see also* Auxin
  biosynthesis, 125, 135
Insertional mutagenesis, 98, 129-131, 182, 189
Invertase, 65, 67, 72, 73
Ipomoea, 211
Isoflavone reductase, 180
Isoflavonoids, 42, 136, 188
Isopentenyl diphosphate (IPP), 149, 151, 154-156
Isoprenoids, 82, 84, 89, 119, 148, 151, 154, 158
  biosynthesis, 151, 154, 158
  pathway, 82, 84, 89
Isoquinoline alkaloid biosynthesis, 171, 176
Isothiocyanates, 224

Jasmonates, 126, 138
  biosynthesis, 131
  methyl, 169-170, 187

Kaempferol, 115
Kinases, 3, 154
Knockout mutants, 233, 238, 241

Lanosterol, 82, 84, 86-87, 89
  lanosterol synthase (LS), 86-87, 89, 98
Lavender, 146
LC-MS, 156, 158
Legume, 8, 32, 44, 76, 82, 180-181, 183, 190, 192, 199
Leguminosae, 180
Leguminous, 136-137, 190
Leucoanthocyanidin dioxygenase (LDOX), 98-99, 102
Leucoanthocyanidin reductase (LCR), 99, 102, 106

Licorice, 136; *see also Glycerrhiza, Glycyrrhiza*
Lignin, 16, 180, 185
Limonene, 150-151
    limonene synthase, 151
Liquiritigenin, 136
Loss-of-function mutation, 112, 119-120, 132
Lupane, 82, 84

Maize, 19, 25-26, 96, 98, 101, 106, 116, 118, 139, 231
Manihot esculenta, 227
Mass spectrometry (MS), 4, 7, 31, 34, 36-39, 41, 44, 47-49, 52, 70-73, 129, 146, 155-156, 187-188, 232, 236
Matrix-assisted laser desorption ionization time-of-flight mass spectrometry (MALDI-TOFMS), 48
Matthiola incana, 99
Meadow rue, 170
Medicagenic acid, 44
Medicago,
    Medicago Genome Initiative (MGI), 32, 181, 183, 192
    M. sativa, 32, 44, 199
    M. truncatula, 8, 31-32, 39-40, 42, 44, 48, 50, 52-53, 76, 179-181, 183-185, 187-190, 192
Medicarpin, 43, 136
Metabolic engineering, 34, 64-65, 69, 71, 73, 75-76, 105, 158, 218, 242-243
Metabolic profiling, 34, 38, 47, 50, 53, 64, 71-74, 76, 119, 155-156, 189; *see also Profiling*
Metabolic stress, 241
Metabolomics, 7, 8, 10, 34, 36, 55
Methionine metabolism, 26
Methionine *S*-methyltransferase (MMT), 24-26

Methyl jasmonate, 169-170, 187
Methylated flavonoids, 146
Methylation reactions, 16
Methylenetetrahydrofolate reductase (MTHFR), 15, 19, 21, 26
Methylsalicylic acid synthase (MSAS), 198
Methyltransferases, 24, 26, 170-171
Mevalonates, 149-151, 153-156, 158-159
    pathway, 149
Mevalonate-independent pathway, 149, 151, 153-154, 156, 158
Mevalonic acid, 88
Microarray, 4, 34, 72, 130-131, 185
Mint, 146
Misexpression, 125, 130
Mixtures, 38-39, 48, 187, 237
Molecular plant physiology, 64, 76
Monocots, 25, 88, 181
Monoterpenes, 146, 148, 149
    biosynthesis, 146, 154, 157
Morphine, 126, 164, 169, 172-174
    biosynthetic pathway, 173
Multidimensional chromatography, 34
Multidisciplinary databases, 181
Multienzyme complex, 96, 103
Multifunctional, 84, 198, 227
Mutants, 3, 5-6, 16, 53, 74, 76, 84, 88-89, 98, 101, 103-104, 106, 112, 114-116, 119-120, 129-132, 135, 138, 140, 153, 187, 189, 190, 192, 205, 208-210, 214, 217-218, 232, 233, 237-239, 241
    Arabidopsis mutants, 63, 73
    flavonoid mutants, 98, 101, 103
    gene knock-outs, 131, 141, 181, 190, 233, 238, 241
    P450 mutants, 129
    pap1-D mutant, 115-116
    sad1 mutants, 88
Mutational approaches, 96, 105, 140

gain-of-function mutations, 112, 119
gene knock-outs, 131, 141, 181, 190, 233, 238, 241
  insertional mutagenesis, 98, 129-131, 182, 189
  loss-of-function mutation, 112, 119-120, 132
  PCR-based reverse genetics, 125, 130
  reverse genetics, 129-131, 135, 138
  site-directed mutagenesis, 209, 214
MYB genes, 116, 119
MYB transcription factors, 119
MYB-like transcription factors, 116
Mycobacterium tuberculosis, 203
Myrosinases, 224, 234

N-hydroxylases, 237
Naringenin, 103, 136-137, 203, 210-212
Nematode, 105
Nitriles, 224
Nitrogen fixation, 96
Nodulation gene inducers, 180
Norcoclaurine 6-*O*-methyltransferase, 171
Nutraceuticals, 192, 198
Nutritional enhancement, 99, 243

Oat, 84-85, 88-89; *see also Avena*
Olea europaea, 84
Oleanane, 82
Oligodeoxynucleotides, 165, 167, 169
Oligonucleotides, 185, 227
Oligosaccharides, 36
One-carbon (C_1) reactions, 16
Open reading frames (ORFs), 2, 32, 186-187
Opium poppy, 164, 170, 172
Oryza sativa, 86
OSC superfamily, 84, 86, 88

Oxazolidine-2-thiones, 224
Oxidoreductases, 151, 172
Oxidosqualene cyclase (OSC), 84, 86-89
Oximes, 224, 226-227, 232-234, 237, 240, 242
Oxocampestanol, 132
Oxylipins, 138
Oxidosqualene, 81
Oxytetracycline, 198

Panax ginseng, 84
PAP1, 99, 111, 114-116, 118-120
pap1-D mutant, 115-116
Papaver
  P. bracteatum, 173-174
  P. orientale, 173-174
  P. rhoeas, 173
  P. somniferum, 164, 170, 172-173
Parallel analyses, 37, 55
PCR-based reverse genetics, 125, 130
Pea, 70, 133, 180, 231
Pentose phosphate pathway, 151
Peppermint, 146, 147, 149, 151, 154, 156, 158
Peppermint essential oils, 149, 151
Peppermint oil gland secretory cell, 149, 154, 156, 158
Petunia, 96, 99, 101, 105-106, 116, 119
  P. hybrida, 211
PHA polymers, 68
Pharmacologically active, 164, 198
Phenanthrene alkaloids, 173
Phenolase, 169
Phenolic, 42, 115, 148
Phenylalanine ammonia lyase (PAL), 118
Phenylpropanoids, 188
  biosynthesis, 171
  metabolism, 101, 103
  pathway, 96, 103, 114-115, 136-137

Phosphofructokinase, 68
Photodiode array (PDA), 41-42, 47
Photorespiration, 18
Phylogenetic, 54, 187, 235, 237
Phylogenetically related, 198, 232, 236; see also Evolutionary relationship
Phytoalexins, 136, 180, 198, 218
Phytohormones, 126
Phytoprotectant, 81, 84; see also Chemical defenses
Pine, 16, 215, 217
Pinus strobus, 210
Pinus sylvestris, 215
Piper methylsticum, 210
Pisum sativum, 84; see also Pea
Plant defense, 82, 85, 187; see also Chemical defense
Plant development, 126
Plant-insect-interactions, 224, 242
Pleiotropic effects, 70, 99
Polycyclic aromatic products, 198
Polyhydroxyalkanoate (PHA), 68
Polyhydroxybutyrate (PHB), 63, 68-71, 74
Polyketides,
  biosynthesis, 204
  biosynthetic diversity, 218
  tetraketides, 199, 204-205, 208-211, 214, 218
  triketides, 210-211, 214
Polyketide synthases (PKSs), 198-199, 203-204, 208-209, 211, 214, 218
  type III polyketide synthases, 198, 202-203, 208, 215, 218-219
Positional cloning, 99, 106
Potato, 64-65, 67, 72-73, 89
Prairie gentian, 137
Prenyl transferases, 151
Principal component analysis (PCA), 31, 53
Proanthocyanidins, 98

Profiling, 4, 7, 32, 34-39, 42, 44, 47-52, 55, 64, 71-72, 74, 76, 159, 181, 185, 187-188; see also Metabolic profiling
Protection, 89, 96, 105, 126
  against pathogens, 82, 84; see also Chemical defenses
  against stress, 82
  against biotic stress, 126
  against UV radiation, 96
  plants, 89
Proteomics, 7, 8, 10, 34, 55, 68, 75-76, 176, 187-189
Pseudomonas saccharophila, 73
Pterocarpan phytoalexins, 180
Puntranjiva roxburghii, 234
Pyrone synthase, 198

Quercetin, 115

Ralstonia eutropha, 68-70
Rapid amplification of cDNA ends (RACE), 146, 174
Rauwolfia serpentina, 164-165, 173
Reaction mechanism, 199, 204
Reducing cancer risk and coronary heart disease, 136
Relational database management systems, 7
Resveratrol, 199, 203, 214-215, 217-218
Reverse genetics, 129-131, 135, 138
Rhizobium, 180, 198
  nodulation gene inducers, 180
  nodulation genes, 198
Root pathogen, 85
Rosemary, 146
Ruta graveolens, 208

(S)-coclaurine, 171
(S)-norcoclaurine, 171

Saccharomyces cerevisiae, 2, 19, 86, 126, 167; *see also* Yeast
sad1 mutants, 81, 88
S-adenosylmethionine, 16, 19, 21, 24, 26
Sage, 146
Salutaridinol, 173-174
　salutaridinol 7-*O*-Acetyltransferase, 174
Sanguinarine, 119
Saponaria officinalis, 84
Saponins
　biosynthesis, 82, 84, 88-89, 189
　glucosyltransferases, 89
　pathway, 189
Sarcosine oxidase, 22, 24, 26
scFv antibodies, 96, 105
Sclerotinia sclerotiorum, 137
Secologanin, 165
Seed dormancy, 98, 106
Self organizing maps (SOMs), 31, 53
Sequence analysis, 88, 90, 181
Sequence-based approaches, 164
Sequential extraction, 37, 188
Serial analysis of gene expression (SAGE), 34, 185-187
Serine hydroxymethyltransferase (SHMT), 18, 22
Sesquiterpenes, 146, 149
S-formylglutathione hydrolase, 15, 23
Shikimate, 73
Signal, 3, 22, 51, 68, 75, 101, 133, 138, 155, 190
Signal transduction, 3, 101, 133
Silencing effects, 70
Sinapic acid, 115
Sinapis alba, 227, 234-236
Site-directed mutagenesis, 209, 214
S-methylmethionine (SMM) cycle, 15-16, 18, 24, 26
Snapdragon, 96, 106, 118-119

Sorghum, 168, 227, 229, 234-235, 239, 241
S. bicolor, 227, 238
Soybean, 136-137, 172, 180, 190, 192
Starch biosynthesis, 64-65, 67
Sterility phenotype, 138
Steroids, 82, 132
　6-oxidases, 132
　hormones, 132
　alkaloids, 82, 89
Sterols, 82, 84, 87-89, 127, 149
　biosynthesis, 84, 88, 127
Stilbene, 197-199, 214-215
　stilbene synthase (STS), 198-199, 203, 214-215, 217-218
Streptomyces coelicolor, 203
Stress, 74-75, 185, 187, 241
　biotic, 115
Strictosidine, 164-165
　strictosidine synthase, 164-165
Structure-function relationship, 218
Substrate specificity, 132, 171, 229, 233
Sucrose metabolism, 67, 73
Sucrose phosphorylase, 73
Sulphoraphane (4-Methylsulfinyl-butylisothiocyanate), 224
Sunflower, 231
Symbiotic relationship, 32, 136, 180
Systems biology, 3-4, 10, 32

Tabernaemontana divaricata, 84
Tandem mass spectrometry (MS/MS), 31, 34, 41, 44, 47, 188
Taraxacum officinale, 84
Taxol, 126
T-DNA insertion, 112, 115, 129, 131, 141
T-DNA tagging, 99, 106, 112; *see also* Activation tagging
Teasterone, 132
Terpenoids, 126, 188

diterpenes, 149
sesquiterpenes, 146, 149
synthases, 150-151
tetraterpenes, 149
triterpenoids, 44, 48, 82, 84, 86-89, 189
Tetrahydrofolate, 15-16, 19, 21, 22-23
Tetraketide, 199, 204-205, 208-211, 214, 218
Tetraterpenes, 149
Thalictrum tuberosum, 170-171
Thebaine, 173, 175
Thiocyanates, 224
Three-dimensional structure, 209-210
Thyme, 146
Tobacco, 112, 118-120, 137, 184, 231-232, 234
Tomato, 132, 154
Transcription factors, 99, 106, 114, 116, 118-119
Transcriptomics, 7-8, 10, 34, 55
Transformed populations, 131
Transgenes, 70
Transgenic, 34, 64-65, 67, 69-74, 76, 89, 96, 105, 112, 114, 116, 118-120, 137, 158, 198, 229-230, 233, 240-241
potato, 34, 65, 72
Transketolase, 151
Transparent testa (*tt*) phenotype, 98-99

Transport, 19, 24, 35, 65, 96, 101, 103, 148
Trichome, 99
Triglochin maritima, 227
Triketides, 210-211, 214
Triketide styrylpyrones, 210
Triterpenoids, 44, 48, 82, 84, 86-89, 189
triterpene cyclase, 189
triterpene synthases, 84, 88
triterpenoid saponins, 82, 84
Tropaeolaceae, 234-235
Tropaeolum majus, 227, 234
Two-dimensional polyacrylamide gel electrophoresis (2-D PAGE), 34, 187, 188
Type III polyketide synthases, 198, 202, 203, 208, 215, 218, 219

UV radiation, 185
UV-inducible, 187

Vinblastine, 173
Virus induced gene silencing (VIGS), 190

Yeast, 19, 24, 65, 67, 72, 86, 132-133, 135, 137, 167, 187, 231, 237; *see also Saccharomyces cerevisiae*
genome, 187